Bioaugmentation Techniques and Applications in Remediation

Bioaugmentation Techniques and Applications in Remediation

Edited by
Inamuddin
Charles Oluwaseun Adetunji
Mohd Imran Ahamed
Tariq Altalhi

CRC Press
Taylor & Francis Group
Boca Raton London New York

CRC Press is an imprint of the
Taylor & Francis Group, an **informa** business

First edition published 2022
by CRC Press
6000 Broken Sound Parkway NW, Suite 300, Boca Raton, FL 33487-2742

and by CRC Press
4 Park Square, Milton Park, Abingdon, Oxon, OX14 4RN

CRC Press is an imprint of Taylor & Francis Group, LLC

Library of Congress Cataloging-in-Publication Data
Names: Inamuddin, 1980- editor. | Adetunji, Charles Oluwaseun, editor. |
Ahamed, Mohd Imran, editor. | Altalhi, Tariq, editor.
Title: Bioaugmentation techniques and applications in remediation /
[edited by] Inamuddin, Charles Oluwaseun Adetunji, Mohd Imran Ahamed, Tariq A. Altalhi.
Description: First edition. | Boca Raton, FL : CRC Press, 2022. | Includes
bibliographical references and index.
Identifiers: LCCN 2021061043 | ISBN 9781032034997 (hbk) |
ISBN 9781032035017 (pbk) | ISBN 9781003187622 (ebk)
Subjects: LCSH: Bioremediation.
Classification: LCC TD192.5 .B5433 2022 | DDC 628.5—dc23/eng/20220223
LC record available at https://lccn.loc.gov/2021061043

ISBN: 978-1-032-03499-7 (hbk)
ISBN: 978-1-032-03501-7 (pbk)
ISBN: 978-1-003-18762-2 (ebk)

DOI: 10.1201/9781003187622

Typeset in Times
by codeMantra

Contents

Preface

It has been observed that rapid population expansion has raised the amount of anthropogenic activity, resulting in high levels of pollution in water, air, and solid waste, as well as an increase in the pressure placed on agricultural lands. Furthermore, a high rate of mortality has resulted in a drop in the number of endangered species as well as an increase in the number of human and environmental threats. The global population has been forecasted to increase tremendously to 9 billion by the year 2050. This is likely to put a burden on economic growth, particularly considering the low level of natural resources available in emerging nations, when compared to developed countries. Moreover, emerging nations have a higher tendency of population expansion, which is likely related to the high rate of rural-to-urban migration. Management of increased solid waste, air, and water pollution will be a challenge as well. Therefore, there is a need to search for sustainable and environmental techniques that could help in the mitigation of all these aforementioned challenges. The utilization of biotechnology using bioaugmentation techniques which entail the release of consortia of microorganisms or specific competent strains into a contaminated environment or the augmentation of the catabolic potential at a contaminated site to enhance bioremediation of contaminants is the need of the hour.

Bioaugmentation techniques are intended to provide detailed information on the application of these techniques using beneficial microorganisms for the rejuvenation of a contaminated environment. Detailed information on the utilization of bioaugmentation approaches for adequate remediation of the environment such as sediments, water, and soil polluted with organic and inorganic contaminants is necessary for developing new techniques. Special highlights are provided on practical applications of bioaugmentation techniques. Detailed information on the commercially available products as well as abiotic and biotic factors that could enhance and facilitate the high rate of environmental rejuvenation of contaminants using bioaugmentation techniques is also provided. Moreover, distinct emphasis is given to the applications of nanomaterials which on combining with bioaugmentation techniques could facilitate an enhanced level of bioremediation efficiency as well as prevention of several challenges encountered during remediation processes. Due to their large surface area, enhanced adsorption features facilitate the attachment of functional groups to enhance affinities towards target molecules. It is an invaluable guide to planners, governmental and non-governmental organizations, environmentalists, biotechnologists, engineers, students, professors, scientists, and R&D industrial experts working in the field of bioaugmentation techniques. The summaries of the work reported in the following 11 chapters are as follows.

Chapter 1 explores the challenges and benefits of using bioaugmentation in the removal of chemically stable compounds, highlighting the fact that the technique is more suitable for short-term treatments, as well as being associated with other methodologies.

Chapter 2 describes structural and functional properties of lignin that make it a promising candidate as a renewable chemical feedstock. Additionally, various methods employed for the extraction of lignin are also discussed with a particular emphasis on ionic liquid-based methods.

Chapter 3 addresses the pesticide evolution and its devastating effects on the environment. Furthermore, the biological way of eliminating pesticide compounds through microorganisms from the polluted site is detailed. Methods involved in bioaugmentation and the key role of biotic and abiotic variables in the establishment of heirloom/introduced microorganisms are also elaborated in this chapter.

Chapter 4 elaborates bioaugmentation process for amending microbes in petroleum product-contaminated sites/sludge. It summarizes the details of common microbes used in bioremediation. It also discusses the factors affecting the bioaugmentation process and the different bioaugmentation techniques. Bioremediation of petroleum product-contaminated water/soil/oil and sludge by implementing various bioaugmentation techniques is also presented.

Chapter 5 overviews different bioaugmentation methods used in bioremediation of cyanide. A limited proportion of the cyanide present in the environment is created naturally by some living species, whereas the majority comes from industrial waste. Enzymes responsible for various cyanide breakdown pathways are frequently produced by microorganisms.

Chapter 6 discusses the various biological methods reported for the removal of recalcitrant pollutants in industrial wastewater. Major importance is given to the methods used for the elimination of hydrocarbons, pesticides, polychlorinated biphenyls, and phthalates. Recent reports on bioaugmentation of pharmaceutical byproducts are also discussed.

Chapter 7 captures pollution issues, environmental pollutants, and the concept of bioaugmentation. Methods of bioaugmentation, their pros and cons, and applications of bioaugmentation are also discussed. The utilization of nanomaterials to bioaugmentation and their future prospects are detailed.

Chapter 8 discusses how plasmid-mediated bioaugmentation approaches are used to improve the cleanup of contaminated sites. Additionally, it gives an insight into determining factors and constraints for assessing the success of plasmid bioaugmentation in the contaminated environment of concern and also intends to identify knowledge gaps for future research efforts.

Chapter 9 deliberates the prospects of bioaugmentation in rhizoengineering as well as rhizosphere engineering processes in bioaugmentation for degradation of xenobiotic compounds. The different types of remediation such as phytoremediation and rhizoremediation are explained, in view of rhizoengineering. Furthermore, factors affecting bioaugmentation and the different approaches are also discussed.

Chapter 10 reviews the recovery of metals from electronic wastes (e-wastes), an example of the municipal wastes of major concern through the bioaugmentation process. The extraction mechanisms of gold and other base metals from e-wastes are further discussed. Microorganisms involved in the recovery of these metals are also introduced.

Chapter 11 touches on bioremediation techniques with a focus on in situ and ex situ approaches. The effect of heavy metals and radionuclides on the biological system is evaluated. The counter mechanisms achieved through microbial activity are presented. The factors affecting bioaugmentation are analyzed. In addition, the recent trends using nanotechnology are summarized.

Editors

Inamuddin, PhD is working as Assistant Professor at the Department of Applied Chemistry, Aligarh Muslim University, Aligarh, India. He obtained Master of Science degree in Organic Chemistry from Chaudhary Charan Singh (CCS) University, Meerut, India, in 2002. He received his Master of Philosophy and Doctor of Philosophy degrees in Applied Chemistry from Aligarh Muslim University (AMU), India, in 2004 and 2007, respectively. He has extensive research experience in multidisciplinary fields of Analytical Chemistry, Materials Chemistry, and Electrochemistry and, more specifically, Renewable Energy and Environment. He has worked on different research projects as project fellow and senior research fellowship funded by the University Grants Commission (UGC), Government of India, and Council of Scientific and Industrial Research (CSIR), Government of India. He has received Fast Track Young Scientist Award from the Department of Science and Technology, India, to work in the area of bending actuators and artificial muscles. He has completed four major research projects sanctioned by the University Grant Commission, Department of Science and Technology, Council of Scientific and Industrial Research, and Council of Science and Technology, India. He has published 196 research articles in international journals of repute and nineteen book chapters in knowledge-based book editions published by renowned international publishers. He has published 150 edited books with Springer (UK), Elsevier, Nova Science Publishers, Inc. (USA), CRC Press Taylor & Francis Asia Pacific, Trans Tech Publications Ltd. (Switzerland), IntechOpen Limited (UK), Wiley-Scrivener, (USA), and Materials Research Forum LLC (USA). He is a member of various journals' editorial boards. He is also serving as Associate Editor for journals (*Environmental Chemistry Letter, Applied Water Science and Euro-Mediterranean Journal for Environmental Integration*, and *Springer-Nature*), Frontiers Section Editor (Current Analytical Chemistry, Bentham Science Publishers), Editorial Board Member (*Scientific Reports-Nature*), Editor (*Eurasian Journal of Analytical Chemistry*), and Review Editor (Frontiers in Chemistry, Frontiers, UK). He is also guest-editing various special thematic special issues to the journals of Elsevier, Bentham Science Publishers, and John Wiley & Sons, Inc. He has attended as well as chaired sessions in various international and national conferences. He has worked as a Postdoctoral Fellow, leading a research team at the Creative Research Initiative Center for Bio-Artificial Muscle, Hanyang University, South Korea, in the field of renewable energy, especially biofuel cells. He has also worked as a Postdoctoral Fellow at the Center of Research Excellence in Renewable Energy, King Fahd University of Petroleum and Minerals, Saudi Arabia, in the field of polymer electrolyte membrane fuel cells and computational fluid dynamics of polymer electrolyte membrane fuel cells. He is a life member of the Journal of the Indian Chemical Society. His research interest includes ion exchange materials, a sensor for heavy metal ions, biofuel cells, super capacitors, and bending actuators.

Charles Oluwaseun Adetunji is presently a faculty member at the Microbiology Department, Faculty of Sciences, Edo State University Uzairue (EDSU), Edo State, Nigeria, where he utilized the application of biological techniques and microbial bioprocesses for the actualization of sustainable development goals and agrarian revolution, through quality teaching, research, and community development. He is currently the Acting Director of Intellectual Property and Technology Transfer, Ag Dean for Faculty of Science, and the Head of department of Microbiology at EDSU. He is a Visiting Professor and the Executive Director for the Center of Biotechnology, Precious Cornerstone University, Ibadan. He has won several scientific awards and grants from renowned academic bodies like Council of Scientific and Industrial Research (CSIR) India, Department of Biotechnology (DBT) India, The World Academy of Science (TWAS) Italy, Netherlands Fellowship Programme (NPF) Netherlands, The Agency for International Development Cooperation; Israel, Royal Academy of Engineering, UK among many others. He have published many scientific journal articles and conference proceedings in refereed national and international journals with over 370 manuscripts. He was ranked among the top 500 prolific authors in Nigeria between 2019 till date by SciVal/ SCOPUS. His research interests include Microbiology, Biotechnology, Post-harvest management, and Nanotechnology. He was recently appointed as the President and Chairman Governing Council of the Nigerian Bioinformatics and Genomics Network Society. He is presently a series editor with Taylor and Francis, USA editing several textbooks on Agricultural Biotechnology, Nanotechnology, Pharmafoods, and Environmental Sciences. He is an editorial board member of many international journals and serves as a reviewer to many double-blind peer review journals like Elsevier, Springer, Francis and Taylor, Wiley, PLOS One, Nature, American Chemistry Society, Bentham Science Publishers etc. He is a member of many scientific and professional including bodies like American Society for Microbiology, Biotechnology Society of Nigeria, Nigerian Society for Microbiology, and he is presently the General/Executive Secretary of Nigerian Young Academy. He has won a lot of international recognition and also acted as a keynote speaker delivering invited talk/position paper at various Universities, research institutes and several centers of excellence which span across several continent of the globe. He has over the last fifteen years built strong working collaborations with reputable research groups in numerous and leading Universities across the globe. He is the convener for Recent Advances in Biotechnology, which is an annual international conference where renown Microbiologist and Biotechnologist come together to share their latest discoveries. He is the president and founder of the Nigerian Post-Harvest and Food Biotechnology Society.

Mohd Imran Ahamed, PhD is working as Research Associate at the Department of Chemistry, Aligarh Muslim University (AMU), Aligarh, India. He received his B.Sc. (Hons) Chemistry and Ph.D. (Chemistry) degrees from AMU. He has completed his M.Sc. (Organic Chemistry) from Dr. Bhimrao Ambedkar University, Agra, India. He has published several research and review articles in various international scientific journals. He has co-edited 57 books with Springer (UK), Elsevier, CRC Press – Taylor & Francis Asia Pacific, Materials Research Forum LLC (USA), and Wiley-Scrivener (USA). His research work includes ion-exchange chromatography, wastewater treatment, and analysis, bending actuator and electrospinning.

Tariq Altalhi, PhD, joined the Department of Chemistry at Taif University, Saudi Arabia as Assistant Professor in 2014. He received his doctorate degree from the University of Adelaide, Australia in the year 2014 with Dean's Commendation for Doctoral Thesis Excellence. He was promoted to the position of the head of Chemistry Department at Taif University in 2017 and Vice Dean of Science College in 2019 currently. In 2015, one of his works was nominated for Green Tech awards from Germany, Europe's largest environmental and business prize, amongst top 10 entries. He has co-edited various scientific books. His group is involved in fundamental multidisciplinary research in nanomaterial synthesis and engineering, characterization, and their application in molecular separation, desalination, membrane systems, drug delivery, and biosensing. In addition, he has established key contacts with major industries in Kingdom of Saudi Arabia.

Contributors

V.C. Akubude
Department of Agricultural and
    Bioresource Engineering
Federal University of Technology
Owerri, Nigeria

L.P. Ananthalekshmi
School of Biosciences
Mahatma Gandhi University
Kottayam, India

Siti Khodijah Chaerun
Department of Metallurgical Engineering
Faculty of Mining and Petroleum Engineering
and
Geomicrobiology-Biomining and
    Biocorrosion Laboratory
Microbial Culture Collection Laboratory
Biosciences and Biotechnology Research
    Center (BBRC)
Institut Teknologi Bandung
Bandung, Indonesia

M. Chandrasekaran
Horticultural College and
    Research Institute for Women
Tamil Nadu Agricultural University
Tiruchirappalli, India

Aniruddha Chatterjee
Centre for Advanced Materials Research and
    Technology
Plastic and Polymer Engineering Department
Maharashtra Institute of Technology
Aurangabad, India

S.Z.Z. Cobongela
Nanotechnology Innovation Centre
Advanced Materials Division
Mintek, South Africa

Eduardo Santos da Silva
Departamento de Biotecnologia
Centro de Biotecnologia
Universidade Federal da Paraíba
João Pessoa, Brasil

Deviany Deviany
Department of Chemical Engineering
Institut Teknologi Sumatera
Lampung, Indonesia

N.D. Dhanraj
School of Biosciences
Mahatma Gandhi University
Kottayam, India

Uroosa Ejaz
Department of Microbiology
University of Karachi
and
Department of Biosciences
Shaheed Zulfikar Ali Bhutto Institute of
    Science and Technology (SZABIST)
Karachi, Pakistan

G. Gokulapriya
Anbil Dharmalingam Agricultural College and
    Research Institute
Tiruchirappalli, India

A.O. Igwe
Department of Chemistry
Michael Okpara University of Agriculture
Umudike, Nigeria

K. Jayachandran
School of Biosciences
Mahatma Gandhi University
Kottayam, India

M.S. Jisha
School of Biosciences
Mahatma Gandhi University
Kottayam, India

Elizabeth Mary John
School of Biosciences
Mahatma Gandhi University
Kottayam, India

Indu C. Nair
Department of Biotechnology
SAS SNDP Yogam College
Konni, India

P.C. Obumseli
Department of Agricultural and
    Bioresource Engineering
Federal University of Technology
Owerri, Nigeria

Greeshma Odukkathil
Centre for Environmental Studies
Anna University
Chennai, India

V.C. Okafor
Department of Agricultural and Bioresource
    Engineering
Federal University of Technology
Owerri, Nigeria

T.F Oyewusi
Department of Agricultural Engineering
Adeleke University
Ede, Nigeria

Luiz Gustavo Pragana
Departamento de Biotecnologia
Centro de Biotecnologia
Universidade Federal da Paraíba
João Pessoa, Brasil

Ajinkya Satdive
Centre for Advanced Materials Research
    and Technology
Plastic and Polymer Engineering Department
Maharashtra Institute of Technology
Aurangabad, India

Muhammad Sohail
Department of Microbiology
University of Karachi
Karachi, Pakistan

R.P. Soundararajan
Horticultural College and
    Research Institute for Women
Tamil Nadu Agricultural University
Tiruchirappalli, India

Saurabh Tayde
Centre for Advanced Materials Research
    and Technology
Plastic and Polymer Engineering Department
Maharashtra Institute of Technology
Aurangabad, India

Bhagwan Toksha
Basic Science and Humanity Department
Maharashtra Institute of Technology
Aurangabad, India

Shyam Tonde
Centre for Advanced Materials Research
    and Technology
Plastic and Polymer Engineering Department
Maharashtra Institute of Technology
Aurangabad, India

Edna Mary Varghese
School of Biosciences
Mahatma Gandhi University
Kottayam, India

Ulrich Vasconcelos
Departamento de Biotecnologia
Centro de Biotecnologia
Universidade Federal da Paraíba
João Pessoa, Brasil

Namasivayam Vasudevan
Centre for Environmental Studies
Anna University
Chennai, India

1 Bioaugmentation Techniques for Removal of Heterocyclic Compounds and Polycyclic Aromatic Hydrocarbons

Ulrich Vasconcelos, Luiz Gustavo Pragana,
and Eduardo Santos da Silva
Universidade Federal da Paraíba

CONTENTS

1.1 INTRODUCTION

Approximately 90% of the discharge of contaminants into the environment originates from human activity; of this total, about 70% of the contaminants are oil hydrocarbons (Cavalcanti et al., 2016), highly hydrophobic compounds with low solubility (Guo et al., 2011). The weathering of these oil hydrocarbons in sediment decreases the bioavailability of contaminant uptake by microorganisms due to different factors such as soil texture, partitioning, sorption rate, pH, organic matter content, and aging (Aljerf and Al Masri, 2018). This weathering also includes processes of adsorption, dissolution, volatilization, biotransformation, oxidation, and photolysis (Oualha et al., 2019).

Hydrocarbons composed of cyclic, heterocyclic, aromatic, and polycyclic aromatic molecules (PAHs) are chemically more stable molecules when compared to oleophins (Poater et al., 2018). This characteristic is due to the organization of the carbonic skeleton, composed of at least two aromatic rings and condensed cyclopentanes, arranged in lines, angles, or groups (Vo-Dinh et al., 1998). Some of these are associated with nitrogen, oxygen, or sulfur atoms (Asif and Wenger, 2019). As a result, these hydrocarbons show greater recalcitrance when disposed of in the environment and, therefore, less bioavailability (Lladó et al., 2013; Huesemann et al., 2004).

Heterocyclic compounds and, mainly, PAHs exhibit greater complexity in terms of recalcitrance. In addition, PAHs comprise a family of hundreds of organic substances, formed by incomplete combustion, considered to be one of the most frequent classes of molecules in the environment (Pope et al., 2000). There are 16 priority PAHs listed by the United States Environmental Protection Agency (USEPA) (Table 1.1). These are representative compounds of the class in terms of toxicity, as well as occurring in concentrations higher than the other PAHs (Ravindra et al., 2008; Thomas and Wornat, 2008).

DOI: 10.1201/9781003187622-1

TABLE 1.1

Physicochemical and Genotoxic Characteristics of the 16 Priority PAHs Listed by the United States Environmental Protection Agency

PAHs	R	MW	G[a]	S	VP	K_{ow}	FP	BP	D
Naphthalene	2	128.17	2B	31.0	8.9×10^{-2}	3.4	80	218	1.14
Acenaphthene	3	154.21	3	3.8	3.8×10^{-3}	3.9	94	280	1.02
Acenaphthylene	3	152.20	3	16.1	2.9×10^{-2}	4.1	91	280	1.01
Fluorene	3	166.22	3	1.9	3.2×10^{-3}	4.2	116	295	1.20
Phenanthrene	3	178.23	3	1.1	6.8×10^{-4}	4.6	101	339	1.25
Anthracene	3	178.23	3	0.05	2.6×10^{-5}	4.5	218	340	1.25
Fluoranthene	4	202.26	3	0.26	8.1×10^{-6}	5.2	111	375	1.25
Pyrene	4	202.26	3	0.13	4.3×10^{-6}	5.2	146	404	1.27
Benzo[a]anthracene	4	228.29	2B	0.01	1.5×10^{-7}	5.6	162	435	1.27
Chrysene	4	228.29	2B	0.002	7.8×10^{-9}	5.9	254	448	1.27
Benzo[b]fluoranthene	5	252.32	2B	0.002	8.1×10^{-8}	6.1	168	481	1.24
Benzo[k]fluoranthene	5	252.32	2B	0.001	9.6×10^{-11}	6.8	217	480	1.24
Benzo[a]pyrene	5	252.32	1	0.004	4.9×10^{-9}	6.5	179	495	1.24
Dibenzo[ah]anthracene	5	276.34	2A	0.005	2.1×10^{-11}	6.5	267	324	1.28
Benzo[ghi]perylene	6	278.35	3	3×10^{-4}	1.0×10^{-10}	6.6	278	550	1.30
Indeno[1,2,3-cd]pyrene	6	278.35	2B	0.062	1.4×10^{-10}	7.1	164	536	1.30

Source: International Agency for Research on Cancer (2021), Bojes and Pope (2007), Cai et al. (2007), Eom et al. (2007), Zhang et al. (2006), Daugulis and McCracken (2003), Pope et al. (2000).

R, rings; MW, molecular weight; G, genotoxicity ([a]classes: 1, Carcinogenic to humans; 2A, Probably carcinogenic to humans; 2B, Possibly carcinogenic to humans; 3, Not classifiable as to its carcinogenicity to humans); S, solubility; VP, vapor pressure (mmHg); K_{ow}, octanol-water partition coefficient (log); FP, fusion point (°C); BP, boiling point (°C); D, density (g/L).

The hydrophobic properties of PAHs result in high affinity with many substrates, including soil particles. Soil is considered the main environment susceptible to oil hydrocarbons contamination (Gong et al. 2007; Mater et al., 2006; Enell et al., 2005; Watanabe, 2001). Soil is a very complex biological system, and extremely biodiverse. In terms of contamination by complex hydrocarbons, microbial reactions play a key role in its recovery (Orgiazzi et al., 2015; Watanabe and Hamamura, 2003). Exposure to PAHs and/or heterocyclic hydrocarbons promotes negative impacts on all layers of the soil, causing the reduction of fertility as well as imbalance in the trophic chain (Herwijnen et al., 2003).

Additionally, the exposure of the microbiota to hydrocarbons causes drastic changes, requiring rapid adaptation and tolerance to the contaminant (Sarkar et al., 2016). The stress caused to the ecosystem promotes selective pressures, allowing generations of hydrocarbonoclastic individuals to establish themselves (Yang et al., 2016; Röling et al., 2002). This involves two important factors: hydrocarbon toxicity on cells and abundance of carbon from the hydrocarbons on the carbon content in the environmental organic matter, resulting in an imbalance of the C:N:P ratios (Sutton et al., 2013).

The use of hydrocarbons as a source of carbon and energy by some microbes has been known since the beginning of the 20th century. Starting in the 1940s, the mechanisms began to be unveiled (Bushnell and Haas, 1941). Microbes that exhibit the ability to assimilate hydrocarbons are highly distributed in the environment, although they do not use these compounds as a preferred source of carbon and energy (Dashti et al., 2015). After exposure to hydrocarbons, however, the population of hydrocarbonoclastic microbes becomes more prevalent (Teramoto et al., 2013). The process of hydrocarbon mineralization by microorganisms is divided into two distinct stages. The first one, more accelerated, is mediated by the bioavailability of the contaminant. The second slower stage is controlled by the hydrocarbon sorption/desorption ratio (Kaplan and Kitts, 2004).

Hydrocarbonoclastic microbes exhibit different mechanisms to promote the assimilation of hydrocarbons such as synthesis of nonspecific enzymes that recognize cyclic compounds (De Boer et al., 2005), expression of oxidoreductases (Limongi et al., 2020), synthesis of biosurfactants (Martínez-Toledo and Rodríguez-Vasquez, 2013), biofilm formation (Nie et al., 2016), and production or assimilation of compatible solutes (Welsh, 2000). It is noteworthy that the process of biological removal of recalcitrant hydrocarbons is more effective when simple molecules, with greater bioavailability, are present assuming the role of cosubstrates (Nascimento et al., 2013).

In the case of environments containing high concentrations of recalcitrant hydrocarbons, the indigenous microbiota may be moderate or highly inhibited (Xu et al., 2018). This results in changes in population dynamics and the mineralization process of these compounds, which can occur via metabolic cooperation or by cometabolism (Johnsen et al., 2005; Aldén et al., 2001), in which cosubstrates are crucial (Brown et al., 2017). This occurs where bioaugmentation is the strategy of choice for interventions applied in environments contaminated by recalcitrant hydrocarbons (Cavalcanti et al., 2019). While in cometabolism, a microbe or a microbial group transforms a particular compound into a metabolite without any direct benefits for the cell (Arp et al., 2001), cosubstrates are assimilable organic compounds that, when used by microbes, favor the degradation of more complex molecules (Baboshin et al., 2003).

1.2 PRINCIPLES OF BIOAUGMENTATION STRATEGIES FOR THE REMOVAL OF RECALCITRANT HYDROCARBONS

Bioremediation is a sustainable, eco-friendly, attractive, and low-cost technique to recover a contaminated site, speeding up the process of mineralization of organic compounds, in less time than the natural attenuation process (Atlas, 1981). When microbes, for example, bacteria and fungi, are employed in bioremediation, the removal of the contaminant is based on stimulating the microbiota's metabolic activity, which converts the contaminant into biomass and metabolites (Abatenh et al., 2017). The use of bioaugmentation as a mineralization strategy for recalcitrant hydrocarbons is based on two basic principles: (i) when the microbial density present is not sufficient to maintain degradation efficiently, requiring the addition of competent cells (Jacques et al., 2007), and (ii) when the C:N:P ratio is unbalanced, failing to stimulate the microbiota by adding nutrients and cosubstrates (Mehrzad et al., 2015).

The removal of recalcitrant hydrocarbons in the soil, using bioaugmentation, requires more time because it is more complex (Yuste et al., 2000). The greatest difficulty comes from the fact that a significant part of the heterocyclic hydrocarbons and PAHs, as well as their metabolites, may be incorporated into the humic matter, changing the sorption/desorption ratio of these compounds, thus limiting the process (Barrios, 2007). Other intrinsic and extrinsic factors are related to the bioavailability of heterocyclic hydrocarbons and PAHs during removal using bioaugmentation. These are summarized in Table 1.2.

The use of bioaugmentation to remove recalcitrant hydrocarbons in the soil is a controlled process that requires conditions that are often not found naturally. The type and structure of the soil directly affect the bioavailability of the contaminant, as well as the organization and density of the degrading communities (Scherr et al., 2007). Generally, the treatment of sandy soils produces better results than clay soils (Haghollahi et al., 2016).

Soil revolving is one of the not natural primary conditions, allowing homogenization, while promoting aeration, breakdown of aggregates, elimination of volatile compounds, and increased contact of microorganisms with nutrients and substrates (Joner et al., 2004). The presence of oxygen is crucial for the beginning of the degradative process because several enzymes involved in the cleavage of the molecules are oxidoreductases (Shome, 2020).

TABLE 1.2

Limiting Factors in the Removal of Recalcitrant Hydrocarbons in the Soil Using Bioaugmentation

Factor		References
Intrinsic	**Microbiota**	
	Composition	Øvreås (2000)
	Size and diversity	Liu et al. (2006)
	Metabolic versatility	Marín et al. (2005)
	Increased enzyme activity	Löser et al. (1998)
	Production of biosurfactants	Singh et al. (2007)
	Viability of electron acceptors	Mulligan and Yong (2004)
	Ecological interactions	Tilman (2000)
	Contaminant	
	Type	Atlas (1981)
	Molecule size and structure	Bossert et al. (1984)
	Bioavailability	Semple et al. (2007)
	Toxicity of metabolites	Boopathy (2000)
	Contamination history	Płaza et al. (2005)
Extrinsic	**Environment**	
	Temperature	Dibble and Bartha (1979)
	Time	Rivas (2006)
	Seasonal changes	Marschner and Kalbitz (2003)
	Moisture	Ettema and Wardle (2002)
	Organic matter	Amellal et al. (2001)
	Essential nutrients	Čejková et al. (1997)
	pH	Rahman et al. (2002)
	Soil properties	Margesin et al. (2007)
	Contamination depth	Vasudevan and Rajaram (2001)
	Presence of cosubstrate	Romantchuk et al. (2000)

The concentration of nutrients, mainly represented by nitrogen and phosphorus, is the most common limiting factor in bioaugmentation (Mariano et al., 2009). An abundance, rather than an excess, of nutrients favors microbial growth and increases the rate of degradation (Karamalidis et al., 2010).

The optimum temperature ranges between 30°C and 40°C. Under these conditions, greater mobilization of hydrocarbons occurs, increasing solubility (Perfumo et al., 2007; Leahy and Colwell, 1990). In addition, higher temperatures accelerate microbial activity (Popp et al., 2006). Regarding pH, the ideal range for biodegradation of oil hydrocarbons is 7.4–7.8. Values equal to or greater than 8.5 are inhibitory (Wang et al., 2011; Rowland et al., 2000).

The ideal water content in soil ranges from 15% to 20%. In this range, a favorable scenario may establish competition between water molecules for the same mineral sites of the hydrocarbon sorption in soil aggregates. Thus, a greater number of hydrocarbons may be found free, increasing the bioaccessibility (Kottler et al., 2001).

In addition, there is an interaction between abiotic factors, establishing a favorable scenario for microorganisms. This positively impacts the removal rate of recalcitrant hydrocarbons (Jia et al., 2004). The availability of nutrients and the ability of the microbiota to tolerate and initially assimilate the lower carbon chain compounds involves successions in the dominant microbiota (Hua and Wang, 2004), conditioned on the bioaccess and bioavailability of carbon sources (Jiao et al., 2017), as well as the class of hydrocarbon (Manse et al., 2020).

In the bioaugmentation, the origin of the microbes introduced in the matrix to be treated may be from the contaminated environment (Dejonghe et al., 2001) or a foreign site (Al-Mailem et al., 2019). The usual way of introducing the inoculum is through bacterial consortia or commercial cocktails, to the detriment of axenic cultures (Radwan et al., 2019). As consortia, the inoculants promote more efficiency to the process, even if a certain group is not ecologically the main degrader, in terms of prevalence under conditions without the contaminant (Festa et al., 2017). Microbial consortia can be composed exclusively of bacteria (Ayotamuno et al., 2006), fungi (Meysami and Baheri, 2003), or mixed (Boonchan et al., 2000).

The introduction of pollutants induces an adaptive response of the indigenous microbiota (Varjani and Upasani, 2017). An environment with a population of hydrocarbonoclastic microbes equal to or less than 10^5 CFU/g of sediment has been found to limit the natural recovery capacity of the site (Hammershøj et al., 2019). When this happens, the number of cells is not suitable to effect metabolic reactions efficiently (Singkran, 2013). Thus, the cell density of the inoculum, composed of competent microbes, needs to vary between 10^7 and 10^{12} CFU/g of soil (MacNaughton et al., 1999). However, for soils contaminated by heterocyclic hydrocarbons and PAHs, it is believed that the inoculum needs to be greater than or equal to the microbial population naturally found in the soil to be treated. This may prevent important obstacles to the effects expected after the inoculation, such as adaptation, predation, competition, immobility, and inhibition, depending on the presence of toxic substances and metabolites or the imbalance of nutrients (Abdulsalam and Omale, 2009).

Certain characteristics of the consortia microbes may contribute negatively to the low yield of the bioaugmentation, such as a prolonged lag phase (Bouchez et al., 2000), presence of compounds that inhibit microbial growth (Dua et al., 2002), motility (McClure et al., 1991) and/or negative ecological interactions among indigenous microbiota and the inoculum (Mille-Lindblom et al., 2006). In these cases, competition between the two populations can trigger die-off in the added cells, i.e., a process that culminates in cell death without the generation of descendants (Jung et al., 2005; Potin et al., 2004). Ecological interactions between different microbial groups, however, guarantee the balance of a microbial population (Arruda et al., 2020), and may intensify the process of removing the contaminant (Seneviratne et al., 2008; Chaîneau et al., 2005).

1.3 ON METHODS AND PERSPECTIVES OF BIOAUGMENTATION IN THE REMOVAL OF RECALCITRANT HYDROCARBONS

After inoculation of the consortium, it is expected that the additional population will increase logarithmically in a short time, depending on the rate of contaminant degradation enhancement (Zhu et al., 2015). In terms of the removal of recalcitrant hydrocarbons from the soil and water, bioaugmentation is considered a safe technique and has been used for several decades. Success on a laboratory scale, however, is often not reproducible under the conditions of an *in situ* decontamination (Mohan et al., 2006; Ringelberg et al., 2001). The results are considered unpredictable because they involve the survival of the hydrocarbonoclastic microorganisms introduced (Hamdi et al., 2007; Nikolopoulou et al., 2006; Márquez-Rocha et al., 2005). This indicates that little is still known about the changes in microbial activity suffered by the population introduced into the matrix to be treated (Li et al., 2018), creating a controversy and a certain skepticism regarding bioaugmentation (Azubuike et al., 2016).

As stated earlier, bioaugmentation is based on the survival and maintenance of the catabolic activity of inoculated microorganisms whose degradation pathways for target contaminants accelerate the removal process (Nowak and Mrozik, 2016). If compared with different bioremediation methodologies, for the same time-span, bioaugmentation is more efficient than monitored natural attenuation, while biostimulation is better than bioaugmentation, making the association of these two techniques to be seen as a good alternative (Vasconcelos et al., 2013). It should also be noted that each history of contamination is unique, and it is not enough to indicate the best treatment before a previous investigation (Mancera-López et al., 2008; Bento et al., 2005).

For the removal of recalcitrant hydrocarbons, it is possible to introduce an enriched and adapted indigenous biomass or hydrocarbonoclastic microbes from another location (Ruberto et al., 2009). However, rational selection of dominant species from contaminated sites is recommended to obtain suitable degraders for specific contaminants (Thompson et al., 2005). Adaptation to the contaminant is crucial for increasing the survival of the microorganism in the environment. The homogeneous dispersion of added microbes and competition with indigenous organisms, however, can lead to unsatisfactory results (Adams et al., 2015).

Regarding the use of indigenous microbes, Ueno et al. (2007) proposed the term autochthonous bioaugmentation. The method selects hydrocarbonoclastic organisms from a contaminated site; this may solve the problem of introducing foreign species that need adaptation to the environment (Bosco et al., 2020). The advantages of employing native inhabitants from the contaminated site to the detriment of allochthonous microbes, contribute significantly to the yield of the process because it avoids competitive stress with the indigenous microbiota and promotes metabolic cooperation during the removal of the hydrocarbons (Ali et al., 2016). The selection of native microbes may also be carried out randomly (Vasconcelos et al., 2011). However, many important microbes in a hydrocarbon removal process are not cultivable (Amann et al., 1995) or retrieved by traditional methods of isolation (Breznak, 2002).

On the other hand, allochthonous bioaugmentation is more commonly practiced. In allochthonous bioaugmentation, hydrocarbonoclastic microbes are isolated from environments historically contaminated by oil. These are known to be metabolically active against these compounds (Fodelianakis et al., 2015). Preferably, hydrocarbonoclastic strains that encode key genes in the degradation of hydrocarbons are introduced. The alpha subunit of dioxygenase that promotes ring hydroxylation is usually used as a biomarker to quantify and select populations capable of degrading hydrocarbons (Cebron et al., 2008). In addition, the expression of other genes is decisive in the selection of strains, such as alkane hydroxylases, monooxygenases, and the P450 oxygenase system (Das and Chandran, 2010).

Some cocktails from recognized hydrocarbonoclastic organisms can be purchased and used as an inoculum; however, the addition of commercial cocktails is not a guarantee of efficiency in all cases (Ali et al., 2016). Allochthonous microbes may or may not exhibit their metabolic abilities when transferred from one environment to another. They need to identify and overcome limitations in the removal process to improve the activity of the consortium strains (Oualha et al., 2019).

Table 1.3 summarizes some bioaugmentation strategies in the removal of PAHs and heterocyclic hydrocarbons. Some are innovative and aim to minimize the limiting factors.

The activity of the indigenous microbiota alone or combined with foreign microorganisms may increase the degradation of recalcitrant hydrocarbons, but these are negatively influenced by numerous biotic and abiotic factors that can limit the process, namely: inoculum size, increased lag phase, antagonistic association, physiological stresses, high mortality of added cells, disturbed cell adhesion, inappropriate biological availability, bioavailability, limitation of dispersion, pH variations, oxygen status, barriers in the mass transfer between the pollutant and the microorganism, absence of nutrients, and the risk of introducing pathogenic microbes into the environment (Steliga et al., 2020; Borah and Yadav, 2017; Azubuike et al., 2016; Vasconcelos et al., 2013; Quan et al., 2010; Pepper et al., 2002).

Bacteria are recognized as the most effective microbes for removing complex hydrocarbons in the environment (Abena et al., 2019), and *Achromobacter* sp., *Sphingomonas* sp., *Rhodococcus* sp., *Acinetobacter* sp., *Pseudomonas* sp., *Burkholderia* sp., *Exiguobacterium* sp., and *Bacillus* sp. are the most promising bacteria (Bosco et al., 2020; Muangchinda et al., 2020; Uwadiae and Omoayena, 2017; Castiglione et al., 2016). In addition, Proteobacteria comprise microbes with the greatest potential for degradation of recalcitrant compounds; for this reason, they are preferred in the composition of consortia (Li et al., 2020). These bacteria exhibit extracellular enzymes that attack polymers. The products of enzymatic reactions passively diffuse across the membrane into the cytoplasm where they undergo cleavage of the rings and are used in central metabolism (Morya et al., 2020).

The use of consortia to the detriment of axenic cultures has benefits. The metabolic differences between the isolates allow better results due to metabolic cooperation (Viñas et al., 2002). However, the desired microbial activity depends on the place where these microbes were

TABLE 1.3

Main Bioaugmentation Strategies and Techniques to Improve the Inoculum Survival Aiming the Removal of Heterocyclic and Polycyclic Aromatic Hydrocarbons

Technique	Reference
Autochthonous bioaugmentation	Ueno et al. (2007)
Randomly selected indigenous consortia	Vasconcelos et al. (2011)
Allochthonous consortia	Chen et al. (2019)
Use of commercial cocktails	Ali et al. (2016)
Encompassed inoculation	Innemanová et al. (2018)
Inoculum volume 5% of the size of the matrix to be treated	Kuppusamy et al. (2016)
Correction of the substrate: inoculum ratio	Wong et al. (2014)
Use of Genetically Engineered Microorganisms	Garbisu et al. (2017)
Use of vector with biodegradation-relevant genes	El faltrousi and Agathos (2005)
Inoculum with microbes producing biosurfactants	Roy et al. (2018)
Application of inoculum with dispersants	Al Kharusi et al. (2016)
Use of cell carriers	Tyagi et al. (2011)
Cell immobilization	Zhang et al. (2019)
Nanodispersion	Horváthová et al (2019)
Association with phytoremediation	Kong et al. (2018)
Association with biostimulation	Gomez and Sartaj (2013)

recovered, as well as on the process of adaptation of the isolates and the tolerance of the cells to the pollutant. This promotes synergistic effects, which may lead to complete and rapid degradation (Bhattacharya et al., 2015).

It is mandatory that the size of the inoculum must be high (Papadopoulou et al., 2018). A cell concentration above 10^8 CFU/g exhibits a greater potential for degradation, even when the consortium is formed by allochthonous bacteria (Chen et al., 2019). Part of the cells can proliferate and play a key role in removing contaminants (Feng et al., 2015). In contrast, the added microbiota may lead to the dominance of other groups, which can eliminate the inoculum, without prejudice to the indigenous microbiota (Liu et al., 2015).

New bioaugmentation models can avoid many variations in the structure of bacterial communities by introducing organisms belonging to the same core indigenous rate (Fernandez et al., 2019). The addition of a new consortium, however, always changes the composition of the indigenous microbial community. This may benefit the native microbiota by promoting changes to other catabolic degradation pathways. Thus, bioaugmentation may reinforce the development of a highly competent indigenous dominant microbiota (Yang et al., 2016). On the other hand, a foreign microbial population disturbs the native microbiota, which may reduce the efficiency of bioaugmentation (Fernandez et al., 2019).

The maintenance of the inoculum over the long term is not achieved under laboratory conditions because the stationary phase lasts for a few weeks and then the population begins to decline due to nutrient depletion and metabolite accumulation. This behavior can be even more relevant under field conditions. Due to this unpredictability, bioaugmentation is only indicated for cleaning a recalcitrant hydrocarbons-contaminated site over a relatively short time (Uwadiae and Omoayena, 2017).

It is believed that percentages between 70% and 75% removal of recalcitrant hydrocarbons can be achieved between 30 and 100 days (Cavalcanti et al., 2019). Above this, the limiting factors start to be more significant (Mrozika and Piotrowska-Seget, 2010). Recently, a field study lasting more than 60 weeks, aimed at removing a diesel/biodiesel mixture from the soil, found that the dynamics of the indigenous microbial diversity was not altered by the addition consortium in the long term, suggesting that the inoculum did not survive. In conclusion, the authors reinforce the premise that bioaugmentation may not be indicated for prolonged treatments, since it is effective only in the initial stages (Woźniak-Karczewska et al., 2019).

In contrast, the increase in CO_2 emissions throughout the entire process described above particularly, in its final half, may be indicative of maintaining a viable microbial density. This could be positive in terms of the rate of biodegradation and mineralization of hydrocarbons (Sabaté et al., 2004). In addition, the presence of essential nutrients ensures the maintenance and abundance of classes of microorganisms, inducing dynamics in populations, without changing the community and metabolic interplay that might be involved in the acceleration of bioremediation (Roy et al., 2018). However, the seasonal variation of the community must also be considered (Fernandez et al., 2019).

The efficiency of bioaugmentation may be increased by 40% when encompassed inoculation is applied. The technique uses foreign microbes inoculated into small portions of the matrix to be treated. After the propagation time of the inoculum, the small portions are transferred to larger areas, and percentages of PAHs removal of about 60%–70% can be achieved, with preferential degradation of the high molecular weight compounds (Innemanová et al., 2018).

The preference for higher molecular weight compounds can also be observed when indigenous hydrocarbonoclastic microorganisms are chosen at random; under this condition, the removal percentages also reach around 70% (Brzeszcz et al., 2020; Vasconcelos et al., 2011). In addition, in associations of bioaugmentation with biostimulation, by adding N and P (Lang et al., 2016) or bioaugmentation with phytoremediation, by using rhizobacteria (Mesa et al., 2015), the preferential degradation of compounds with high molecular weight can vary between 30% and 45% (Kong et al., 2018). It is important to note that the microbial preference for high molecular weight hydrocarbons occurs because most low molecular weight compounds suffer photo-oxidation, photolysis, volatilization, and bioaccumulation (Besaltatpour et al., 2011).

The success of bioaugmentation may also depend on a better understanding of donor cells that improve the ecological fitness of the host cells (Bosco et al., 2020). To avoid the reduction of viability and abundance after the addition of the inoculum, some bacteria require biodegradation-relevant genes that can be horizontally transferred by means of vectors, through conjugation with indigenous microorganisms, ecologically important to the process of removing the contaminant (El faltrousi and Agathos, 2005). Thus, the use of genetically engineered microorganisms (GEM) may be a suitable strategy; the risks of using GEM can be minimized with the destruction of the organism after the treatment (Mustafa et al., 2015). The success of bioaugmentation using GEM is dependent on factors such as phylogenetic distance, ratio of donor and recipient cells, efficient transfer of plasmids, and expression of the transferred genes. In addition, the morphological, physiological, and biochemical characteristics of cells are also relevant (Garbisu et al., 2017).

In order to guarantee an inoculum's survival for a longer time, several strategies may be tried. Dispersers increase the desorption of aggregated hydrocarbons and improve the diffusion in the aqueous phase of certain matrices, such as in the soil and improve bioavailability and bioaccess (Yang et al., 2001). Some dispersers can be applied directly to the site (Al Kharusi et al., 2016), while others are introduced via nanosystems (Horváthová et al., 2019). In order to lower costs, the bioaugmentation may be carried out with an inoculum of bacteria that produces biosurfactants. *Pseudomonas aeruginosa* proved to be a promising alternative. In a recent study, the addition of the bacteria increased the bioavailability of the contaminants, by promoting changes in the hydrocarbonoclastic bacterial community, without the need for aeration under anoxic conditions (Zhao et al., 2018a).

Saponins, natural biodegradable compounds, can also be used as dispersants. Saponins reduce the hydrophobicity of contaminants, improving the bioavailability for bacteria (Zhao et al., 2018b). This phenomenon is explained by Derjaguin–Landau and Verwey–Overbeek theory, which involves the estimation of the magnitude and variation of the inter-particular distance of a London-Van der Waals attraction energy between two surfaces and the repulsive energy resulting from the overlapping of ionic atmospheres between them (Oliveira, 1997).

More recently, the use of cell immobilizers is seen as a safe alternative that has been proposed to enhance the survival of the inoculum in the soil. Cell immobilizers can resist harsh soil conditions and can be prepared with low-cost materials, such as agro-industrial waste (Xue et al., 2017) or polymers with cell-carrying properties, such as agar, alginate, gelatin, and gums (Tyagi et al., 2011). Cell immobilizers absorb pollutants more, contributing to greater degradation, based on two mechanisms: improving the interaction and adsorption of bacteria (Ullah et al., 2017) and micropore volume filling (Brányik et al., 2006).

The immobilization of bacterial consortia in soil produces a humid and aerated microenvironment favorable to the development of microorganisms, allowing acclimatization *in situ*, as well as increasing cell density and stability. In addition, it contributes to the synergism between allochthonous and indigenous microbes, making the latter the dominant degrading organism. Cell immobilizers also increase soil porosity and permeability, favoring microbial access to hydrocarbons and contributing to increased enzyme activity (Zhang et al., 2019).

1.4 CONCLUSION

Bioaugmentation is a cost-effective alternative to soil remediation when used to remove recalcitrant molecules, such as PAHs and heterocyclic hydrocarbons. It requires prior evaluation, however, before the interventions are carried out, in order to know the degree of contamination, as well as the viability of the indigenous microbiota. In these cases, bioaugmentation is more efficient in the short term, although some strategies are being developed to remove the contaminants. Simple associations with biostimulation or phytoremediation can be tested, as well as recently developed and very promising approaches involving cellular immobilizers, carriers, and GEM inoculation. The maximum biological removal of these compounds, however, only accounts for about 70% efficiency, which invites further investigation.

REFERENCES

Abatenh, E.; Gizaw, B.; Tsegaye, Z.; Wassie, M. 2017. The role of microorganisms in bioremediation – a review. *Open J Environ Biol* 2:38–46.

Abdulsalam, S.; Omale, A.B. 2009. Comparison of biostimulation and bioaugmentation techniques for the remediation of used motor oil contaminated soil. *Braz Arch Bio Technol* 52: 747–754.

Abena, M.T.B.; Li, T.; Shah, M.N.; Zhong, W. 2019. Biodegradation of total petroleum hydrocarbons (TPH) in highly contaminated soils by natural attenuation and bioaugmentation. *Chemosphere* 234: 864–874.

Adams, G.O.; Fufeyin, P.T.; Okoro, S.E.; Ehinomen I. 2015. Bioremediation, biostimulation and bioaugmentation: a review. *Int J Environ Bioremed Biodegrad* 3:28–39.

Al Kharusi, S.; Abed, R.M.M.; Dobretsov, S. 2016. EDTA addition enhances bacterial respiration activities and hydrocarbon degradation in bioaugmented and non-bioaugmented oil-contaminated desert soils. *Chemosphere* 147: 279–286.

Aldén, L.; Demoling, F.; Bääth, E. 2001. Rapid method of determining factors limiting bacterial growth in soil. *Appl Environ Microbiol* 67: 1830–1838.

Ali, N.; Dashti, N.; Salamah, S.; Al-Awadhi, H.; Sorkhoh, N.; Radwan, S. 2016. Autochthonous bioaugmentation with environmental samples rich in hydrocarbonoclastic bacteria for bench-scale bioremediation of oily seawater and desert soil. *Environ Pollut Sci Res* 23: 8686–8698.

Aljerf, L.; Al Masri, N. 2018. Persistence and bioaccumulation of persistent organic pollutants (POPs) in the soil and aquatic ecosystems: Syrian frontiers in ecology and environment. *Sustain Environ* 3: 358–386.

Al-Mailem, D.M.; Kansour, M.K.; Radwan, S.S. 2019. Cross-bioaugmentation among four remote soil samples contaminated with oil exerted just inconsistent effects on oil-bioremediation. *Front Microbiol* 10: 2827. doi: 10.3389/fmicb.2019.02827.

Amann, R.I.; Ludwig, W.; Schleifer, K.H. 1995. Phylogenetic identification and *in situ* detection of individual microbial cells without cultivation. *Microbiol Rev* 59: 143–169.

Amellal, N.; Portal, J-M.; Berthelin, J. 2001. Effect of soil structure on the bioavailability of polycyclic aromatic hydrocarbons within aggregates of a contaminated soil. *Appl Geochem* 16: 1611–1619.

Arp, D.J.; Yeager, C.M.; Hyman, M.R. 2001. Molecular and cellular fundamentals of aerobic cometabolism of trichloroethylene. *Biodegradation* 12: 81–103.

Arruda, R.R.A.; Bonifácio, T.T.C.; Oliveira, B.T.M.; Silva, J.E.G.; Vasconcelos, U. 2020. Assessment of indole and pyocyanin in the relationship of *Pseudomonas aeruginosa* to *Escherichia coli*. *Int J Develop Res* 10: 34122–34128.

Asif, M.; Wenger, L.M. 2019. Heterocyclic aromatic hydrocarbon distributions in petroleum: A source facies assessment tool. *Org Geochem* 137: doi: 10.1016/j.orggeochem.2019.07.005.

Atlas, R.M. 1981. Microbial degradation of petroleum hydrocarbons: an environment perspective. *Microbiol Rev* 45: 180–209.

Ayotamuno, M.J.; Kogbara, K.B.; Ogaji, S.O.T.; Probert, S.P. 2006. Bioremediation of a crude oil polluted agricultural-soil at Port Harcourt, Nigeria. *Appl Energy* 83: 1249–1247.

Azubuike, C.C.; Chikere, C.B.; Okpokwasili, G.C. 2016. Bioremediation techniques–classification based on site of application: principles, advantages, limitations and prospects. *World J Microbiol Biotechnol* 32: 180. doi: 10.1007/s11274-016-2137-x.

Baboshin, M.; Finklstein, Z.I.; Golovleva, L.A. 2003. Fluorene cometabolism by *Rhodococcus rhodochrous* and *Pseudomonas fluorescens* cultures. *Microbiology* 72: 194–198.

Barrios, E. 2007. Soil biota ecosystems services and land productivity. *Ecol Econ* 64: 283–296.

Bhattacharya, M.; Guchhait, S.; Biswas, D.; Datta, S. 2015. Waste lubricating oil removal in a batch reactor by mixed bacterial consortium: a kinetic study. *Bioprocess Biosyst Eng* 38: 2095–2106.

Bento, F.M.; Camargo, F.A.O.; Okeke, B.C.; Frankenberger, W.T. 2005. Comparative of bioremediation of soils contaminated with diesel oil by natural attenuation, biostimulation and bioaugmentation. *Bioresour Technol* 96: 1049–1055.

Besaltatpour, A.; Hajabbasi, M.; Khoshgoftarmanesh, A.; Dorostkar, V. 2011. Landfarming process effects on biochemical properties of petroleum-contaminated soils. *Soil Sediment Contam Int J* 20:234–248. doi: 10.1080/15320383.2011.546447.

Bojes, H.K.; Pope, P.G. 2007. Characterization of EPA's 16 priority pollutant polycyclic aromatic hydrocarbons (PAHs) in tank bottom solids and associated contaminated soils at an exploration and production sites in Texas. *Regul Toxicol Pharmacol* 47: 288–295.

Boonchan, S.; Britz, M.L.; Stanley, G.A. 2000. Degradation and mineralization of high-molecular-weight polycyclic aromatic hydrocarbons by defined fungal-bacterial cocultures. *Appl Environ Microbiol* 66: 1007–1019.

Boopathy, R. 2000. Factors limiting bioremediation technologies. *Bioresour Technol* 74: 63–67.

Borah, D.; Yadav, R.N.S. 2017. Bioremediation of petroleum based contaminants with biosurfactant produced by a newly isolated petroleum oil degrading bacterial strain. *Egypt J Pet* 26: 181–188.

Bosco, F.; Casale, A.; Mazzarino, I.; Godio, A.; Ruffino, B.; Mollea, C.; Chiampo, F. 2020. Microcosm evaluation of bioaugmentation and biostimulation efficacy on diesel-contaminated soil. *J Chem Technol Biotechnol* 95: 904–912.

Bossert, I.; Kachel, W.M.; Bartha, R. 1984. Fate of hydrocarbons during oily sludge disposal in soil. *Appl Environ Microbiol* 47: 763–767.

Bouchez, T.; Patureau, D.; Dabert, P.; Juretschko, S.; Doré, J.; Delgenès, P.; Moletta, R.; Wagner, M. 2000. Ecological study of bioaugmentation failure. *Environ Microbiol* 2: 179–190.

Brányik, T.; Silva, D.P.; Vicente, A.A.; Lehnert, R.; Silva, J.B.A.; Dostálek, P.; Teixeira, J.A. 2006. Continuous immobilized yeast reactor system for complete beer fermentation using spent grains and corncobs as carrier materials. *J Ind Microbiol Biotechnol* 33: 1010–1018.

Breznak, J.A. 2002. A need to retrieve the not-yet-cultured majority. *Environ Microbiol* 4: 4–5.

Brown, D.M.; Bonte, M.; Gill, R.; Dawick, J.; Boogaard, P.J. 2017. Heavy hydrocarbon fate and transport in the environment. *Q J Eng Geol* 50: 333–346.

Brzeszcz, J.; Kapusta, P.; Steliga, T.; Turkiewicz. 2020. Hydrocarbon removal by two differently developed microbial inoculants and comparing their actions with biostimulation treatment. *Molecules* 25: 661. doi:10.3390/molecules25030661.

Bushnell, L.D.; Haas, H.F. 1941. The utilization of certain hydrocarbons by microorganisms. *J Bacteriol* 41: 653–673.

Cai, Q-Y.; Mo, C-H.; Wu, Q-T.; Zeng, Q-Y.; Katsoyiannis, A. 2007. Bioremediation of polycyclic aromatic hydrocarbons contaminated sewage sludge by different composing process. *J Harzard Mater* 142: 535–542.

Castiglione, M.R.; Giorgetti, L.; Becarelli, S.; Siracusa, G.; Lorenzi, R.; Di Gregorio, S. 2016. Polycyclic aromatic hydrocarbon-contaminated soils: bioaugmentation of autochthonous bactéria and toxicological assessment of the bioremediation process by means of *Vicia faba* L. *Environ Sci Pollut Res* 23: 7930–7941.

Cavalcanti, T.G., Souza, A.F., Ferreira, G.F., Dias, D.S.B., Severino, L.S., Morais, J.P.S., Sousa, K.A., Vasconcelos, U. 2019. Use of agro-industrial waste in the removal of phenanthrene and pyrene by microbial consortia in soil. *Waste Biomass Valor* 10: 205–214.

Cavalcanti, T.G.; Viana, A.A.G.; Guedes, T.P.G.; Freire, A.S.; Travassos, R.A.; Vasconcelos, U. 2016. Seed options for toxicity tests in soil contaminated with oil. *Can J Pure Appl Sci* 10: 4039–4045.

Cebron, A.; Norini, M.P.; Beguiristain, T.; Leyval, C. 2008. Real-Time PCR quantification of PAH-ring hydroxylating dioxygenase (PAH-RHDalpha) genes from Gram positive and Gram negative bacteria in soil and sediment samples. *J Microbiol Methods* 73: 148–159.

Čejková, A.; Masák, J.; Jirků, V. 1997. Use of mineral nutrients and surface-active substances in a biodegradation process modulation. *Folia Microbiol* 42: 513–516.

Chaîneau, C.H.; Rougeux, G.; Yéprémian, C.; Oudot, J. 2005. Effect of nutrient concentration on the biodegradation of crude oil and associated microbial populations in soil. *Soil Biol Biochem* 37: 1490–1497.

Chen, Y-A.; Liu, P-W.G.; Whang, L-M.; Wu, Y-J.; Cheng, S-S. 2019. Biodegradability and microbial community investigation for soil contaminated with diesel blending with biodiesel. *Process Saf Environ Protect* 130: 115–125.

Das, N.; Chandran, P. 2010. Microbial degradation of petroleum hydrocarbon contaminants: an overview. *Biotechnol Res Int* 2011: 1–13.

Dashti, N., Ali, N., Eliyas, M., Khanafer, M., Sorkhoh, N.A., Radwan, S.S. Most hydrocarbonoclastic bacteria in the total environment are diazotrophic, which highlights their value in the bioremediation of hydrocarbon contaminants. 2015. *Microbes Environ* 30: 70–75.

Daugulis, A.J.; McCracken, C. M. 2003. Microbial degradation of high and low molecular weight polyaromatic hydrocarbons in a two-phase partitioning bioreactor by two strains of *Sphingomonas* sp. *Biotechnol Lett* 25: 1441–1444.

De Boer, W.; Folman, L.B.; Summerbell, R.C.; Boddy, L. 2005. Living in a fungal world: impact of fungi on soil bacterial niche development. *FEMS Microbiol Rev* 29: 795–811.

Dejonghe, N.; Boon, N.; Seghers, D.; Top, E.M.; Verstraete, W. 2001. Bioaugmentation of soils by increasing microbial richness: missing links. *Environ Microbiol* 3: 649–657.

Dibble, J.T.; Bartha, R. 1979. Effect of environmental parameters on the biodegradation of oil sludge. *Appl Environ Microbiol* 37: 729–739.

Dua, M.; Singh, A.; Sethunathan, N.; Johri, A.K. 2002. Biotechnology and bioremediation: success and limitations. *Appl Microbiol Biotechnolol* 59: 143–152.

El Faltrousi, S.; Agathos, S.N. 2005. Is bioaugmentation a feasible strategy for pollutant removal and site remediation? *Curr Opin Microbiol* 8: 268–275.

Enell, A.; Reichenberg, F.; Ewald, G.; Warfvinge, P. 2005. Desorption kinetics studies on PAH-contaminated soil under varying temperatures. *Chemosphere* 61: 1529–1538.

Eom, I.C.; Rast, C.; Veber, A.M.; Vasseur, P. 2007. Ecotoxicity of polycyclic aromatic hydrocarbons (PAH)-contaminated soil. *Ecotoxicol Environ Saf* 67: 190–205.

Ettema, C.H.; Wardle, D.A. 2002. Spatial soil ecology. *Trends Ecol Evol* 17: 177–183.

Feng, Y.Z.; Chen, R.R.; Hu, J.L.; Zhao, F.; Wang, J.H.; Chu, H.Y.; Zhang, J.; Dolfing, J.; Lin, X. 2015. *Bacillus asahii* comes to the fore in organic manure fertilized alkaline soils. *Soil Biol Biochem* 81: 186–194.

Fernandez, M.; Pereira, P.P.; Agostini, E.; González, P.S. 2019. How the bacterial community of a tannery effluent responds to bioaugmentation with the consortium SFC 500–1. Impact of environmental variables. *J Environ Manage* 247: 46–56.

Festa, S.; Coppotelli, B.M.; Madueño, L.; Loviso, C.L.; Macchi, M.; Tauil, R.M.N.; Valacco, M.P.; Morelli, I.S. 2017. Assigning ecological roles to the populations belonging to a phenanthrene-degrading bacterial consortium using omic approaches. *PLoS One* 12: e0184505. doi: 10.1371/journal.pone.0184505.

Fodelianakis, S.; Antoniou, E.; Mapelli, F.; Magagnini, M.; Nikolopoulou, M.; Marasco, R.; Barbato M.; Tsiola, A.; Tsikopoulou, I.; Giaccaglia, L.; Mahjoubi, M.; Jaouani, A.; Amer, R.; Hussein, E.; Al-Horani, F.A.; Benzha, F.; Blaghen, M.; Malkawi, H.I.; Abdel-Fattah, Y.; Cherif, A.; Daffonchio, D.; Kalogerakis, N. 2015. Allochthonous bioaugmentation in *ex situ* treatment of crude oil-polluted sediments in the presence of an effective degrading indigenous microbiome. *J Hazard Mater* 287: 78–86.

Garbisu, C.; Garaiyurrebaso, O.; Epelde, L.; Grohmann, E.; Alkorta, I. 2017. Plasmid-Mediated Bioaugmentation for the bioremediation of contaminated soils. *Front Microbiol* 8: 1966. doi: 10.3389/fmicb.2017.01966.

Gomez, F.; Sartaj, M. 2013. Field scale *ex situ* bioremediation of petroleum contaminated soil under cold climate conditions. *Int Biodeter Biodegrad* 85: 375–382.

Gong, Z.; Alef, K.; Wilke, B-M.; Li. P. 2007. Activated carbon adsorption of PAHs from vegetable oil used in soil remediation. *J Hazard Mater* 143: 372–378.

Guo, Y.; Wu, K.; Huo, X.; Xu, X. 2011. Sources, distribution, and toxicity of polycyclic aromatic hydrocarbons. *J Env Health* 73: 22–25.

Haghollahi, A.; Fazaelipoor, M.H.; Schaffie, M. 2016. The effect of soil type on the bioremediation of petroleum contaminated soils. *J Environ Manage* 15: 197–201.

Hamdi, H.; Benzarti, S.; Manusadžianas, L.; Aoyama, I.; Jedidi, N. 2007. Bioaugmentation and biostimulation effects on PAH dissipation and soil ecotoxicity under controlled conditions. *Soil Biol Biochem* 39: 1926–1935.

Hammershøj, R.; Birch, H.; Redman, A.D.; Mayer, P. 2019. Mixture effects on biodegradation kinetics of hydrocarbons in surface water: increasing concentrations inhibited degradation whereas multiple substrates did not. *Environ Sci Technol* 53: 3087–3094.

Herwijnen, R.V.; Van de Sande, B.F.; Van Der Wielen, F.W.M.; Springael, D.; Govers, H.A.J.; Parsons, J.R. 2003. Influence of phenanthrene and flouranthene on the degradation of fluorene and glucose by *Sphingomonas* sp strain LB126 in chemostat cultures. *FEMS Microbiol Ecol* 46: 105–111.

Horváthová, H.; Lászlová, K.; Dercová, K. 2019. Bioremediation vs. nanoremediation: degradation of polychlorinated biphenyls (PCBS) using integrated remediation approaches. *Water Air Soil Pollut* 230: 204. doi: 10.1007/s11270-019-4259-x.

Hua, F., Wang, H.Q. 2004. Uptake and trans-membrane transport of petroleum hydrocarbons by microorganisms. *Biotechnol Biorechnol Equip* 28: 165–175.

Huesemann, M.H.; Hausmann, T.S.; Fortman, T.J. 2004. Does bioavailability limit biodegradation? A comparison of hydrocarbon biodegradation and desorption rates in aged soils. *Biodegradation* 15: 261–274.

Innemanová, P.; Filipová, A.; Michalíková, K.; Wimmerová, L.; Cajthaml, T. 2018. Bioaugmentation of PAH-contaminated soils: A novel procedure for introduction of bacterial degraders into contaminated soil. *Ecol Eng* 118: 93–96.

International Agency for Research on Cancer. 2021. IARC monographs on the identification of carcinogenic hazards to humans. Available at: https://monographs.iarc.who.int/list-of-classifications.

Jacques, R.J.S.; Bento, F.M.; Antoniolli, Z.I.; Camargo, F.A.O. 2007. Biorremediação de solos contaminados com hidrocarbonetos aromáticos policíclicos. *Cienc Rural* 37: 1192–1201.

Jia, J-L., Li, G-H., Zhong, Y. 2004. The relationship between abiotic factors and microbial activities of microbial eco-system in contaminated soil with petroleum hydrocarbons. *Huang Jing Ke Xue* 25: 110–114.

Jiao, S.; Luo, Y.; Lu, M.; Xiao, X.; Lin, Y.; Chen, W.; Wei, G. 2017. Distinct succession patterns of abundant and rare bacteria in temporal microcosms with pollutants. *Environ Pollut* 225: 497–505.

Johnsen, A.R.; Wick, L.Y.; Harms, H. 2005. Principals of microbial PAH-degradation in soil. *Environ Pollut* 133: 71–84.

Joner, E.J.; Hirmann, D.; Szolar, O.H.J.; Todorovic, D.; Leyval, C.; Loibner, A.P. 2004. Priming effects on PAH degradation and ecotoxicity during a phytoremediation experiment. *Environ Pollut* 128: 429–435.

Jung, H.; Ahn, Y.; Choi, H.; Kim, I.S. 2005. Effects of *in situ* ozonation on indigenous microorganisms in diesel contaminated soil: survival and regrowth. *Chemosphere* 61: 923–932.

Kaplan, C.W.; Kitts, C.L. 2004. Bacterial succession in a petroleum land treatment unit. *Appl Environ Microbiol* 70: 1777–1786.

Karamalidis; A.K.; Evangelou, A.C.; Karabika, E.; Koukkou, A.I.; Drainas, C.; Voudrias, E.A. 2010. Laboratory scale bioremediation of petroleum-contaminated soil by indigenous microorganisms and added *Pseudomonas aeruginosa* strain Spet. *Bioresour Technol* 11: 6545–6552.

Kong, F-X.; Sun, G-D.; Liu, Z-P. 2018. Degradation of polycyclic aromatic hydrocarbons in soil mesocosms by microbial/plant bioaugmentation: Performance and mechanism. *Chemosphere* 198: 83–91.

Kottler, B.D.; White, J.C.; Kelsey, J.W. 2001. Influence of soil mixture on the sequestration of organic compounds in soil. *Chemosphere* 42: 893–898.

Kuppusamy, S.; Thavamani, P.; Megharaj, M.; Naidu, R. 2016. Bioaugmentation with novel microbial formula vs. natural attenuation of a long-term mixed contaminated soil-treatability studies in solid- and slurry-phase microcosms. *Water Air Soil Pollut* 227: 25. doi: 10.1007/s11270-015-2709-7.

Lang, F.S.; Destain, J.; Delvigne, F.; Druart, P.; Ongena, M.; Thonart, P. 2016. Biodegradation of Polycyclic Aromatic Hydrocarbons in mangrove sediments under different strategies: Natural attenuation, biostimulation, and bioaugmentation with *Rhodococcus erythropolis* T902.1. *Water Air Soil Pollut* 227: 297. doi: 10.1007/s11270-016-2999-4.

Leahy, J.G.; Colwell, R.R. 1990. Microbial degradation of hydrocarbons in the environment. *Microbial Rev* 54: 305–315.

Li, J.; Luo, C.; Zhang, D.; Song, M.; Cai, X.; Jiang, L.; Zhang, G. 2018. Autochthonous bioaugmentation-modified bacterial diversity of phenanthrene degraders in PAH-contaminated wastewater as revealed by DNA-stable isotope probing. *Environ Sci Technol* 52: 2934–2944.

Li, J.; Wu, C.; Chen, S.; Lu, Q.; Shim, H.; Huang, X.; Jia, C.; Wang, S. 2020. Enriching indigenous microbial consortia as a promising strategy for xenobiotics' cleanup. *J Cleaner Product* 261: 121234. doi: 10.1016/j.jclepro.2020.121234.

Limongi, R.; Oliveira, B.T.M.; Gervazio, K.Y.; Morais, V.C.; Barbosa, P.S.Z.; Cavalcanti, T.G.; Vasconcelos, U.; Amaral, I.P.G. 2020. Biodegradation of pyrene and anthracene by *Pseudomonas aeruginosa* TGC-02 in submerged culture. *Int J Eng Res Appl* 10: 12–20.

Liu, B-R.; Jia, G-M.; Chen, J.; Wang, G.A 2006. Review of methods for studying microbial diversity in soils. *Pedosphere* 16: 18–24.

Liu, X.; Chen, Y.; Zhang, X.; Jiang, X.; Wu, S.; Shen, J.; Sun, X.; Li, J.; Lu, L.; Wang, L.; 2015. Aerobic granulation strategy for bioaugmentation of a sequencing batch reactor (SBR) treating high strength pyridine wastewater. *J. Hazard Mater* 295: 153–160.

Lladó, S.; Covino, S.; Solanas, A.M.; Viñas, M.; Petrucciolli, M.; D'Aniballe, A. 2013. Comparative assessment of bioremediation approaches to highly recalcitrant PAH degradation in a real industrial polluted soil. *J Hazard Mater* 248–249: 407–414.

Löser, C.; Seidel, H.; Zehnsdorf, A.; Stottmeister, U. 1998. Microbial degradation of hydrocarbons in soil during aerobic/anaerobic changes under purely aerobic conditions. *Appl Microbiol Biotechnol* 49: 631–636.

MacNaughton, S.J.; Stephen, J.R.; Venosa, A.D.; Davis, G.A.; Chang, Y-J.; White, D. C. 1999. Microbial population changes during bioremediation of an experimental oil spill. *Appl Environ Microbiol* 65: 3566–3574.

Mancera-López, M.G.; Esparza-García, F.; Chávez-Gomez, B.; Rodriguéz-Vázquez, R.; Saucedo-Castañeda, G.; Barrera-Cortés, J. 2008. Bioremediation of an aged hydrocarbon-contaminated soil by a combined system of biostimulation-bioaugmentation with filamentous fungi. *Int Biotederior Biodegradation* 61: 151–160.

Manse, G., Werner, D., Meynet, P., Ogbaga, C.C. 2020. Microbial community responses to different volatile petroleum hydrocarbon class mixtures in an aerobic sandy soil. *Environ Pollut* 264: 114738. doi: 10.1016/j.envpol.2020.114738.

Margesin, R.; Hämmerle, M; Tscherko, D. 2007. Microbial activity and community composition during bioremediation of diesel-oil contaminated soil: effects of hydrocarbons concentration, fertilizers and incubation time. *Microb Ecol* 53: 259–269.

Mariano, A.P.; de Angelis, D.F.; Pirôllo, M.P.S.; Contiero, J.; Bonotto, D.M. 2009. Investigation about the efficiency of the bioaugmentation technique when applied to diesel oil contaminated soils. *Braz Arch Biol Technol* 52: 1297–1312.

Marín, J.A.; Hernandez, T.; Garcia, C. 2005. Bioremediation of oil refinery sludge by landfarming in semiarid conditions: influence on soil microbial activity. *Environ Res* 98: 185–195.

Marschner, B.; Kalbitz, K. 2003. Controls of bioavailability and biodegradability of dissolved organic matter in soils. *Geoderma* 113: 211–235.

Martínez-Toledo, A.; Rodríguez-Vasquez, R. 2013. In situ biosurfactant production and hydrocarbon removal by *Pseudomonas putida* CB-100 in bioaugmented and biostimulated oil-contaminated soil. *Braz J Microbiol* 44: 595–605.

Mater, L.; Sperb, R.M.; Madureira, L.A.S.; Rosin, A.P.; Correa, A.X.R.; Radetski, C. M. 2006. Proposal of a sequential treatment methodology for the safe reuse of oil-sludge contaminated soil. *J Hazard Mater* 136: 967–971.

Márquez-Rocha, F.J.; Olmos-Soto, J.; Rosano-Hernández, M.C.; Muriel-García, M. 2005. Determination of hydrocarbon-degrading metabolic capabilities of tropical bacterial isolates. *Int Biodeterior Biodegrad* 55: 17–23.

McClure, N.C.; Fry, J.C.; Weightman, A.J. 1991. Survival and catabolic activity of natural and genetically engineered bacteria in a laboratory-scale activated-sludge unit. *Appl Environ Microbiol* 57: 366–373.

Mehrzad, F.; Fataei, E.; Rad, S.N.; Imani, A.A. 2015. The investigation of nutrient addition impact on bioremediation capability of gasoil by *Alcaligenes faecalis*. *J Pure Appl Microbiol* 9: 2185–2191.

Mesa, J.; Rodríguez-Llorente, J.D.; Pajuelo, E.; Piedras, J.M.B.; Caviedes, M.A.; Redondo-Gómez, S.; Mateos-Naranjo, E. 2015. Moving closer towards restoration of contaminated. *J Hazard Mater* 300: 263–271.

Meysami, P.; Baheri, H. 2003. Pre-screening of fungi and bulking agents for contaminated soil bioremediation. *Adv Environ Res* 7: 881–887.

Mille-Lindblom, C.; Fischer, H.; Tranvik, L.T. 2006. Antagonism between bacteria and fungi: substrate competition and a possible tradeoff between fungal growth and tolerance towards bacteria. *Oikos* 113: 233–242.

Mohan, S.V.; Kisa, T.; Ohkuma, T.; Kanaly, R.A.; Shimizu, Y. 2006. Bioremediation technologies for treatment of PAH-contaminated soil and strategies to enhance process efficiency. *Rev Environ Sci Biotechnol* 5: 347–374.

Morya, R.; Salvachúa, D.; Thakur, I.S. 2020. *Burkholderia*: an untapped but promising bacterial genus for the conversion of aromatic compounds. *Trends Biotechnol* 9: 963–975.

Mrozika, A.; Piotrowska-Seget, Z. 2010. Bioaugmentation as a strategy for cleaning up of soils contaminated with aromatic compounds. *Microbiol Res* 165: 363–375.

Muangchinda, C.; Srisuwankarn, P.; Boubpha, S.; Chavanich, S.; Pinyakong, O. 2020. The effect of bioaugmentation with *Exiguobacterium* sp. AO-11 on crude oil removal and the bacterial community in sediment microcosms, and the development of a liquid ready-to-use inoculum. *Chemosphere* 250: 126303. doi: 10.1016/j.chemosphere.2020.126303.

Mulligan, C.N.; Yong, R.N. 2004. Natural attenuation of contaminated soils. *Environ Int* 30: 587–601.

Mustafa, Y.A.; Abdul-Hameed, H.M.; Razak, Z.A. 2015. Biodegradation of 2,4-dichlorophenoxyacetic acid contaminated soil in a roller slurry bioreactor. *Clean-Soil Air Water* 43: 1115–1266.

Nascimento, T.C.F.; Oliveira, F.J.S.; França, F.P. 2013. Biorremediación de um suelo tropical contaminado com resíduos aceitosos intemperizados. *Rev Int Contam Ambie* 29: 21–28.

Nie, M.; He, M.; Lin, Y.; Wang, L.; Jin, P.; Zhang, S.Y. 2016. Immobilization of biofilms of *Pseudomonas aeruginosa* NY3 and their application in the removal of hydrocarbons from highly concentrated oil-containing wastewater on the laboratory scale. *J Environ Manage* 173: 34–40.

Nikolopoulou, M.; Pasadakis, N.; Kalogerakis, N. 2006. Enhanced bioremediation of crude oil utilizing lipophylic fertilizers. *Desalination* 211: 286–295.

Nowak, A.; Mrozik, A. 2016. Facilitation of co-metabolic transformation and degradation of monochlorophenols by *Pseudomonas* sp. CF600 and changes in its fatty acid composition. *Water Air Soil Pollut* 227: 83. doi: 10.1007/s11270-016-2775-5.

Oliveira, R. 1997. Understanding adhesion: A means for preventing fouling. *Exp Thermal Fluid Sci* 14: 316–322.

Orgiazzi, A.; Dunbar, M.B.; Panagos, P.; de Groot, G.A.; Lemanceau, P. 2015. Soil biodiversity and DNA barcodes: opportunities and challenges. *Soil Biolol Biochem* 80: 244–250.

Oualha, M.; Al-Kaabi, N.; Al-Ghouti, M.; Zouari, N. Identification and overcome of limitations of weathered oil hydrocarbons bioremediation by an adapted Bacillus sorensis strain. *J Environ Manage* 250: 109455. doi: 10.1016/j.jenvman.2019.109455.

Øvreås, L. 2000. Population and community level approaches for analyzing microbial diversity in natural environments. *Ecol Lett* 3: 236–251.

Papadopoulou, E.S.; Genitsaris, S.; Omirou, M.; Perruchon, C.; Stamatopoulou, A.; Ioannides, I.; Karpouzas, D.G. 2018. Bioaugmentation of thiabendazole-contaminated soils from a wastewater disposal site: factors driving the efficacy of this strategy and the diversity of the indigenous soil bacterial community. *Environ Pollut* 233: 16–25.

Pepper, I.L.; Gentry, T.J.; Newby, D.T.; Roane, T.M.; Josephson, K.L. 2002. The role of cell bioaugmentation and gene bioaugmentation in the remediation of co-contaminated soils. *Environ Health Perspect* 110: 943–946.

Perfumo, A.; Banat, I.M.; Marchant, R.; Vezzulli, L. 2007. Thermally enhanced approaches for bioremediation of hydrocarbon-contaminated soils. *Chemosphere* 66: 179–184.

Płaza, G.; Nałęcz-Jawecki, G.; Ulfig, K.; Brigmon, R.L. 2005. Assessment of genotoxic activity of petroleum hydrocarbon-bioremediated soil. *Ecotoxicol Environ Saf* 62: 415–420.

Poater, J.; Duran, M.; Solà, M. 2018. Aromaticity determines the relative stability of kinked vs. straight topologies in Polycyclic Aromatic Hydrocarbons. *Front Chem* 6: 561. doi: 10.3389/fchem.2018.00561.

Pope, C.J.; Peters, W.A.; Howard, J. B. 2000. Thermodynamic driving forces for PAH isomerization and growth during thermal treatment of polluted soils. *J Hazard Mater* 79: 189–208.

Popp, N.; Schlömann, M.; Mau, M. 2006. Bacterial diversity in active stage of a bioremediation system for mineral oil hydrocarbon-contaminated soils. *Microbiology* 152: 3291–3304.

Potin, O.; Veignie, E.; Rafin, C. 2004. Biodegradation of polycyclic aromatic hydrocarbons (PAHs) by *Cladosporium sphaerospermum* isolated from an aged PAH contaminated soil. *FEMS Microbiol Ecol* 51: 71–78.

Quan, X.; Tang, H.; Xiong, W.; Yang, Z. 2010. Bioaugmentation of aerobic sludge granules with a plasmid donor strain for enhanced degradation of 2,4-dichlorophenoxyacetic acid. *J Hazard Mater* 179: 1136–1142.

Radwan, S.S.; Al-Mailem, D.M.; Kansour, M.K. 2019. Bioaugmentation failed to enhance oil bioremediation in three soil samples from three different continents. *Sci Rep* 9: 19508. doi: 10.1038/s41598-019-56099-2.

Rahman, K.S.M.; Thahira-Rahman, J.; Lakshmanaperumalsamy, P.; Banat, I.M. 2002. Towards efficient crude oil degradation by a mixed bacterial consortium. *Bioresour Technol* 85: 257–261.

Ravindra, K.; Sokhi, R.; Van Grieken, R. 2008. Atmospheric polycyclic aromatic hydrocarbons: source attribution, emission factors and regulation. *Atmos Environ* 42: 2895–2921.

Ringelberg, D.B.; Talley, J.W.; Perkins, G.J.; Tucker, S.G.; Luthy, R.G.; Bouwer, E.J.; Fredrickson, H.L. 2001. Succession of phenotypic, genotypic and metabolic community characteristics during *in vitro* bioslurry treatment of polycyclic aromatic hydrocarbon-contaminated sediments. *Appl Environ Microbiol* 67: 1542–1550.

Rivas, R.F. 2006. Polycyclic aromatic hydrocarbons sorbed on soils: a short review of chemical oxidation based treatments. *J Hazard Mater* 138: 234–251.

Röling, W.F.M.; Milner, M.G.; Jones, D.M.; Lee, K.; Daniel, F.; Swannell, R.J.P.; Head, I.M. 2002. Robust hydrocarbon degradation and dynamics of bacterial communities during nutrient-enhanced oil spill bioremediation. *Appl Environ Microbiol* 68: 5537–5548.

Romantchuk, M.; Sarand, I.; Petänen, T.; Peltola, R.; Jonsson-Vihanne, M.; Koivula, T; Yrjälä, K.; Haahtela, K. 2000. Means to improve the effect of *in situ* bioremediation of contaminated soil: an overview of novel approaches. *Environ Pollut* 107: 179–185.

Rowland, A.P.; Lindley, D.K.; Hall, E.H.; Rossall, M.J.; Wilson, D.R.; Benham, D.G.; Harrison, A.F.; Daniels, R.E. 2000. Effects of beach sand properties, temperature, rainfall on the degradation rates of oil in buried oil/beach sand mixtures. *Environ Pollut* 109: 109–118.

Roy, A.; Dutta, A.; Pal, S.; Gupta, A.; Sarkar, J.; Chatterjee, A.; Saha, A.; Sarkar, P.; Sar, P.; Kazy, S.K. 2018. Biostimulation and bioaugmentation of native microbial community accelerated bioremediation of oil refinery sludge. *Biores Technol* 253: 22–32.

Ruberto, L.; Dias, R.L.; Lo Balbo, A.; Vazquez, S.C.; Hernandez, E.; Mac Cormack, W.P. 2009. Influence of nutrients addition and bioaugmentation on the hydrocarbon biodegradation of a chronically contaminated Antarctic soil. *J Appl Microbiol* 106: 1101–1110.

Sabaté, J.; Viñas, M.; Solanas, A.M. 2004. Laboratory-scale bioremediation experiments on hydrocarbon-contaminated soils. *Int Biodegrad Biodeterior* 54: 19–25.

Sarkar, J.; Kazy, S.K.; Gupta, A.; Dutta, A.; Mohapatra, B.; Roy, A.; Bera, P.; Mitra, A.; Sar, P. 2016. Biostimulation of indigenous microbial community for bioremediation of petroleum refinery sludge. *Front Microbiol* 7: doi: 10.3389/fmicb.2016.01407.

Scherr, K.; Aichberger, H.; Braun, R.; Loibner, A.P. 2007. Influence of soil fractions on microbial degradation behavior of mineral hydrocarbons. *Eur J Soil Biol* 43: 341–350.

Semple, K.J.; Doick, K.J.; Wick, L.Y.; Harms, H. 2007. Microbial interactions with organic contaminant in soil: definitions processes and measurement. *Environ Pollut* 150: 166–176.

Seneviratne, G.; Zavahir, J.S.; Bandara, W.M.M.S. 2008. Fungal-bacterial biofilms: their development for a novel biotechnological applications. *World J Microbiol Biotechnolol* 24: 739–743.

Shome, R. 2020. Role of microbial enzyme in bioremediation. *eLifePress* 1: 15–20.

Singh, A.; Van Hamme, J.D.; Ward, O.P. 2007. Surfactant in microbiology and biotechnology part 2: application aspects. *Biotechnol Adv* 25: 99–121.

Singkran, N. 2013. Classifying risk zones by the impacts of oil spills in the coastal waters of Thailand. *Mar Pollut Bull* 70: 34–43.

Sutton, N.B.; Maphosa, F.; Morillo, J.A.; Al Soud, W.A.; Langenhoff, A.A.M.; Grotenhuis, T.; Rijnaarts, H.H.M.; Smidt, H. 2013. Impact of long-term diesel contamination on soil microbial community structure. *Appl Environ Microbiol* 79: 619–630.

Steliga, T.; Wojtowicz, K.; Kapusta, P.; Brzeszcz, J. 2020. Assessment of biodegradation efficiency of Polychlorinated Biphenyls (PCBs) and Petroleum Hydrocarbons (TPH) in soil using three individual bacterial strains and their mixed culture. *Molecules* 25: 709. doi: 10.3390/molecules25030709.

Teramoto, M.; Queck, S.Y.; Ohnishi, K. 2013. Specialized hydrocarbonoclastic bacteria prevailing in seawater around a port in the Strait of Malacca. *PLoS One* 8: e66594. doi: 10.1371/journal.pone.0066594.

Thomas, S.; Wornat, M.J. 2008. The effects of oxygen on the yields of polycyclic aromatic hydrocarbons formed during the pyrolysis and fuel-rich oxidation of cathecol. *Fuel* 87: 768–781.

Thompson, I.P.; van der Gast, C.J.; Ciric, L.; Singer, A.C. 2005. Bioaugmentation for bioremediation: the challenge of strain selection. *Environ Microbiol* 7:909–915.

Tilman, D. 2000. Causes, consequences and ethics of biodiversity. *Nature* 405: 208–211.

Tyagi, M.; da Fonseca, M.M.R.; de Carvalho, C.C.C.R. 2011. Bioaugmentation and biostimulation strategies to improve the effectiveness of bioremediation processes. *Biodegradation* 22:231–241.

Ueno, A.; Ito, Y.; Yumoto, I.; Okuyama, H. 2007. Isolation and characterization of bacteria from soil contamination with diesel oil and the possible use of these in autochthonous bioaugmentation. *World J Microbiol Biotechnol* 23: 1739–1745.

Ullah, M.W.; Shi, Z.; Shi, X.; Zeng, D.; Li, S.; Yang, G. 2017. Microbes as structural templates in biofabrication: study of surface chemistry and applications. *ACS Sustain Chem Eng* 5: 11163–11175.

Uwadiae, S.; Omoayena, E. 2017. Induced degradation of crude oil polluted soil by microbial augmentation. *J Eng Stu Res* 23: 37–44.

Varjani, S.J.; Upasani, V.N. 2017. A new look on factors affecting microbial degradation of petroleum hydrocarbon pollutants. *Int Biodeterior Biodegrad* 120: 71–83.

Vasconcelos, U.; França, F.P.; Oliveira, F.J.S. 2011. Removal of high-molecular weight polycyclic aromatic hydrocarbons. *Quim Nova* 34: 218–221.

Vasconcelos, U.; Oliveira, F.J.S.; França, F.P. 2013. Raw glycerol as cosubstrate on the PAHs biodegradation in soil. *Can J Pure Appl Sci* 7: 2203–2209.

Vasudevan, N.; Rajaram, P. 2001. Bioremediation of oil-sludge contaminated soil. *Environ Int* 26: 409–411.

Viñas, M.; Grifoll, M.; Sabaté, J.; Solanas, A.M. 2002. Biodegradation of a crude oil by three microbial consortia of different origins and metabolic capabilities. *J Ind Microbiol Biotechnol* 28: 252–260.

Vo-Dinh, T.; Fetzer, J.; Campiglia, A.D. 1998. Monitoring and characterization of polyaromatic compounds in the environment. *Talanta* 47: 943–969.

Wang, Q.; Zhang, S.; Li, Y.; Klassen, W. 2011. Potential approaches to improving biodegradation of hydrocarbons for bioremediation of crude oil pollution. *Environ Protection J* 2: 47–55.

Watanabe, K. 2001. Microorganisms relevant in bioremediation. *Curr Opin Biotechnol* 12: 237–241.

Watanabe, K.; Hamamura, N. 2003. Molecular and physiological approaches to understanding the ecology of pollutant degradation. *Curr Opin Biotechnol* 14: 289–295.

Welsh, D.T. 2000. Ecological significance of compatible solute accumulation by micro-organisms: from single cells to global climate. *FEMS Microbiol Rev* 24: 263–290.

Wong, Y.M.; Wu, T.Y.; Juan, J.C. 2014. A review of sustainable hydrogen production using seed sludge via dark fermentation. *Renew Sustain Energy Rev.* 34: 471–482.

Woźniak-Karczewska, M.; Lisiecki, P.; Białas, W.; Owsianiak, M.; Piotrowska-Cyplik, A.; Wolko, Ł.; Ławniczak, Ł; Heipieper, H.J.; Gutierrez, T.; Chrzanowski, Ł. 2019. Effect of bioaugmentation on long-term biodegradation of diesel/biodiesel blends in soil microcosms. *Sci Total Environ* 671: 948–958.

Xu, X.; Liu, W; Tian, S.; Wang, W.; Qi, Q.; Jiang, P.; Gao, X.; Li, F.; Li, H.; Yu, H. 2018. Petroleum hydrocarbon-degrading bacteria for the remediation of oil pollution under aerobic conditions: a perspective analysis. *Front Microbiol* 9: 2885. doi: 10.3389/fmicb.2018.02885.

Xue, J.; Wu, Y.; Shi, K.; Xiao, X.; Gao, Y.; Lin, L.; Qiao, Y. 2017. Study on the degradation performance and kinetics of immobilized cells in straw-alginate beads in marine environment. *Bioresour Technol* 280: 88–94.

Yang, S.; Wen, X.; Shi, Y.; Liebner, S.; Jin, H.; Perfumo, A. 2016. Hydrocarbon degraders establish at the costs of microbial richness, abundance and keystone taxa after crude oil contamination in permafrost environments. *Sci Rep* 6: 37473. doi: 10.1038/srep37473.

Yang, Y.; Katte, D.; Smets, B.F. Pignatello, J.J.; Grasso, D. 2001. Mobilization of oil organic matter by complexing agents and implications for polycyclic aromatic hydrocarbon desorption. *Chemosphere* 43: 1013–1021.

Yang, Z.; Guo, R.; Shi, X.; He, S.; Wang, L.; Dai, M.; Qiu, Y.; Dang, X. 2016. Bioaugmentation of *Hydrogenispora ethanolica* LX-B affects hydrogen production through altering indigenous bacterial community structure. *Bioresour Technol* 211: 319–326.

Yuste, L.; Corbella, M.E.; Turiégano, M.J.; Karlson, U.; Poyet, A.; Rojo, F. 2000. Characterization of bacterial strains able to grow on high molecular mass residues from crude oil processing. *FEMS Microbiol Ecol* 32: 69–75.

Zhang, B.; Zhang, L.; Zhang, X. 2019. Bioremediation of petroleum hydrocarbon contaminated soil by petroleum-degrading bacteria immobilized on biochar. *RSC Adv* 9: 35304–35311.

Zhang, X.-X.; Cheng, S.-P.; Zhu, C.-J.; Sun, S-L. 2006. Microbial PAH-degradation in soil: degradation pathways and contributing factors. *Pedosphere* 16: 555–565.

Zhao, F.; Li, P.; Guo, C.; Shi, R.-J.; Zhang, Y. 2018a. Bioaugmentation of oil reservoir indigenous *Pseudomonas aeruginosa* to enhance oil recovery through *in-situ* biosurfactant production without air injection. *Bioresour Technol* 251: 295–302.

Zhao, Y; Qu, D.; Zhou, R.; Yang, X.; Kong, W.; Ren, H. 2018b. Enhancing bacterial transport with saponins in saturated porous media for the bioaugmentation of groundwater: visual investigation and surface interactions. *Environ Sci Pollut Res* 25:26539–26549.

Zhu, X.; Chen, M.; He, X.; Xiao, Z.; Zhou, H.; Tan, Z. 2015. Bioaugmentation treatment of PV wafer manufacturing wastewater by microbial culture. *Water Sci Technol* 72: 754–761.

2 Bioaugmentation for Lignin Removal from the Paper Industry

Uroosa Ejaz

University of Karachi

Shaheed Zulfikar Ali Bhutto Institute of
Science and Technology (SZABIST)

Muhammad Sohail

University of Karachi

CONTENTS

2.1 INTRODUCTION

Rapid increase in industrialization, urbanization, and population is the main cause of damage to the environment and ecological balance (Tribedi et al. 2018). On the one hand, industries play an imperative role in shaping the economy of the country and improving the quality of people (Murillo-Luna, Garcés-Ayerbe, and Rivera-Torres 2011). On the other hand, the discharge of pollutants to land, water, and air has caused virtually irreversible loss to the environment. The paper and pulp industry provides many economic benefits and fulfills the demands of modern society; so, it is one of the most important industrial sectors worldwide (Hossain and Ismail 2015). This industry consumes a huge amount of electricity, fossil fuels, and natural resources, i.e., water and wood

DOI: 10.1201/9781003187622-2

(Nataraj et al. 2007; Wang, Gu, and Ma 2007; Savant, Abdul-Rahman, and Ranade 2006). The paper industry is considered the sixth largest sector that pollutes the environment, following other industrial sectors of steel, textile, leather, cement, and oil mills (Haq, Mazumder, and Kalamdhad 2020). It releases numerous solid, liquid, and gaseous wastes into the environment which results in different characteristics of wastewater (Table 2.1). Lignin is the major by-product waste from the paper industry, which imparts dark brown color and causes an increase in biological oxygen demand (BOD) and chemical oxygen demand (COD), and toxicity to the exposed communities (Sen et al. 2020). Gaete et al. (2000), observed the deleterious effects of lignin at the planktonic level and in the benthic zones which led to the reduced diversity of phyto-zooplankton and zoobenthos (Sobral et al. 1998) and also disturbed the benthic invertebrate and algal communities (Larsson and Förlin 2002). Therefore, it is important to find eco-friendly and efficient techniques to treat the effluent of paper.

2.2 LIGNIN

Lignin is an aromatic biopolymer which is the second most abundant biopolymer and is formed by the polymerization of tyrosine and phenylalanine to form three types of phenylpropane units, i.e., p-hydroxyphenyl, syringyl, and guaiacyl, collectively known as monolignols (Rashid, Ejaz, and Sohail 2021). It is the second most sustainable and abundant component in wood and carbon source in nature after cellulose (Pin et al. 2020). Cellulose and hemicellulose form a framework in the plant cell wall in which lignin is embedded as a connector which strengthens the cell wall by solidifying it like a tar (Khan et al. 2020; Ejaz and Sohail 2020a). The presence of a highly cross-linked structure consisting of oxyphenyl propanoid units makes the degradation of lignin difficult (Hossain and Ismail 2015).

2.2.1 LIGNOCELLULLOSIC STREAM

Lignocellulose residues include grasses, woods, and a variety of forestry products (mill waste, sawdust), agricultural residues (wheat straw, corn stover), and municipal waste (Sánchez 2009). Industries which utilize the lignocellulosic biomass as the feedstock produce lignocellulosic streams. Hemicellulose and cellulose are utilized in industrial processes for manufacturing of paper or are degraded into sugars for biofuel production. Industries employ the treatment process to remove lignin from the plant biomass (Mathews, Pawlak, and Grunden 2015). For paper production, the presence of lignin reduces the quality of the product. Lignin decreases brightness and also leads to decreased

TABLE 2.1
Characteristics of Wastewater from Different Studies

Suspended Solids (mg/L)	COD (mg/L)	BOD (mg/L)	Lignin (mg/L)	pH	Color (Pt–Co)	References
-	2585.6	-	500.58	6.3	1,640	Sen et al. (2020)
3300	716	155	-	8.5–9.5	-	Chaudhry and Paliwal (2019)
-	774	426	529	8.1	1065	Zainith et al. (2019)
-	218	-	-	8.2	330	Sudarshan et al. (2017)
-	1,441	-	21	6.9	-	Hay et al. (2016)
841	2,716	3285.61	2,584	10.2	9732.8	Paliwal et al. (2016)
1900–3138	3380–4930	1650–2565	-	6.2–7.8	-	Zwain et al. (2013)
981	2,420	185	-	7.4	1,761	Garg et al. (2012)
5,240	1,170–1,510	142–221	133–265	6.4–7.3	354–563	Eskelinen et al. (2010)
91,500	70,000	503	-	9	109,284	De los Santos Ramos et al. (2009)
7,150	1,275	556	-	7	-	Avşar and Demirer (2008)

paper strength; therefore, its removal is necessary. Depending on the methods adopted for extraction, several lignin preparations have been reported including steam-exploded lignin, organosolv lignin, ligninsulfonates, and Kraft lignin (Kumar et al. 2020). Most of the lignin is discharged in the form of effluent during the paper manufacturing process (Haq, Mazumder, and Kalamdhad 2020). It is reported that paper and pulp mills produce 50–70 million tons of lignin worldwide per year (Mandlekar et al. 2018).

2.3 PRETREATMENT METHODS TO TREAT PAPER INDUSTRY WASTE

Mostly pollutants of organic nature, lignin, and lignin-derived products are found in paper mill effluent. Lignin is usually present as a co-product in the effluent of the paper industry (Haq, Mazumder, and Kalamdhad 2020). Therefore, contaminated effluent needs treatment(s) prior to its discharge to the environment or its reuse. Various procedures have been reported for the recovery and removal of lignin.

2.3.1 Physicochemical Methods

Physicochemical treatment processes include removal of toxic compounds, colors, floating matters, colloidal particles, and suspended solids by non-conventional and conventional methods like nanofiltration, ultrafiltration, reverse osmosis, electrolysis, ozonation, oxidation, coagulation, adsorption, screening, flotation, and sedimentation (Dey, Choudhury, and Das 2013; Sharma, Goel, and Capalash 2007; Singh et al. 2011).

2.3.1.1 Ozonation

Ozonation can disinfect and degrade inorganic and organic pollutants present in contaminated wastewater. Various reports describe the advantages of this method in decreasing COD and removal of toxic and color compounds from industrial effluent (Haq, Mazumder, and Kalamdhad 2020). Mänttäri, Viitikko, and Nyström (2006) observed that a high dose of ozone (~ 800, 900, and 1,100 mg/L) resulted in >50% lignin and color removal and decrease in turbidity. In another study, a removal by 46% of lignin and a decrease in 40% and 11% BOD and COD, respectively, were found in ozone-treated pulp mill effluent (Ruas et al. 2007). A decrease by 40%–96.6% in lignin content of paper mill effluent has also been reported by using ozone (Michniewicz, Stufka-Olczyk, and Milczarek 2012).

2.3.1.2 Membrane Technologies

The use of membrane technique has been extensively studied in the last few decades to treat wastewater. Gönder, Arayici, and Barlas (2012) used ultrafiltration membrane to treat industrial effluent. But, high cost and technical limitations render its use at the commercial level (Greenlee et al. 2010).

2.3.1.3 Adsorption

Adsorption is a widely used technique to remove color and other organic compounds from industrial effluent. Fuller's earth, activated carbon, silica, and coal ash are used as an adsorptive matrix to remove lignin and color from the wastewater of the paper industry (Pokhrel and Viraraghavan 2004; Kamali et al. 2019). For instance, Das and Patnaik (2000) observed 61% and 80.4% removal of lignin by slag and furnace dust, respectively.

2.3.1.4 Coagulation and Precipitation

In the tertiary treatment process, coagulation and precipitation methods are used. In this method, larger flocks are produced from smaller particles in industrial effluent by the addition of metal salts. Wang et al. (2011) used $AlCl_3$ and starch-graft-polyacrylamide as coagulants and flocculants, respectively, to treat wastewater. In another study, corn stover was used for lignin removal with metals viz.

ZnO, Fe$_2$SO$_4$, CuO, MgO, and NiO (Li and Zhang 2011). MgO was proved as the best catalyst as compared to other metals; this process did not generate the by-product (acids, 5-hydroxymethylfurfural, and furfural). Moreover, the liquor obtained after the pretreatment process could be directly subjected to fermentation or hydrolysis (Li and Zhang 2011).

2.3.2 Disadvantages of Physicochemical Pretreatment

The high cost of physicochemical pretreatment methods is the major impediment in its application on a large scale (Subba Rao and Venkatarangaiah 2014; Bagal and Gogate 2014; Siegrist and Joss 2012). Furthermore, toxic by-products are formed by some of the physicochemical techniques. For instance, chlorine-based bleaches are used in physicochemical processes which emit toxic products to air, soil, and water (Gutiérrez, Rodríguez, and Del Río 2006).

2.3.3 Biological Pretreatment

Microorganisms including algae, bacteria, fungi, and their enzymes are used in the biological treatment process. Biological pretreatment methods are more eco-friendly, cost-effective, and efficient to reduce COD and BOD from industrial wastewater (Hossain and Ismail 2015). However, many complex pollutants are not degraded by microorganisms and consequently persist in the wastewater. Therefore, the process of bioaugmentation is preferred.

2.3.4 Bioaugmentation

Bioaugmentation is the addition of cultured microorganisms into a contaminated site to enhance bioremediation of pollutants (Tribedi et al. 2018). Specific pollutant which is dominant in the environment can be degraded by the process of bioaugmentation.

2.3.4.1 Principle of Bioaugmentation
The rationale of bioaugmentation technique is to improve the rate of removal of pollutants by adding specific microorganisms (Leahy and Colwell 1990). The microbes that are added to the site augment the existing biomass by increasing its ability to produce an acceptable effluent. Microbial species chosen for the process of bioaugmentation are carefully screened to eliminate pathogenic organisms. Microbes are natural forms selected from the environment and then grown in bulk quantities. Bioaugmentation not only eliminates the pollutants but also results in increased microbial and genetic diversity present at that site (Shukla, Singh, and Sharma 2010; Dejonghe et al. 2001).

2.3.4.2 Factors Influencing Bioaugmentation
Adaptation of microbes to a contaminated site is the most important factor which influences the process of bioaugmentation (Tribedi et al. 2018). Furthermore, it also depends on the ability of the newly introduced microbial consortia to compete with the predators, abiotic factors, and indigenous microorganisms (Omokhagbor Adams et al. 2020). Other factors include nutrient content, aeration, presence of organic matter, moisture, temperature, and pH (Tribedi et al. 2018).

2.3.4.3 Lignin Removal by Bioaugmentation
Addition of lignin-biodegrading microorganisms into wastewater can result in lignin removal. Various microbial strains have been reported for this purpose either as mono- or co-culture. Zheng et al. (2013) reported 50% lignin removal by using a consortium of *Pandoraea* B-6 and *Comamonas* B-9 (bacteria), and *Aspergillus* F-1 (fungus). In another study, Hailei et al. (2006) used *Azotobacter* sp., *Coriolus versicolor,* and *Phanerochate chrysosporium* for the treatment of paper mill effluent. Chen et al. (2012) reported that *Gordonia* strain JW8 can efficiently degrade alkaline lignin.

Many studies particularly indicate the potential of white-rot fungi, Basidimycotina, and Ascomycotina in breakdown of lignin (Haq et al. 2016) as compared to other microorganisms

(Haq, Mazumder, and Kalamdhad 2020). The white-rot fungi utilize lignin for their metabolism and growth (Asina et al. 2016). Some widely known white-rot/wood decay fungi (Table 2.2) are *Myrothecium verrucaria*, *Stereum hirsutum*, *Pleurotus ostreatus*, *Daedalea flavida*, *Coriolus versicolor*, and *Trametes versicolor* (Mishra and Thakur 2010; Wang et al. 2017). *Phanerochaete chrysosporium*, *Schizophyllum commune*, and *Tinctoria borbonica* have also been applied for the removal of lignin from the effluent of the paper mill (Haq, Mazumder, and Kalamdhad 2020). These fungi produce extracellular enzymes, viz. manganese (MnP), laccase (Lac), and lignin peroxidases (LiP), which play a significant role in the degradation of lignin (Blanchette 1995). The enzyme laccase has also been used to remove lignin from wastewater (Nzila, Razzak, and Zhu 2016). It was found that *Aspergillus flavus* can reduce 45% COD and 94% lignin in pulp industry wastewater (Haq, Mazumder, and Kalamdhad 2020).

Several workers also utilized "fungi imperfecti" along with basidiomycetes for this purpose. For example, *Trametes versicolor*, *Trichoderma* spp., and *Aspergillus niger* were used in a consortium to reduce color and lignin from the hardwood pulp bleach wastewater (Kamali and Khodaparast 2015; Dashtban et al. 2010). However, the fungal ligninolytic system can be affected by extreme environmental conditions which include high pH and temperature and the presence of toxic material in the effluent. Abd-elsalam and Hafez (2008) reported that the fungal ligninolytic system is considered less effective to treat wastewater of paper mills due to the structural hindrance caused by a fungal filament. Moreover, fungi are not stable for *in situ* treatment under extreme substrate and environmental conditions such as high lignin concentration, oxygen limitation, and higher pH. Therefore, the interest was diverted in the utilization of bacterial cultures for lignin removal in industrial effluent. Various bacterial treatment methods, like biological filters, activated sludge, aeration lagoon, anaerobic and aerobic bioreactors are used to treat paper industry effluent (Haq, Mazumder, and Kalamdhad 2020). However, these conventional methods are not efficient to be used on commercial level due to their high maintenance requirements, length of operation, and higher operation cost (Raj et al. 2014). Consequently, there is a need for the search for bacteria that can efficiently degrade lignin. It is reported that *Streptomyces*, *Arthrobacter*, *Bacillus*, *Nocardia*, *Alcaligenes*, *Flavibacterium*, *Xanthomonas*, and *Pseudomonas* sp. can degrade lignin (Chandra and Singh 2012; Wang et al. 2013; Xu et al. 2018). Bacteria grow faster as compared to fungi, and the bacterial ligninolytic system can survive in harsh environmental conditions and are more thermostable, salt, and pH tolerant (Harms, Schlosser, and Wick 2011). Degradation of monomeric lignin by bacteria has been widely reported (Table 2.3), whereas only a few bacterial strains can break the complex derivatives of lignin which are generated in pulping processes (Chandra et al. 2007). *Pseudomonas aeruginosa*, *Bacillus* sp., *B. subtilis* and *B. megaterium* can degrade 50%–97% lignin from paper mill effluent (Haq, Mazumder, and Kalamdhad 2020; Raj et al. 2014; Tiku et al. 2010). In another study, *Bacillus* sp., *B. endo-phyticus*, and *B. subtilis* removed 16.87%, 17.9%, and 17.8% lignin, respectively, whereas a consortium of these three bacterial strains removed 40.19% lignin in paper mill effluent (Ojha and Tiwari 2016). Moreover, a consortium of bacteria and protozoa has also been reported for lignin degradation in paper industry effluent (Hossain and Ismail 2015).

TABLE 2.2
Fungal Species Involve in the Degradation of Lignin

Fungal Specie	Lignin Removal (%)	References
Sordaria macrospora k-hell and *Myceliophthora thermophile*	15–20	Yang, Gu, and Lin (2020)
Echinodontium taxodii	66.7–73.3	Shi et al. (2016)
Aspergillus flavus strain F10	39–61	Barapatre and Jha (2016)
Merulius aureus and *Fusariumsam bucinum*	79	Malaviya and Rathore (2007)
Gliocladium virens	52	Kamali and Khodaparast (2015)

TABLE 2.3

Bacterial Species Involve in the Degradation of Lignin

Bacterial Specie	Lignin Removal (%)	References
Arthrobacter sp. C2	40.1	Abdel-Hamid, Solbiati, and Cann (2013)
Pseudomonas fluorescens DSM 50090 and *Rhodococcus opacus* DSM1069	80	Ravi et al. (2019)
Serratia liquefaciens	58	Haq et al. (2016)
Citrobacter freundii (FJ581026) and *Citrobacter* sp. (FJ581023)	71	Chandra and Bharagava (2013)
Bacillus sp.	37	Chandra and Purohit (2007)
Bacillus sp.	28–53	Mishra and Thakur (2010)
Citrobacter freundii and *Serratia marcescens*	65	Abhishek et al. (2015)

2.3.4.4 Limitations of Bioaugmentation Technologies

It is seen that the number of introduced microorganisms decreases shortly after the addition to a treatment site. Both biotic and abiotic stresses can result in death of exogenous microorganism. These stresses occur due to factors associated with quorum sensing, grazing by protozoa, shock of pollutant load, phage infections, competition between indigenous and introduced microorganisms, nutrient limitations, pH and temperature changes, and insufficient substrates (Bouchez et al. 2000).

2.3.4.5 Methods to Improve Process of Bioaugmentation

2.3.4.5.1 Immobilization

Immobilization (encapsulation or entrapment) of microorganisms offers several advantages as compared to free cell bioaugmentation, i.e., it enhances physical and biological stabilities, it provides protection against bacteriophage infections and protozoal grazing. Enhanced cell survival and high biomass concentration can be achieved by immobilization technique (Ejaz and Sohail 2020; Ejaz, Ahmed, and Sohail 2018). It is reported that immobilized *P. chrysosporium* reduced 83% lignin in paper mill effluent (Gomathi et al. 2012). However, the immobilization process remains costly, especially when this technique has to be employed for the treatment of huge volumes of wastewaters.

2.3.4.5.2 Genetically Modified Microorganisms

Bioaugmentation can be improved by using genetically modified microorganisms (GMM) harboring genes encoding catabolic enzymes which are involved in the biodegradation of pollutants are incorporated (Nzila, Razzak, and Zhu 2016). Genetically modified *Pseudomonas putida* and *Pseudomonas* sp. strains have been reported for the degradation of monoaromatic compounds (Lyon et al. 2013; Shankar et al. 2011). However, the use of GMM is associated with serious concerns regarding their long-term environmental effects (Nzila, Razzak, and Zhu 2016).

2.3.4.5.3 Quorum Sensing

Bioaugmented bacteria can produce a hydrated polymeric matrix which can form biofilms. The observation showed that quorum sensing leads to better biodegradation of pollutants (Wang et al. 2013). However, more research is needed to establish the link between quorum sensing and bioaugmentation (Nzila, Razzak, and Zhu 2016).

2.3.4.5.4 Nanotechnology

Nanotechnology can tremendously improve bioaugmentation. Zhang et al. (2015) reported that carbon-nanotubes at the concentration >25 mg/L resulted in increased biodegradation of atrazine by *Arthrobacter* sp. Reduced bioavailability of pollutants is also one of the limitations of bioaugmentation.

To counteract this limitation, bacteria can be functionalized by fixing "thermal responsive nanomaterial" on their surface (Nzila, Razzak, and Zhu 2016). The use of nanomaterials in bioaugmentation, nonetheless, is still in an early stage of investigation.

2.4 CONCLUSION

The paper industry is responsible for polluting the environment, particularly water reservoirs. Conventional physicochemical pretreatment methods are available for lignin removal from industrial effluent. But release of toxic by-products by physicochemical methods is a concern for the environment. Therefore, biological pretreatment methods can be employed for lignin degradation. Particularly, bioaugmentation is an attractive strategy for the removal of lignin from paper industry effluent. Bioaugmentation can be improved by utilizing immobilization, genetic engineering, and nanobiotechnological techniques. Opportunities exist to improve the biodegradation of pollutants in contaminated wastewater.

REFERENCES

Abd-elsalam, H.E., and E.E., Hafez, Genetic Engineering, and Technology Application. 2008. "Molecular Characterization of Two Native Egyptian Ligninolytic Bacterial Strains." *Journal of Applied Sciences* 4 (10): 1291–96.

Abdel-Hamid, A.M., J.O. Solbiati, and, I.K.O. Cann 2013. "Insights into Lignin Degradation and Its Potential Industrial Applications." In *Advances in Applied Microbiology*. doi: 10.1016/B978-0-12-407679-2.00001-6.

Abhishek, A., A., Dwivedi, N., Tandan, and U Kumar. 2015. "Comparative Bacterial Degradation and Detoxification of Model and Kraft Lignin from Pulp Paper Wastewater and Its Metabolites." *Applied Water Science* 7 (2): 757–67. doi: 10.1007/s13201-015-0288-9.

Asina, F., I., Brzonova, K., Voeller, E., Kozliak, A., Kubátová, B. Yao, and Y. Ji, 2016. "Biodegradation of Lignin by Fungi, Bacteria and Laccases." *Bioresource Technology* 220: 414–24. doi: 10.1016/j.biortech.2016.08.016.

Avşar, E., and G.N. Demirer, 2008. "Cleaner Production Opportunity Assessment Study in SEKA Balikesir Pulp and Paper Mill." *Journal of Cleaner Production* 16 (4): 422–31. doi: 10.1016/j.jclepro.2006.07.042.

Bagal, M.V., and P.R. Gogate, 2014. "Wastewater Treatment Using Hybrid Treatment Schemes Based on Cavitation and Fenton Chemistry: A Review." *Ultrasonics Sonochemistry* 21 (1): 1–14. doi: 10.1016/j.ultsonch.2013.07.009.

Barapatre, A., and H. Jha. 2016. "Decolourization and Biological Treatment of Pulp and Paper Mill Effluent by Lignin-Degrading Fungus Aspergillus Flavus Strain F10." *International Journal of Current Microbiology and Applied Sciences* 5 (5): 19–32.

Blanchette, R.A. 1995. "Degradation of the Lignocellulose Complex in Wood." *Canadian Journal of Botany* 73 (S1): 999–1010. doi: 10.1139/b95-350.

Bouchez, T., D. Patureau, P. Dabert, S. Juretschko, J. Doré, P. Delgenès, R. Moletta, and M. Wagner. 2000. "Ecological Study of a Bioaugmentation Failure." Environmental Microbiology 2 (2): 179–90. doi: 10.1046/j.1462-2920.2000.00091.x.

Chandra, R., A. Raj, H.J. Purohit, and A. Kapley. 2007. "Characterisation and Optimisation of Three Potential Aerobic Bacterial Strains for Kraft Lignin Degradation from Pulp Paper Waste." Chemosphere 67 (4): 839–46. doi:10.1016/j.chemosphere.2006.10.011.

Chandra, R., and R.N. Bharagava. 2013. "Bacterial Degradation of Synthetic and Kraft Lignin by Axenic and Mixed Culture and Their Methabolic Products." *Journal of Environmental Biology* 34: 991–99.

Chandra, R., and R. Singh. 2012. "Decolourisation and Detoxification of Rayon Grade Pulp Paper Mill Effluent by Mixed Bacterial Culture Isolated from Pulp Paper Mill Effluent Polluted Site." *Biochemical Engineering Journal* 61: 49–58. doi:10.1016/j.bej.2011.12.004.

Chaudhry, S., and R. Paliwal. 2019. "Techniques for Remediation of Paper and Pulp Mill Effluents: Processes and Constraints." *Handbook of Environmental Materials Management*, 1747–65. doi:10.1007/978-3-319-73645-7_134.

Chen, J., P. Zhan, B. Koopman, G. Fang, and Y. Shi. 2012. "Bioaugmentation with Gordonia Strain JW8 in Treatment of Pulp and Paper Wastewater." *Clean Technologies and Environmental Policy* 14 (5): 899–904. doi:10.1007/s10098-012-0459-4.

Das, C.P., and L.N. Patnaik. 2000. "Removal of Lignin by Industrial Solid Wastes." *Practice Periodical of Hazardous, Toxic, and Radioactive Waste Management* 4: 156–61.

Dashtban, M., H. Schraft, T.A. Syed, and W. Qin. 2010. "Fungal Biodegradation and Enzymatic Modification of Lignin." *International Journal of Biochemistry and Molecular Biology* 1 (1): 36–50.

Dejonghe, W., N. Boon, D. Seghers, E. M. Top, and W. Verstraete. 2001. "Bioaugmentation of Soils by Increasing Microbial Richness: Missing Links." Environmental Microbiology 3 (10): 649–57. doi:10.1046/j.1462-2920.2001.00236.x.

De los Santos Ramos, W., T. Poznyak, I. Chairez, and R. Córdova 2009. "Remediation of Lignin and Its Derivatives from Pulp and Paper Industry Wastewater by the Combination of Chemical Precipitation and Ozonation." Journal of Hazardous Materials 169 (1–3): 428–34. doi:10.1016/j.jhazmat.2009.03.152.

Dey, S., M. Dutta Choudhury, and S. Das. 2013. "A Review on Toxicity of Paper Mill Effluent on Fish." Bulletin of Environment, Pharmacology and Life Sciences 2 (February): 17–23. http://bepls.com/feb_2013/4.pdf.

Ejaz, U., A. Ahmed, and M. Sohail. 2018. "Statistical Optimization of Immobilization of Yeast Cells on Corncob for Pectinase Production." *Biocatalysis and Agricultural Biotechnology* 14: 450–56. doi:10.1016/j.bcab.2018.04.011.

Ejaz, U., and M. Sohail. 2020. "Supporting Role of Lignin in Immobilization of Yeast on Sugarcane Bagasse for Continuous Pectinase Production." *Journal of the Science of Food and Agriculture*. doi:10.1002/jsfa.10764.

Eskelinen, K., H. Särkkä, T. Agustiono Kurniawan, and M.E.T. Sillanpää. 2010. "Removal of Recalcitrant Contaminants from Bleaching Effluents in Pulp and Paper Mills Using Ultrasonic Irradiation and Fenton-like Oxidation, Electrochemical Treatment, and/or Chemical Precipitation: A Comparative Study." *Desalination* 255 (1–3): 179–87. doi:10.1016/j.desal.2009.12.024.

Gaete, H., A. Larrain, E. Bay-Schmith, J. Baeza, and J. Rodriguez. 2000. "Ecotoxicological Assessment of Two Pulp Mill Effluent, Biobio River Basin, Chile." Bulletin of Environmental Contamination and Toxicology 65 (2): 183–89. doi:10.1007/s001280000113.

Garg, S.K., M. Tripathi, S. Kumar, S.K. Singh, and S.K. Singh. 2012. "Microbial Dechlorination of Chloro-organics and Simultaneous Decolorization of Pulp-Paper Mill Effluent by Pseudomonas Putida MTCC 10510 Augmentation." *Environmental Monitoring and Assessment* 184 (9): 5533–44. doi:10.1007/s10661-011-2359-1.

Gomathi, V., B. Cibichakravarthy, A. Ramanathan, S. Nallapeta, V. Ramanjaneya, R. Mula, and J. Rayalu. 2012. "Decolourization of Paper Mill Effluent By Immobilized Cells." *International Journal of Plant, Animal and Environmental Sciences* 2 (1): 141–46.

Gönder, Z.B., S. Arayici, and H. Barlas. 2012. "Treatment of Pulp and Paper Mill Wastewater Using Utrafiltration Process: Optimization of the Fouling and Rejections." *Industrial and Engineering Chemistry Research* 51 (17): 6184–95. doi:10.1021/ie2024504.

Greenlee, L.F., F. Testa, D.F. Lawler, B.D. Freeman, and P. Moulin. 2010. "Effect of Antiscalants on Precipitation of an RO Concentrate: Metals Precipitated and Particle Characteristics for Several Water Compositions." *Water Research* 44 (8): 2672–84. doi:10.1016/j.watres.2010.01.034.

Gutiérrez, A., I.M. Rodríguez, and J.C. Del Río. 2006. "Chemical Characterization of Lignin and Lipid Fractions in Industrial Hemp Bast Fibers Used for Manufacturing High-Quality Paper Pulps." *Journal of Agricultural and Food Chemistry* 54 (6): 2138–44. doi:10.1021/jf052935a.

Hailei, W., L. Guosheng, L. Ping, and P. Feng. 2006. "The Effect of Bioaugmentation on the Performance of Sequencing Batch Reactor and Sludge Characteristics in the Treatment Process of Papermaking Wastewater." *Bioprocess and Biosystems Engineering* 29 (5–6): 283–89. doi:10.1007/s00449-006-0077-9.

Haq, I., P. Mazumder, and A.S. Kalamdhad. 2020. "Recent Advances in Removal of Lignin from Paper Industry Wastewater and Its Industrial Applications – A Review." *Bioresource Technology* 312: 123636. doi:10.1016/j.biortech.2020.123636.

Haq, I., S. Kumar, V. Kumari, S.K. Singh, and A. Raj. 2016. "Evaluation of Bioremediation Potentiality of Ligninolytic Serratia Liquefaciens for Detoxification of Pulp and Paper Mill Effluent." *Journal of Hazardous Materials* 305: 190–99. doi:10.1016/j.jhazmat.2015.11.046.

Harms, H., D. Schlosser, and L.Y. Wick. 2011. "Untapped Potential: Exploiting Fungi in Bioremediation of Hazardous Chemicals." *Nature Reviews Microbiology* 9 (3): 177–92. doi:10.1038/nrmicro2519.

Hay, J.X.W., T. Yeong Wu, B. Junn Ng, J. Ching Juan, and J.Md. Jahim. 2016. "Reusing Pulp and Paper Mill Effluent as a Bioresource to Produce Biohydrogen through Ultrasonicated Rhodobacter Sphaeroides." Energy Conversion and Management 113: 273–80. doi:10.1016/j.enconman.2015.12.041.

Hossain, K., and N. Ismail. 2015. "Bioremediation and Detoxification of Pulp and Paper Mill Effluent: A Review." *Research Journal of Environmental Toxicology* 9 (3): 113–34. doi:10.3923/rjet.2015.113.134.

Kamali, M., D.P. Suhas, M.E. Costa, I. Capela, and T.M. Aminabhavi. 2019. "Sustainability Considerations in Membrane-Based Technologies for Industrial Effluents Treatment." *Chemical Engineering Journal* 368 (February): 474–94. doi:10.1016/j.cej.2019.02.075.

Kamali, M., and Z. Khodaparast. 2015. "Review on Recent Developments on Pulp and Paper Mill Wastewater Treatment." Ecotoxicology and Environmental Safety 114: 326–42. doi:10.1016/j.ecoenv.2014.05.005.

Khan, M.T., U. Ejaz, and M. Sohail. 2020. "Evaluation of Factors Affecting Saccharification of Sugarcane Bagasse Using Cellulase Preparation from a Thermophilic Strain of Brevibacillus." *Current Microbiology* 77: 2422–29. doi:10.1007/s00284-020-02059-3.

Kumar, A., Anushree, J. Kumar, and T. Bhaskar. 2020. "Utilization of Lignin: A Sustainable and Eco-Friendly Approach." *Journal of the Energy Institute* 93 (1): 235–71. doi:10.1016/j.joei.2019.03.005.

Larsson, D., G. Joakim, and L. Förlin. 2002. "Male-Biased Sex Ratios of Fish Embryos near a Pulp Mill: Temporary Recovery after a Short-Term Shutdown." *Environmental Health Perspectives* 110 (8): 739–42. doi:10.1289/ehp.02110739.

Leahy, J.G., and R.R. Colwell. 1990. "Microbial Degradation of Hydrocarbons in the Environment." *Microbiological Reviews* 54 (3): 305–15. doi:10.1128/mmbr.54.3.305-315.1990.

Li, S., and X. Zhang. 2011. "The Study of PAFSSB on RO Pre-Treatment in Pulp and Paper Wastewater." Procedia Environmental Sciences 8: 4–10. doi:10.1016/j.proenv.2011.10.003.

Lyon, D., Y., Timothy M. Vogel, De Lyon, E.C. De Lyon, and E. Cedex. 2013. "Chapter 1 Bioaugmentation for Groundwater Remediation : An Overview." In Bioaugmentation for Groundwater Remediation; Ward, C.H., Ed.; Springer Science and Business Media: New York, 1–37. doi:10.1007/978-1-4614-4115-1.

Malaviya, P. and V. S. Rathore. 2007. "Bioremediation of Pulp and Paper Mill e Z Uent by a Novel Fungal Consortium Isolated from Polluted Soil." Bioresource Technology 98: 3647–3651. doi:10.1016/j.biortech.2006.11.021.

Mandlekar, N., A. Cayla, F. Rault, S. Giraud, F. Salaün, G. Malucelli, and J.-P. Guan. 2018. "An Overview on the Use of Lignin and Its Derivatives in Fire Retardant Polymer Systems." In Lignin - Trends and Applications. doi:10.5772/intechopen.72963.

Mänttäri, M., K. Viitikko, and M. Nyström. 2006. "Nanofiltration of Biologically Treated Effluents from the Pulp and Paper Industry." Journal of Membrane Science 272 (1–2): 152–160. doi:10.1016/j.memsci.2005.07.031.

Mathews, S.L., J. Pawlak, and A.M. Grunden. 2015. "Bacterial Biodegradation and Bioconversion of Industrial Lignocellulosic Streams." *Applied Microbiology and Biotechnology* 99 (7): 2939–2954. doi: 10.1007/s00253-015-6471-y.

Michniewicz, M., J. Stufka-Olczyk, and A. Milczarek. 2012. "Ozone Degradation of Lignin; Its Impact upon the Subsequent Biodegradation." *Fibres and Textiles in Eastern Europe* 96 (6 B): 191–96.

Mishra, M., and I. Shekhar Thakur. 2010. "Isolation and Characterization of Alkalotolerant Bacteria and Optimization of Process Parameters for Decolorization and Detoxification of Pulp and Paper Mill Effluent by Taguchi Approach." *Biodegradation* 21 (6): 967–78. doi: 10.1007/s10532-010-9356-x.

Murillo-Luna, J.L., C. Garcés-Ayerbe, and P. Rivera-Torres. 2011. "Barriers to the Adoption of Proactive Environmental Strategies." *Journal of Cleaner Production* 19 (13): 1417–25. doi: 10.1016/j.jclepro.2011.05.005.

Nataraj, S. K., S. Sridhar, I. N. Shaikha, D. S. Reddy, and T. M. Aminabhavi. 2007. "Membrane-Based Microfiltration/Electrodialysis Hybrid Process for the Treatment of Paper Industry Wastewater." *Separation and Purification Technology* 57 (1): 185–92. doi: 10.1016/j.seppur.2007.03.014.

Nzila, A., S.A. Razzak, and J. Zhu. 2016. "Bioaugmentation: An Emerging Strategy of Industrial Wastewater Treatment for Reuse and Discharge." *International Journal of Environmental Research and Public Health* 13 (9). doi: 10.3390/ijerph13090846.

Ojha, A.K., and M Tiwari. 2016. "Lignin Decolorization and Degradation of Pulp and Paper Mill Effluent by Ligninolytic Bacteria." *Iranica Journal of Energy and Environment* 7 (3): 282–93. doi: 10.5829/idosi.ijee.2016.07.03.11.

Omokhagbor Adams, G., P.T. Fufeyin, S.E. Okoro, and I. Ehinomen. 2020. "Bioremediation, Biostimulation and Bioaugmention: A Review." *International Journal of Environmental Bioremediation & Biodegradation* 3 (1): 28–39. doi:10.12691/ijebb-3-1-5.

Paliwal, R., S. Uniyal, M. Verma, A. Kumar, and J.P.N. Rai. 2016. "Process Optimization for Biodegradation of Black Liquor by Immobilized Novel Bacterial Consortium." *Desalination and Water Treatment* 57 (40): 18915–26. doi: 10.1080/19443994.2015.1092892.

Pin, T.C., V.M. Nascimento, A.C. Costa, Y. Pu, A.J. Ragauskas, and S.C. Rabelo. 2020. "Structural Characterization of Sugarcane Lignins Extracted from Different Protic Ionic Liquid Pretreatments." *Renewable Energy* 161: 579–92. doi: 10.1016/j.renene.2020.07.078.

Pokhrel, D., and T. Viraraghavan. 2004. "Treatment of Pulp and Paper Mill Wastewater - A Review." *Science of the Total Environment* 333 (1–3): 37–58. doi: 10.1016/j.scitotenv.2004.05.017.

Raj, A., M.M. Krishna Reddy, R., Chandra, H.J. Purohit, and A. Kapley. 2007. "Biodegradation of Kraft-Lignin by *Bacillus* sp. Isolated from Sludge of Pulp and Paper Mill." *Biodegradation* 18: 783–92. doi:10.1007/s10532-007-9107-9.

Raj, A., S. Kumar, I. Haq, and S.K. Singh. 2014. "Bioremediation and Toxicity Reduction in Pulp and Paper Mill Effluent by Newly Isolated Ligninolytic *Paenibacillus* sp." *Ecological Engineering* 71: 355–62. doi: 10.1016/j.ecoleng.2014.07.002.

Rashid, R., Ejaz, U. and Sohail, M. 2021. "Biomass to Xylose." In *Sustainable Bioconversion of Waste to Value Added Products*, 247–65. doi: 10.1007/978-3-030-61837-7_15.

Ravi, K., O.Y. Abdelaziz, M. Nöbel, J.G. Hidalgo, M.F. Gorwa Grauslund, C.P. Hulteberg, and G. Lidén. 2019. "Biotechnology for Biofuels Bacterial Conversion of Depolymerized Kraft Lignin." *Biotechnology for Biofuels*. 1–14. doi: 10.1186/s13068-019-1397-8.

Ruas, D.B., A.H. Mounteer, A.C. Lopes, B.L. Gomes, F.D. Brandão, and L.M. Girondoll. 2007. "Combined Chemical Biological Treatment of Bleached Eucalypt Kraft Pulp Mill Effluent." *Water Science and Technology* 55 (6): 143–50. doi: 10.2166/wst.2007.222.

Sánchez, C. 2009. "Lignocellulosic Residues: Biodegradation and Bioconversion by Fungi." *Biotechnology Advances* 27: 185–94. doi: 10.1016/j.biotechadv.2008.11.001.

Savant, D.V., R. Abdul-Rahman, and D.R. Ranade. 2006. "Anaerobic Degradation of Adsorbable Organic Halides (AOX) from Pulp and Paper Industry Wastewater." *Bioresource Technology* 97 (9): 1092–1104. doi: 10.1016/j.biortech.2004.12.013.

Sen, S.K., S. Raut, M. Gaur, and S. Raut. 2020. "Biodegradation of Lignin from Pulp and Paper Mill Effluent: Optimization and Toxicity Evaluation." *Journal of Hazardous, Toxic, and Radioactive Waste* 24 (4): 04020032. doi: 10.1061/(asce)hz.2153-5515.0000522.

Shankar, J., P.C. Abhilash, H.B. Singh, R.P. Singh, and D.P. Singh. 2011. "Genetically Engineered Bacteria : An Emerging Tool for Environmental Remediation and Future Research Perspectives." *Gene* 480 (1–2): 1–9. doi: 10.1016/j.gene.2011.03.001.

Sharma, P., R. Goel, and N. Capalash. 2007. "Bacterial Laccases." *World Journal of Microbiology and Biotechnology* 23 (6): 823–832. doi: 10.1007/s11274-006-9305-3.

Shi, Y., X. Yan, Q. Li, X. Wang, S. Xie, L. Chai, and J. Yuan. 2016. "Directed Bioconversion of Kraft Lignin to Polyhydroxyalkanoate by *Cupriavidus basilensis* B-8 without Any Pretreatment." *Process Biochemistry* 52: 238–242. doi: 10.1016/j.procbio.2016.10.004.

Shukla, K.P., N.K. Singh, and S. Sharma. 2010. "Bioremediation: Developments, Current Practices and Perspectives." *Genetic Engineering and Biotechnology Journal* 2010: 1–20.

Siegrist, H., and A. Joss. 2012. "Review on the Fate of Organic Micropollutants in Wastewater Treatment and Water Reuse with Membranes." *Water Science and Technology* 66 (6): 1369–76. doi: 10.2166/wst.2012.285.

Singh, Y.P., P. Dhall, R.M. Mathur, R.K. Jain, V.V. Thakur, V. Kumar, R. Kumar, and A. Kumar. 2011. "Bioremediation of Pulp and Paper Mill Effluent by Tannic Acid Degrading *Enterobacter* sp." *Water, Air, and Soil Pollution* 218 (1–4): 693–701. doi: 10.1007/s11270-010-0678-4.

Sobral, O., R. Ribeiro, F. Gonçalves, and A.M.V.M. Soares. 1998. "Ecotoxicity of Pulp Mill Effluents from Different Prebleaching Processes." *Bulletin of Environmental Contamination and Toxicology* 61 (6): 738–45. doi: 10.1007/s001289900823.

Subba Rao, A.N., and V.T. Venkatarangaiah. 2014. "Metal Oxide-Coated Anodes in Wastewater Treatment." *Environmental Science and Pollution Research* 21 (5): 3197–3217. doi: 10.1007/s11356-013-2313-6.

Sudarshan, K., K. Maruthaiya, P. Kotteeswaran, and A. Murugan. 2017. "Reuse the Pulp and Paper Industry Wastewater by Using Fashionable Technology." *Applied Water Science* 7 (6): 3317–22. doi: 10.1007/s13201-016-0477-1.

Tiku, D.K., A. Kumar, R. Chaturvedi, S.D. Makhijani, A. Manoharan, and R. Kumar. 2010. "Holistic Bioremediation of Pulp Mill Effluents Using Autochthonous Bacteria." *International Biodeterioration and Biodegradation* 64 (3): 173–83. doi: 10.1016/j.ibiod.2010.01.001.

Tribedi, P., M. Goswami, P. Chakraborty, K. Mukherjee, G. Mitra, P. Bhattacharyya, and S. Dey. 2018. "Bioaugmentation and Biostimulation: A Potential Strategy for Environmental Remediation." *Journal of Microbiology & Experimentation* 6 (5): 223–31. doi: 10.15406/jmen.2018.06.00219.

Wang, B., L. Gu, and H. Ma. 2007. "Electrochemical Oxidation of Pulp and Paper Making Wastewater Assisted by Transition Metal Modified Kaolin." *Journal of Hazardous Materials* 143 (1–2): 198–205. doi: 10.1016/j.jhazmat.2006.09.013.

Wang, J.P., Y.Z. Chen, Y. Wang, S.J. Yuan, and H.Q. Yu. 2011. "Optimization of the Coagulation-Flocculation Process for Pulp Mill Wastewater Treatment Using a Combination of Uniform Design and Response Surface Methodology." *Water Research* 45 (17): 5633–40. doi: 10.1016/j.watres.2011.08.023.

Wang, J.-h., H.-z. He, M.-z. Wang, S. Wang, J. Zhang, W. Wei, H.-x. Xu, Z.-m. Lv, and D.-s. Shen. 2013. "Bioresource Technology Bioaugmentation of Activated Sludge with *Acinetobacter* sp. TW Enhances Nicotine Degradation in a Synthetic Tobacco Wastewater Treatment System." *Bioresource Technology* 142: 445–53. doi: 10.1016/j.biortech.2013.05.067.

Wang, Q.f., L.l. Niu, J. Jiao, N. Guo, Y.p. Zang, Q.y. Gai, and Y.j. Fu. 2017. "Degradation of Lignin in Birch Sawdust Treated by a Novel Myrothecium Verrucaria Coupled with Ultrasound Assistance." *Bioresource Technology* 244: 969–74. doi: 10.1016/j.biortech.2017.07.164.

Wang, Y., Q. Liu, L. Yan, Y. Gao, Y. Wang, and W. Wang. 2013. "A Novel Lignin Degradation Bacterial Consortium for Efficient Pulping." *Bioresource Technology* 139: 113–19. doi: 10.1016/j.biortech.2013.04.033.

Yang, X., C. Gu, and Y. Lin. 2020. "A Novel Fungal Laccase from Sordaria Macrospora k - Hell : Expression, Characterization, and Application for Lignin Degradation." *Bioprocess and Biosystems Engineering* 43: 1133–1129. doi: 10.1007/s00449-020-02309-5.

Zainith, S., D. Purchase, G.D. Saratale, L.F.R. Ferreira, M. Bilal, and R.N. Bharagava. 2019. "Isolation and Characterization of Lignin-Degrading Bacterium Bacillus Aryabhattai from Pulp and Paper Mill Wastewater and Evaluation of Its Lignin-Degrading Potential." *3 Biotech* 9 (3): 1–11. doi: 10.1007/s13205-019-1631-x.

Zhang, C., M. Li, X. Xu, and N. Liu. 2015. "Effects of Carbon Nanotubes on Atrazine Biodegradation by *Arthrobacter* sp." *Journal of Hazardous Materials* 287: 1–6. doi: 10.1016/j.jhazmat.2015.01.039.

Zheng, Y., L.Y. Chai, Z.H. Yang, C.J. Tang, Y.H. Chen, and Y. Shi. 2013. "Enhanced Remediation of Black Liquor by Activated Sludge Bioaugmented with a Novel Exogenous Microorganism Culture." *Applied Microbiology and Biotechnology* 97 (14): 6525–35. doi: 10.1007/s00253-012-4453-x.

Zwain, H.M., S.R. Hassan, N.Q. Zaman, H.A. Aziz, and I. Dahlan. 2013. "The Start-up Performance of Modified Anaerobic Baffled Reactor (MABR) for the Treatment of Recycled Paper Mill Wastewater." *Journal of Environmental Chemical Engineering* 1 (1–2): 61–64. doi: 10.1016/j.jece.2013.03.007.

3 Bioaugmentation of Pesticides-Contaminated Environment

G. Gokulapriya and M. Chandrasekaran
Anbil Dharmalingam Agricultural College and Research Institute

R.P. Soundararajan
Centre for Plant Protection Studies, Tamil Nadu Agricultural University

CONTENT

3.1 INTRODUCTION

In recent days, pesticides are being a part of human life and the environment. During the last two decades, pesticide usage has increased worldwide dramatically to fulfill the need for food products. Pesticide is a chemical agent used to kill or control pests such as fungus, insects, rodents, or undesired organisms causing harm to crops during various stages, harvest, and storage. It originates from the Latin word "cida" which means to kill (Horsak et al., 1964). The compounds *viz.* Insecticides, fungicides, herbicides, rodenticides, molluscicides, nematicides, plant growth regulators, and others fall under the category of "pesticide." Several thousand years back, Sumerians, Greeks, and Romans used various diverse substances such as sulfur, mercury, arsenic, copper, and a few plant extracts to kill pests. However, the lack of knowledge on chemistry resulted in higher rates of application and phytotoxicity. An era of synthetic pesticides emerged during the late 19th and early 20th centuries, i.e., in 1940 (Abubakar et al., 2020; Ortiz-Hernandez et al., 2013). The synthetic product DDT (dichlorodiphenyltrichloroethane) was the first inorganic pesticide developed and became the most extensively used insecticide for the management of disease-transmitting vectors and several types of insect pests in agriculture crop production. Since DDT is hydrophobic (water-insoluble; lipid-soluble) in nature, it may lodge in fatty tissues and bioaccumulate through the food chain. Rachel Carson, an American marine biologist documented the numerous shortcomings and illnesses of using DDT in her book "Silent Spring" (1962). By 1972, DDT was outlawed in the United States, and worldwide production and use began to decline as well (Rosenfeld and Feng, 2011). In general, pesticide entered into an ecosystem, it passes through many processes, including transformation/degradation, volatilization, plant uptake, sorption–desorption, surface water runoff, and groundwater transport (Chowdhury et al., 2008). One of the primary processes that determine a pesticide's destiny and transit in the environment is transformation or degradation. It encompasses abiotic transformation such as hydrolysis, oxidation, reduction, photolysis, wet deposition, and biotic transformation (degradation by living organisms). A pesticide may either be converted into a degraded substance or totally mineralized into a carbon field during these steps (Karpouzas and Walker, 2000; Singh et al., 2006; Ortiz-Hernandez et al., 2013). Pesticides pose numerous disorders in humans and other living organisms as well. Globally, every year, it is estimated that between 1 and 41 million people suffer from exposure to pesticides (Gyawali, 2018). The epithelial cell T47D developed hormone-dependent breast cancer after being exposed to glyphosate (Thongprakaisang et al., 2013). Several other diseases are brain tumors, prostate cancer, and tumors in the lung, liver, thyroid, uterus, lymphoid glands, and mammary gland. Parkinson's disease in humans is associated with excessive production of Reactive Oxygen Species mediated by many pesticides (Sabarwal et al., 2018). The studies show that carbamate pesticides disturb the immune system resulting in an increased risk of tumor-caused cancer problems (Dhouib et al., 2016; Porto et al., 2011). Most of the pesticides persist in the environment and impose a serious threat on the biotic community. An increase in concern toward the environment has forced us to develop an ecologically sound approach for the effective elimination of synthetic chemical pesticides from a contaminated site. Biological method is an important technology to remediate contaminants. Among that, bioaugmentation enhances the remediation of contaminants effectively (Cycon et al., 2017; Parte et al., 2017).

3.2   IMPACT OF PESTICIDES

Pesticides are inevitable in agriculture is an undeniable truth. Every year millions of tons of pesticides have been exploited in order to sink the yield loss and to produce undamaged products. About 4,122,334 tons of pesticides are used globally in 2018. Meanwhile, pesticide consumption in India is around 58,160 tons (FAO, 2018). Pesticides are advantageous to safeguard crops from biotic stress imposed by pests and diseases. Indiscriminate application of pesticides might be favorable for the resistance development in insects but it is havoc to the public, environment, and other non-target organisms. As organochlorine pesticides are lipophilic, they accumulate in the adipose tissue of animals resulting in biomagnification and remain unaffected over a longer time period (Boudh and Singh, 2019). Owing to the accumulation of pesticides in the environment, the biota of phytoplankton and zooplankton is altered. It also produces carcinogenic, neurotoxic effects on human beings. Underground water quality is also affected by leaching of contaminants (Ortiz-Hernandez et al., 2013). Unintentional and intentional exposure of the public to pesticides poses serious acute and chronic health impacts. Acute effects include irritation of the nose and throat, skin sores, itching of the eyes and skin, nausea, diarrhea, disorientation, myopia, cataracts, neurological and hepatic consequences. Threats associated with chronic pesticidal poisoning are bone marrow infection, neurotoxicity, cancer, male and female infertility, allergies, asthma, etc. (World Health Organization, 1990; Boudh and Singh, 2019).

3.3   SOIL CONTAMINATION

Soil serves as a growing media for microflora and microfauna. In 1 g of soil, there are around 100 million bacteria (5,000–7,000 different species) and over 10,000 fungus colonies (Metting, 1993; Dindal, 1990) were recorded. "If we lose both fungi and bacteria," says Dr. Elaine Ingham, "soil would degrade." Plants growth is supported not only by water, sunlight but also by the soil inhabitant microbes for their nutrient availability. Chlorophenols, excess fertilizers, pesticides, petroleum and associated chemicals, polycyclic aromatic hydrocarbons (PAHs), and heavy metals all disturb the type of soil (Aktar et al., 2009; Gong et al., 2009; Kavamura and Esposito, 2010; Udeigwe et al., 2011; Xu et al., 2012; Hu et al., 2013; Tang et al., 2014). It is possible to determine pesticide persistence, mobility, and transformation products using metrics such as half-life (DT_{50}), octanol/water partition coefficient (K_{ow}), soil-sorption constant (K_{oc}), and water solubility in soil. Soil organic matter and pH play an important factor in the breakdown of pesticides. If the organic matter is less, the adsorption of pesticides with soil will be less, and the availability of pesticides for degradation is more. Soil pH is negatively correlated with adsorption. Pesticides like atrazine, 2,4,5-T, picloram, and 2,4-D have greater adsorption under acidic pH (Andreu and Picó, 2004; Aktar et al., 2009). Some insecticides obstruct the microorganisms participating in the nitrogen mineralization process. Triclopyr, for example, is a herbicide that inhibits the microorganisms that convert ammonia to nitrite (Pell et al., 1998). Similarly, the organophosphorus herbicide glyphosate inhibits nitrogen-fixing bacteria's development and its function (Santos and Flores, 1995). Plant roots have a symbiotic interaction with mycorrhizal fungi, which assist them in the nutrient absorption. Triclopyr, a pyridine group herbicide; oxadiazon, an oxadiazole herbicide and glyphosate, an organophosphate herbicide, have all been reported to disrupt mycorrhizal fungal species (Boudh and Singh, 2019; Zaller et., 2014; Hagner et al., 2019; Aktar et al., 2009; Helander et al., 2018).

3.4 WATER CONTAMINATION

Water is one of the five components needed for living on earth. Pesticide contamination in water is due to the release of persistent pesticides from agricultural fields, urban use for house gardening, and manufacturing factories. Contamination also occurred by drift/runoff from the treated site to rivers, lakes, and other water bodies (Boudh and Singh, 2019; Aktar et al., 2009). According to the National Water Quality Assessment, surface water is more frequently polluted than groundwater due to the direct entry of overland pesticides through runoff (Aktar et al., 2009; Boudh and Singh, 2019). Among 25 pesticides detected, 10% are found in surface water and only 2% in groundwater. Out of it, 6 are insecticides and 11 are herbicides applied for farmlands and in residential gardens (Gilliom et al., 2006). Owing to pesticide toxicity, around 6–14 million fishes died per year. Pesticides that occur in the streams and groundwaters are atrazine, simazine, prometon, tebuthiuron, 2,4-D, diuron, diazinon, chlorpyrifos, and carbaryl. More than 58% of drinking water samples collected from Bhopal were known to be contaminated with organochlorine pesticides. If once the groundwater is contaminated, it is very tedious to clear up the contaminants present in it (Boudh and Singh, 2019; Gilliom et al., 2006). Yamuna River in India is badly polluted due to the increase in population and industrialization. Hexachlorocyclohexane and DDT concentrations in samples taken from various areas on the Yamuna River varied from 12.76 to 593.49 and 66.17 to 722.94 ng/L, respectively (Kaushik et al., 2008; Syafrudin et al., 2021). Similarly, organophosphate herbicides mobilized from agriculture runoff and wastewater pollute Kurose river flows in Hiroshima, affecting aquatic creatures in the Seto Inland Sea. Pesticide concentrations ranged from 2.8 (fenarimol) to 1194 ng/L (diazinon) (Syafrudin et al., 2021; Chidya et al., 2018).

3.5 BIOAUGMENTATION AND ITS CONCEPTS

The process of conversion of noxious pollutants into less toxic or nontoxic form by living organisms like plants, microbes, or its enzymes is bioremediation (Silva et al., 2019; Jobby and Desai, 2017). The incomplete mineralization of toxic compounds by microbes resulted in the formation and accumulation of more toxic compounds than the parent compound. In order to overcome such struggles, the concept of bioaugmentation evolved (Coppola et al., 2011). Bioaugmentation is a method of *in situ* engineered bioremediation (Figure 3.1). Bioaugmentation is a green technology that strengthens the biodegradative potential of contaminated areas by the inoculation of specific strains or

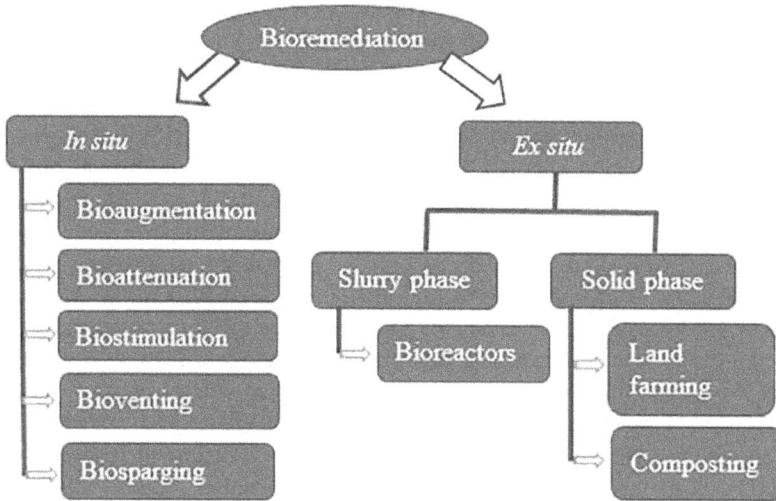

FIGURE 3.1   Bioremediation methods (Tewari et al., 2012).

consortia/group of microorganisms with suitable catalytic properties that can degrade the target molecules (Das and Mukherjee, 2007; Mrozik and Piotrowska-Seget, 2010).

Bioaugmentation helps to avoid the production of toxic metabolites by enhancing the bioremediation process (Castro-Gutierrez et al., 2018). When soil is contaminated with specific pesticide compounds, microbes responsible for such degradation are insufficient. In such cases, bioaugmentation aids to clear up the pollutants by improving specific microbial populations (Tewari et al., 2012). Bioaugmentation is classified as autochthonous, allochthonous, and gene bioaugmentation, depending on the origin of the inoculants. Indigenous microorganisms from the contaminated site are isolated, cultured, and again injected onto the same site in autochthonous bioaugmentation. In the case of allochthonous bioaugmentation, pesticide-degrading microorganisms are isolated from another site, exogenously cultured, and inoculated onto the contaminated site. For gene bioaugmentation microorganisms are encoded with pesticide-degrading genes or enzymes through rDNA technology and such genetically modified microorganisms are incorporated into the pesticide concentrated environment (Pieper and Reineke, 2000; Cycon et al., 2017; Zhang et al., 2012). When the quantity of native microorganisms is insufficient to degrade the native pollutants, bioaugmentation is used (Cycon et al., 2017; Gentry et al., 2004; Mrozik and Piotrowska-Seget, 2010).

3.6   MICROORGANISMS IN BIOAUGMENTATION

In nature, microorganisms are ubiquitous and diverse in the way of antagonistic and synergistic populations in their community. The ability of microorganisms to degrade pesticides through their catabolic activity into an end product like carbon dioxide, water, or intermediate metabolites that are required for its growth has promoted its utilization in the remediation of pollutants (Verma and Kuila, 2019). Microorganisms interact on the substrates chemically and physically, resulting in breakdown or structural alterations in the target molecule. Usually, more than one microorganism is involved in pesticide degradation. The magnitude of microbial population in a contaminated site can be strengthened either by the stimulation of heirloom microbes by supplementing with nutrients or electron acceptors (biostimulation) or certain specific microorganisms are introduced to a localized site (bioaugmentation). The inoculated microbes might be a single strain or consortia. Release of microbial consortia is considered to be more effective than a single strain since the product of one strain may be utilized by another strain for its growth. Many studies have indicated that using bacterial consortia to degrade aromatic chemicals is more successful than deploying a single strain (Mrozik and Piotrowska-Seget, 2010; Ghazali et al., 2004; Goux et al., 2003).

Among the microbial populations, bacteria, fungus, and actinomycetes are the primary and active pesticide degraders (Briceno et al., 2007; Ortiz-Hernandez et al., 2013) (Table 3.1). In general, fungi transform the contaminants by altering the molecular structure of the compound, and the released product is nontoxic. Such biotransformed compound is further degraded by the activity of bacteria present in contaminated sites (Ortiz-Hernandez et al., 2013; Diez, 2010). Bacteria act as a significant and prominent degrader of pollutants (Sylvia et al., 2005). *Aspergillus oryzae* CCMI 125, *Fusarium oxysporum* CCMI866, *Lecanicillium saksenae* PP0011, *Lentinula edodes* EL1, and *Penicillium brevicompactum* PP0021 are fungal strains capable to degrade terbuthylazine, difenoconazole, and pendimethalin pesticides from the liquid culture medium (Pinto et al., 2012; Dey et al., 2021). Besides bacteria, algae also got scant importance in transforming contaminants since it occupies the base of the food chain. *Ankistrodesmus* SI2, *Chlorella vulgaris*, and *Scenedesmus* SI1 are involved in the transformation of naphthalene, dibenzofuran, and dibenzo-p-dioxin compounds respectively through their Cytochrome P450 and oxygenase enzymatic activity (Todd et al., 2002). Degradation of fenamiphos, an organophosphorus pesticide by the five algal species (*Chlamydomonas* sp., *Chlorella* sp., *Scenedesmus* sp. MM1, *Scenedesmus* sp. MM2, *Stichococcus* sp.,) and cyanobacteria species (*Anabaena* sp., *Nostoc* sp. MM1, *Nostoc* sp. MM2, *Nostoc* sp. MM3, *Nostoc muscorum*) into fenamiphos sulfoxide or fenamiphos sulfoxide phenol (Caceres et al., 2008).

3.7   FACTORS LIMITING BIOAUGMENTATION

The outcome of bioaugmentation depends on two factors namely, biotic and abiotic factors. Biotic factors include competition existing between introduced and indigenous microbes on availability of nutrients, infected by the native protozoans, chemical compounds contained in root exudates generated by plants. Temperature, moisture, pH, soil type, organic matter falls under abiotic factors (Gentry et al., 2004; El Fantroussi and Agathos, 2005; Mrozik et al., 2010; Sartoros et al., 2005; Lebeau, 2011).

3.8   TEMPERATURE

Degradation rate of toxic chemicals is mainly driven by changes in temperature. The rate of chemical degradation will double for a rise in temperature of every 10°C (Cress, 1990). Temperature alters the hydrolysis and solubility of pesticide thereby rate of adsorption is affected (Burns, 1975; Racke et al., 1997). As adsorption is an exothermic process and desorption is an endothermic process, a rise in temperature causes less adsorption and more desorption (Perez et al., 2013). Diels and Lookman (2007) reported that the temperature ranges from 5°C to 30°C influence the bioremediation of contaminants. Microorganisms thrive and perform efficiently between 25°C and 35°C, and pesticide breakdown is ideal at 25°C–40°C (Alexander, 1977). Hong et al. (2007), evaluated the optimum temperature and pH for Burkholderia sp. FDS-1 strain growth and activity. The results also suggested that 30°C with slightly alkaline pH was ideal. *Pseudomonas putida* strains ep I and II were also used to degrade organophosphate ethoprophos (Karpouzas and Walker, 2000). The results show that at 25°C and 37°C bacterial strains can degrade ethoprophos completely. 30°C was found to be the appropriate temperature for *Pseudomonas cepacia* to degrade 2, 4, 5-T in soil (Chatterjee et al., 1982).

3.9   SOIL MOISTURE

Its' known phenomena that the soil moisture is essential for the normal growth and development of microorganisms. It indirectly indicates the aeration of soil (Lebeau, 2011). Water serves as a solvent for the mobilization of pesticides and diffusion. Depending upon the pesticide solubility in

TABLE 3.1

Microorganisms Involved in Degradation of Toxic Contaminants

Microbes	Contaminant	Source	Reference(s)
Pseudomonas sp. strain (WBC-3)	Methyl parathion and *para*-nitrophenol	Agricultural loamy soil taken from Wuhan Botanical Garden	Wang et al. (2014)
Pseudomonas stuzeri	Chlorpyriphos	Soil poisoned by pesticide in Egypt	Awad et al. (2011)
Agrobacterium radiobacter (J14a)	Atrazine	Atrazine treated cornfield soil near Sheldon	Struthers et al. (1998)
Desulfitobacterium frappieri PCP-1 (ATC700397)	Pentachlorophenol	Anaerobic sludge from industrial wastewater	Guiot et al. (2002)
			Lanthier et al. (2002)
Mortierella sp. strain W8	Endosulfan	Agricultural soil	Kataoka et al. (2011)
Serratia marcescens (Del-1 and Del-2)	Deltamethrin	Deltamethrin polluted soil, Poland	Cycon et al. (2014)
Bacterial consortium	Chlorpyriphos	Chlorpyriphos tainted soil, India	Lakshmi et al. (2008)
Acremonium sp. (CBMAI 1676)	Fenvalerate	Sponge from unpolluted South Atlantic Ocean, Brazil	Birolli et al. (2016)
Pseudochrobactrum sp. (BSQ1)	Chlorothalonil	Soil from Nanjing, Jiangsu	Xu et al. (2018)
Massilia sp. (BLM18)			
Ralstonia eutropha (JMP134)	2,4-Dichlorophenoxyacetic acid	Contaminated soil	Roane et al. (2001)
Shinella sp. (NJUST26)	2,4-triazole	Triazole-contaminated soil	Wu et al. (2018)
Sphingomonas sp. (NJUST37)	Tricyclazole	Activated sludge	
Sphingobium sp. (JQL4–5)	Bifenthrin	Wastewater sludge	Yuanfan et al. (2010)
	Cypermethrin		
	Cyhalothrin		
	Deltamethrin		
	Fenpropathrin		
Bacillus cereus Y1 strain	Deltamethrin	Soil poisoned with deltamethrin	Zhang et al. (2016)

(Continued)

TABLE 3.1 (*Continued*)
Microorganisms Involved in Degradation of Toxic Contaminants

Microbes	Contaminant	Source	Reference(s)
Achromobacter sp. kg 16 (VKM B-2534D) *Ochrobactrum anthropi* GPK 3 (VKM B-2554D)	Glyphosate	Organophosphate-contaminated podzolic soil	Ermakova et al. (2010)
Actinomycetes consortia *Streptomyces* sp. strain (A2-A5-A11-M7 consortium)	Lindane	Lindane-contaminated soil	Raimondo et al. (2019)
Streptomyces aureus (HP-S-01)	Cypermethrin Deltamethrin	Activated sludge	Chen et al. (2011)
Verticillum sp. (DSP)	Chlorpyriphos	Contaminated soil	Fang et al. (2008)
Lignolytic fungi	Atrazine	Contaminated wastewater	Castro-Gutierrez et al. (2018)
Trametes versicolor	Carbendazim Carbofuran Metalaxyl		
Fusarium solani	γ-HCH	Soil collected from lindane producing pesticide industry	Sagar and Singh (2011)
Chlorella vulgaris	p-Chlorophenol	Liquid wastes contaminated with aromatic compounds	Lima et al. (2004)
Coenochlorispyrenoidosa			

water, the compound will be available to its degraders. About 50% moisture content will give a good rate of degradation (Tewari et al., 2012). If the soil moisture is less, the rate of degradation is slow. Perez et al. (2013) concluded the optimum moisture content needed for the degradation of chlorpyriphos and diazinon in andic soil is between 100% and 80%. Similarly, the efficiency of *Pseudomonas putida* epI to degrade ethoprophos in soil with 60% moisture content was significantly high (Karpouzas and Walker, 2000).

3.10 SOIL PH

Soil pH also has a potent role in pesticide adsorption and degradation. An efficient degradation can be achieved at pH ranges from 5 to 8. If pH of the soil is more, the adsorptive potential of pesticide is low and its availability for microbial degradation will be more (Chowdhury et al., 2008; Lebeau, 2011; Diels and Lookman, 2007; Burns, 1975). Optimum pH is necessary for microorganisms for better growth and development. At lower pH (5.4) degrading potential of *Pseudomonas putida* was lost, while at pH 6.8 and 8.3 the activity of bacteria was observed in soil (Karpouzas and Walker, 2000). Topp et al. (1997) noted that the sorption potential of prometryn herbicide to montmorillonite clay is more at pH 3 than at pH 7. Degradation of atrazine was efficient at pH greater than 7 (Lebeau, 2011).

3.11 ORGANIC MATTER

Bioavailability of contaminants and the survivability of exogenously introduced microbes are much influenced by organic matter (Mrozik and Piotrowska-Seget, 2010). Organic matter in soil can either impart negative effect on degrading microbes by boosting pesticide adsorption mechanism or imparts productive function on microbes by promoting co-metabolism (Chowdhury et al., 2008). Low organic matter in soils had higher 2,4-D mineralization by native microorganisms (Greer and Shelton, 1992). For maintaining the local microbial population it is essential to sustain a minimum level of organic matter (Burns, 1975). Cycon et al. (2014) demonstrated with a laboratory experiment for enhancing deltamethrin degradation by augmenting with *Serratia marcescens* strains. The findings showed that deltamethrin breakdown was slower in silty and silty loam soil, indicating that organic matter plays an important role in pesticide breakdown.

3.12 BIOAUGMENTATION APPROACHES

Efficacy of bioremediation could be attained by maintaining appropriate number and biomass of inoculated strains (Mrozik et al., 2010). Cell bioaugmentation and genetic bioaugmentation are the two approaches to bioaugmentation.

3.13 CELL BIOAUGMENTATION

Cell bioaugmentation emphasizes on survival and catabolic mechanisms of introduced microbial strains. Microbial inoculants can be introduced either as liquid culture or combined with the carrier material. Direct delivery of inoculum could be possible for surface soil through mechanical means into the contaminated site, but in a sub-surface environment direct incorporation could be laborious. Liquid cultures of about 10^9–10^{12} cells/mL of the cell suspension are incorporated with nutrients and then it is injected at targeted sites using liquid delivery pumps (Da Silva et al., 2010). However, microbial affinity to contaminants could not be ascertained while liquid cultures are used. The transport of microbial inoculum can be enhanced by using ultra micro bacteria, bacteria with the altered cell surface, adhesion-deficient bacteria, starved bacteria, surfactants. Starvation alters the adhesion properties of microbial cells to the substrate. Hence, microbial adherence with soil surface can be potentially reduced and inoculum distribution can be increased effectively (Borges et al., 2008).

Distribution of inoculum can also be determined by the type of organism utilized for augmentation (Gentry et al., 2004; Da Silva et al., 2010). Carrier materials safeguard the inoculum from fluctuating external factors and supply nutrients temporarily to introduced microorganisms. Inorganic carrier material includes lignite, biosolids, soil amended with charcoal, manure, lignite, peat, and clay (Rebah et al., 2002; Gentry et al., 2004). Alginate, agar, carrageenan, agarose, and chitosan are naturally derived polymers, and synthetic polymers like acrylamide, polyurethane, polyvinyl, and resins are all act as carrier materials for immobilization of microorganisms. Ideal characteristics of carrier material are: (i) providing a suitable environment for growth and survival of microbial inoculum, (ii) being nontoxic to inoculum and environment, and (iii) introduction of cells targeted as well to encapsulate the microorganisms injected when control is required (Gentry et al., 2004; Van Veen et al., 1997). Studies with phenol and TCE (trichloroethene) degradation with suspended and chitosan-bead immobilized *Pseudomonas putida* show that the immobilized cell culture may tolerate a higher concentration of TCE and its transfer yield confirmed the protective behavior of carrier material to immobilized cells (Chen et al., 2007). Efficacy of bioaugmentation may also be improved by using activated soil. Activated soil naturally contains the contaminant degrading microbial population acquired by continuous exposure of contaminants at a specific site. While applying activated soil, there is no need to culture the inoculum outside the soil and made up of many different groups of microbes which could enhance degradation than inoculated as pure culture (Mrozik et al., 2010; Gentry et al., 2004).

3.14   GENETIC BIOAUGMENTATION

Recently, attempts made on the technology, horizontal gene transfer (HGT) of pollutant degrading genetic traits from donor to native microflora existing in contaminated areas. The main aim of their experiments was to improve the xenobiotic degrading potential into a broader range. Among three mechanisms of gene transfer in bacteria, i.e., transformation, transduction, and conjugation, conjugation is the most efficacious biological process wherein genetic information stored in plasmids is exchanged from donor to recipient bacteria by the direct cell-to-cell contact (Garbisu et al., 2017; Furuya and Lowy, 2006).

In the recent past, the impact of genetically modified bacterial strains on hydrocarbon breakdown in soil has been studied extensively. The effectiveness of two bacterial strains *viz.* natural isolate *Arthrobacter* sp. FG1 and recombinant strain *Pseudomonas putida* PaW340/p DH5 on the breakdown of 4-chlorobenzoic acid were investigated. Dehalogenase genes from *Arthrobacter* sp. FG1 was cloned into *P. putida* PaW340 to construct the recombinant strain. The slurry seeded with the recombinant strain degraded 4-chlorobenzoic acid more quickly than the slurry infected with *Arthrobacter* sp. FG1 (Massa et al., 2009). *Stenotrophomonas* strain YC-1 successfully degraded chlorpyriphos in soil when the *mpd* gene encoding the organophosphorus hydrolase was cloned from a chlorpyriphos-degrading bacterium (strain YC-1) and expressed in *Escherichia coli* (Yang et al., 2006). The most explored technology in genetic bioaugmentation studies is the transfer of genetic materials in plasmid through conjugation, where the survival of the imported host strain for a longer period is not necessary (Lebeau, 2011).

3.15   CONCLUSION

Biological remediation of soil is an environment-sustainable way to solve the contaminants. Over the past few years, many researches toward the environment were on advances in bioaugmentation. Cell bioaugmentation is the most widely prevailing method for delivering microbial inoculum at contaminated sites. To mitigate rapid mortality of cultures in cell bioaugmentation, plasmid-mediated genetic bioaugmentation seems to provide greater opportunity to ameliorate microbial degradation. Though hundreds of studies in bioaugmentation, it is still in laboratory-scale experiments. Besides spatial heterogeneity of soil, variations in contaminant concentration throughout the site limit the bioaugmentation at a small scale. Sustained research works are required, primarily at the ground level, to commercialize the emerging technologies. Types of microorganisms and methods of bioaugmentation technique should be carefully selected to increase the degradation capacity of degraders. Further research is required to gain knowledge on the ecological traits of microflora to avoid abiotic stresses.

REFERENCES

Abubakar, Y., Tijjani, H., Egbuna, C., Adetunji, C.O., Kala, S., Kryeziu, T.L., Ifemeje, J.C. and Patrick-Iwuanyanwu, K.C., 2020. Pesticides, history, and classification. In Natural Remedies for Pest, Disease and Weed Control, pp. 29–42. Academic Press.

Aktar, M.W., Sengupta, D. and Chowdhury, A., 2009. Impact of pesticides use in agriculture: their benefits and hazards. *Interdisciplinary Toxicology*, 2(1), 1–12.

Alexander, M., 1977. *Introduction to Soil Microbiology*. 2nd Ed. Wiley Eastern Limited, New Delhi.

Andreu, V. and Picó, Y., 2004. Determination of pesticides and their degradationproducts in soil: critical review and comparison of methods. *TrAC Trends in Analytical Chemistry*, 23(10–11), 772–789.

Awad, N.S., Sabit, H.H., Abo-Aba, S.E. and Bayoumi, R.A., 2011. Isolation, characterization and fingerprinting of some chlorpyrifos-degrading bacterial strains isolated from Egyptian pesticides-polluted soils. *African Journal of Microbiology Research*, 5(18), 2855–2862.

Birolli, W.G., Alvarenga, N., Seleghim, M.H. and Porto, A.L., 2016. Biodegradation of the pyrethroid pesticidesfenvalerate by marine-derived fungi. *MarineBiotechnology*, 18(4), 511–520.

Borges, M.T., Nascimento, A.G., Rocha, U.N. and Tótola, M.R., 2008. Nitrogenstarvation affects bacterial adhesion to soil. *Brazilian Journal of Microbiology*, 39(3), 457–463.

Boudh, S. and Singh, J.S., 2019. Pesticide contamination: environmental problems and remediation strategies. In *Emerging and Eco-Friendly Approaches for Waste Management*, pp. 245–269. Springer, Singapore.

Briceno, G., Palma, G. and Durán, N., 2007. Influence of organic amendment on the biodegradation and movement of pesticides. *Critical Reviews in Environmental Science and Technology*, 37(3), 233–271.

Burns, R.G., 1975. Factors affecting pesticides loss from soil. *Soil Biochemistry*, 4, 103–141.

Caceres, T.P., Megharaj, M. and Naidu, R., 2008. Biodegradation of the pesticide fenamiphos by ten different species of green algae and cyanobacteria. *Current Microbiology*, 57(6), 643–646.

Carson, R. 1962. Silent spring. Houghton Mufflin, Boston, USA.

Castro-Gutierrez, V., Masis-Mora, M., Carazo-Rojas, E., Mora-Lopez, M. and Rodriguez-Rodriguez, C.E., 2018. Impact of oxytetracycline and bacterial bioaugmentation on the efficiency and microbial community structure of a pesticide-degrading biomixture. *Environmental Science and Pollution Research*, 25(12), 11787–11799.

Chatterjee, D.K., Kilbane, J.J. and Chakrabarty, A.M., 1982. Biodegradation of 2, 4, 5- trichlorophenoxyacetic acid in soil by a pure culture of *Pseudomonas cepacia*. *Applied and Environmental Microbiology*, 44(2), 514.

Chen, S., Lai, K., Li, Y., Hu, M., Zhang, Y. and Zeng, Y., 2011. Biodegradation of deltamethrin and its hydrolysis product 3-phenoxybenzaldehyde by a newly isolated *Streptomyces aureus* strain HP-S-01. *Applied Microbiology and Biotechnology*, 90(4), 1471–1483.

Chen, Y.M., Lin, T.F., Huang, C., Lin, J.C. and Hsieh, F.M., 2007. Degradation of phenol and TCE using suspended and chitosan-bead immobilized *Pseudomonas putida*. *Journal of Hazardous Materials*, 148(3), 660–670.

Chidya, R.C., Abdel-Dayem, S.M., Takeda, K. and Sakugawa, H., 2018. Spatio-temporal variations of selected pesticide residues in the Kurose River in HigashiHiroshima city, Japan. *Journal of Environmental Science and Health, Part B*, 53(9), 602–614.

Chowdhury, A., Pradhan, S., Saha, M., Sanyal, N., 2008. Impact of pesticides on soil microbiologicalparameters and possible bioremediation strategies. *Indian Journal of Microbiology*, 48, 114–127.

Coppola, L., Pilar Castillo, M.D. and Vischetti, C., 2011. Degradation of isoproturon and bentazone in peat and compost based biomixtures. *Pest Management Science*, 67(1), 107–113.

Cress, D., 1990. *Factors Affecting Pesticide Behaviour and Breakdown*. Ph.D. Dissertation Work, Kansas State University.

Cycon, M., Mrozik, A. and Piotrowska-Seget, Z., 2017. Bioaugmentation as a strategy for the remediation of pesticide-polluted soil: A review. *Chemosphere*, 172, 52–71.

Cycon, M., Żmijowska, A. and Piotrowska-Seget, Z., 2014. Enhancement of deltamethrin degradation by soil bioaugmentation with two different strains of *Serratia marcescens*. *International Journal of Environmental Science and Technology*, *11*(5), 1305–1316.

Da Silva, M.L.B. and Alvarez, P.J.J., 2010. Bioaugmentation. In Timmis K.N. (Ed) *Handbook of Hydrocarbon and Lipid Microbiology*. Springer, Berlin, Heidelberg.

Das, K. and Mukherjee, A.K., 2007. Crude petroleum-oil biodegradation efficiency of *Bacillus subtilis* and *Pseudomonas aeruginosa* strains isolated from a petroleum-oil contaminated soil from North-East India. *Bioresource Technology*, *98*(7), 1339–1345.

Dey, P., Gola, D., Chauhan, N., Bharti, R.K. and Malik, A., 2021. Mechanistic insight to bioremediation of hazardous metals and pesticides from water bodies by microbes. In: Maulin P. Shah (Ed) *Removal of Emerging Contaminants Through Microbial Processes*, pp. 467–487. Springer, Singapore.

Dhouib, I., Jallouli, M., Annabi, A., Marzouki, S., Gharbi, N., Elfazaa, S. and Lasram, M.M., 2016. From immunotoxicity to carcinogenicity: the effects of carbamate pesticides on the immune system. *Environmental Science and Pollution Research*, *23*(10), 9448–9458.

Diels, L. and Lookman, R., 2007. Microbial systems for in-situ soil and groundwater remediation. In: Marmiroli, N., Samotokin, B. and Marmiroli, M. (Eds) *Advanced Science and Technology for Biological decontamination of Sites Affected by Chemical and Radiological Nuclear Agents*, pp. 61–77. Springer, Dordrecht.

Diez, M.C., 2010. Biological aspects involved in the degradation of organic pollutants. *Journal of Soil Science and Plant Nutrition*, *10*(3), 244–267.

Dindal, D.L., 1990. *Soil Biology Guide*. Wiley & Sons, New York.

El Fantroussi, S. and Agathos, S.N., 2005. Is bioaugmentation a feasible strategy for pollutant removal and site remediation? *Current Opinion in Microbiology*, *8*(3), 268–275.

Ermakova, I.T., Kiseleva, N.I., Shushkova, T., Zharikov, M., Zharikov, G.A. and Leontievsky, A.A., 2010. Bioremediation of glyphosate-contaminated soils. *Applied Microbiology and Biotechnology*, *88*(2), 585–594.

Fang, H., Xiang, Y.Q., Hao, Y.J., Chu, X.Q., Pan, X.D., Yu, J.Q. and Yu, Y.L., 2008. Fungal degradation of chlorpyrifos by *Verticillium* sp. DSP in pure cultures and its use in bioremediation of contaminated soil and pakchoi. *International Biodeterioration & Biodegradation*, *61*(4), 294–303.

FAO, 2018. Food and Agriculture Organisation of the United Nations. (http://www.fao.org/faostat/en/).

Furuya, E.Y. and Lowy, F.D., 2006. Antimicrobial-resistant bacteria in the community setting. *Nature Reviews Microbiology*, *4*(1), 36–45.

Garbisu, C., Garaiyurrebaso, O., Epelde, L., Grohmann, E. and Alkorta, I., 2017. Plasmid-mediated bioaugmentation for the bioremediation of contaminated soils. *Frontiers in Microbiology*, *8*: 1966.

Gentry, T., Rensing, C. and Pepper, I.A.N., 2004. New approaches for bioaugmentation as a remediation technology. *Critical Reviews in Environmental Science and Technology*, *34*(5), 447–494.

Ghazali, F.M., Rahman, R.N.Z.A., Salleh, A.B. and Basri, M., 2004. Biodegradation of hydrocarbons in soil by microbial consortium. *International Biodeterioration & Biodegradation*, *54*(1), 61–67.

Gilliom, R.J., Barbash, J.E., Crawford, C.G., Hamilton, P.A., Martin, J.D., Nakagaki, N., Nowell, L.H., Scott, J.C., Stackelberg, P.E., Thelin, G.P. and Wolock, D.M., 2006. *Pesticides in the Nation's Streams and Ground Water, 1992–2001(No. 1291)*. US Geological Survey.

Gong, J.L., Wang, B., Zeng, G.M., Yang, C.P., Niu, C.G., Niu, Q.Y., Zhou, W.J. and Liang, Y., 2009. Removal of cationic dyes from aqueous solution using magnetic multi-wall carbon nanotube nanocomposite as adsorbent. *Journal of Hazardous Materials*, *164*(2–3), 1517–1522.

Goux, S., Shapir, N., El Fantroussi, S., Lelong, S., Agathos, S.N. and Pussemier, L., 2003. Long-term maintenance of rapid atrazine degradation in soils inoculated with atrazine degraders. *Water, Air and Soil Pollution: Focus 3*(3), 131–142.

Greer, L.E. and Shelton, D.R., 1992. Effect of inoculant strain and organic mattercontent on kinetics of 2,4-dichlorophenoxyacetic acid degradation in soil. *Applied and Environmental Microbiology*, *58*(5), 1459–1465.

Guiot, S.R., Tartakovsky, B., Lanthier, M., Lévesque, M.J., Manuel, M.F., Beaudet, R., Greer, C.W. and Villemur, R., 2002. Strategies for augmenting the pentachlorophenol degradation potential of UASB anaerobic granules. *Water Science and Technology*, *45*(10), 35–41.

Gyawali, K., 2018. Pesticide uses and its effects on public health and environment. *Journal of Health Promotion*, *6*, 28–36.

Hagner, M., Mikola, J., Saloniemi, I., Saikkonen, K. and Helander, M., 2019. Effects of a glyphosate-based herbicide on soil animal trophic groups and associated ecosystem functioning in a northern agricultural field. *Scientific Reports*, *9*(1), 1–13.

Helander, M., Saloniemi, I., Omacini, M., Druille, M., Salminen, J.P. and Saikkonen, K., 2018. Glyphosate decreases mycorrhizal colonization and affects plant-soil feedback. *Science of the Total Environment,* *642*, 285–291.

Hong, Q., Zhang, Z., Hong, Y. and Li, S., 2007. A microcosm study on bioremediation of fenitrothion-contaminated soil using *Burkholderia* sp. FDS-1. *International Biodeterioration & Biodegradation,* *59*(1), 55–61.

Horsak, R.D., Bedient, B.P., Hamilton, C.M. and Thomas, B.F., 1964. Pesticides. *Environmental Forensics: Contaminant Specific Guide,* 8, 143–165.

Hu, G., Li, J. and Zeng, G., 2013. Recent development in the treatment of oily sludge from petroleum industry: a review. *Journal of Hazardous Materials,* *261*, 470–490.

Jobby, R. and Desai, N., 2017. Bioremediation of heavy metals. In: Kumar, P., Bhola R. Gurjar and Govil, J.N. (eds) *Biodegradation and Bioremediation. Environmental Science and Engineering,* *8*, pp. 201–220. Stadium Press, New Delhi.

Karpouzas, D.G., Walker, A., 2000. Factors influencing the ability of *Pseudomonas putida* epI to degrade-ethoprophos in soil. *Soil Biology and Biochemistry,* *32*, 1753–1762.

Kataoka, R., Takagi, K. and Sakakibara, F., 2011. Biodegradation of endosulfan by *Mortieralla* sp. strain W8 in soil: Influence of different substrates on biodegradation. *Chemosphere,* *85*(3), 548–552.

Kaushik, C.P., Sharma, H.R., Jain, S., Dawra, J. and Kaushik, A., 2008. Pesticideresidues in river Yamuna and its canals in Haryana and Delhi, India. *Environmental Monitoring and Assessment,* *144*(1), 329–340.

Kavamura, V.N. and Esposito, E., 2010. Biotechnological strategies applied to the decontamination of soils polluted with heavy metals. *Biotechnology Advances,* *28*(1), 61–69.

Lakshmi, C.V., Kumar, M. and Khanna, S., 2008. Biotransformation of chlorpyrifos and bioremediation of contaminated soil. *International Biodeterioration & Biodegradation,* *62*(2), 204–209.

Lanthier, M., Tartakovsky, B., Villemur, R., DeLuca, G. and Guiot, S.R., 2002. Microstructure of anaerobic granules bioaugmented with *Desulfitobacterium frappieri* PCP-1. *Applied and Environmental Microbiology,* *68*(8), 4035–4043.

Lebeau, T., 2011. Bioaugmentation for in situ soil remediation: how to ensure the success of such a process. In : Singh, A., et al., (Eds) *Bioaugmentation, Biostimulation and Biocontrol,* pp. 129–186. Springer, Berlin, Heidelberg.

Lima, S.A., Raposo, M.F.J., Castro, P.M. and Morais, R.M., 2004. Biodegradation of p-chlorophenol by a microalgae consortium. *Water Research,* *38*(1), 97–102.

Massa, V., Infantino, A., Radice, F., Orlandi, V., Tavecchio, F., Giudici, R., Conti, F., Urbini, G., Di Guardo, A. and Barbieri, P., 2009. Efficiency of natural and engineered bacterial strains in the degradation of 4-chlorobenzoic acid in soil slurry. *International Biodeterioration & Biodegradation,* *63*(1), 112–115.

Metting, F.B., 1993. Soil Microbial Ecology, Applications in Agriculture and Environmental Management. Marcel Dekker, Inc, New York.

Mrozik, A. and Piotrowska-Seget, Z., 2010. Bioaugmentation as a strategy for cleaning up of soils contaminated with aromatic compounds. *Microbiological Research,* *165*(5), 363–375.

Ortiz-Hernandez, M.L., Sanchez-Salinas, E., Dantan-Gonzalez, E. and Castrejon-Godinez, M.L., 2013. Pesticide biodegradation: mechanisms, genetics and strategies to enhance the process. *Biodegradation-Life of Science,* *10*, 251–287.

Parte, S.G., Mohekar, A.D. and Kharat, A.S., 2017. Microbial degradation of pesticide: a review. *African Journal of Microbiology Research,* *11*(24), 992–1012.

Pell, M., Stenberg, B. and Torstensson, L., 1998. Potential denitrification and nitrification tests for evaluation of pesticide effects in soil. *Ambio,* *27*, 24–28.

Perez, E.H., Paez, M.I. and Figueroa, A., 2013. Effect of humidity and temperature on dissipation of chlorpyrifos and diazinon in andic soils, cauca, Colombia. *Asian Journal of Chemistry,* *25*(16), 9208–9212.

Pieper, D.H. and Reineke, W., 2000. Engineering bacteria for bioremediation. *Current Opinion in Biotechnology,* *11*(3), 262–270.

Pinto, A.P., Serrano, C., Pires, T., Mestrinho, E., Dias, L., Teixeira, D.M. and Caldeira, A.T., 2012. Degradation of terbuthylazine, difenoconazole and pendimethalin pesticides by selected fungi cultures. *Science of the Total Environment,* *435*, 402–410.

Porto, A.L.M., Melgar, G.Z., Kasemodel, M.C. and Nitschke, M., 2011. Biodegradation of pesticides. *Pesticides in the Modern World–Pesticides Use and Management, Stoytcheva M.(Ed.)//Tech,* *1*, 407–438.

Racke, K.D., Skidmore, M.W., Hamilton, D.J., Unsworth, J.B., Miyamoto, J. and Cohen, S.Z., 1997. Pesticides report 38. Pesticide fate in tropical soils(technical report). *Pure and Applied Chemistry,* *69*(6), 1349–1372.

Raimondo, E.E., Aparicio, J.D., Briceño, G.E., Fuentes, M.S. and Benimeli, C.S., 2019. Lindane bioremediation in soils of different textural classes by an *Actinobacteria consortium*. *Journal of Soil Science and Plant Nutrition*, *19*(1), 29–41.

Rebah, F.B., Tyagi, R.D. and Prevost, D., 2002. Wastewater sludge as a substrate for growth and carrier for rhizobia: the effect of storage conditions on survival of *Sinorhizobium meliloti*. *Bioresource Technology*, *83*(2), 145–151.

Roane, T.M., Josephson, K.L. and Pepper, I.L., 2001. Dual-bioaugmentation strategy to enhanceremediation of cocontaminated soil. *Applied and Environmental Microbiology*, *67*(7), 3208–3215.

Rosenfeld, P.E. and Feng, L., 2011. *Risks of Hazardous Wastes*. Oxford: William Andrew, Burlington, pp. 441–454.

Sabarwal, A., Kumar, K. and Singh, R.P., 2018. Hazardous effects of chemical pesticides on human health–Cancer and other associated disorders. *Environmental Toxicology and Pharmacology*, *63*, 103–114.

Sagar, V. and Singh, D.P., 2011. Biodegradation of lindane pesticide by non white-rots soil fungus *Fusarium* sp. *World Journal of Microbiology and Biotechnology*, *27*(8), 1747–1754.

Santos, A. and Flores, M., 1995. Effects of glyphosate on nitrogen fixation of free-living heterotrophic bacteria. *Letters in Applied Microbiology*, *20*(6), 349–352.

Sartoros, C., Yerushalmi, L., Beron, P. and Guiot, S.R., 2005. Effects of surfactant and temperatureonbiotransformationkineticsof anthracene and pyrene. *Chemosphere*, *61*(7), 1042–1050.

Silva, A., Delerue-Matos, C., Figueiredo, S.A. and Freitas, O.M., 2019. The use of algae and fungi for removal of pharmaceuticals by bioremediation and biosorption processes: a review. *Water*, *11*(8), 1555.

Singh, B.K., Walker, A., Wright, D.J., 2006. Bioremedial potential of fenamiphos and chlorpyrifos degradingisolates: influence of different environmental conditions. *Soil Biology and Biochemistry*, *38*, 2682–2693.

Struthers, J.K., Jayachandran, K. and Moorman, T.B., 1998. Biodegradation of atrazine by *Agrobacterium radiobacter* J14a and use of this strain in bioremediation of contaminated soil. *Applied and Environmental Microbiology*, *64*(9), 3368–3375.

Syafrudin, M., Kristanti, R.A., Yuniarto, A., Hadibarata, T., Rhee, J., Al-Onazi, W.A., Algarni, T.S., Almarri, A.H. and Al-Mohaimeed, A.M., 2021. Pesticides in drinking water-a review. *International Journal of Environmental Research and Public Health*, *18*(2), 468.

Sylvia, D.M., Fuhrmann, J.J., Hartel, P.G. and Zuberer, D.A., 2005. *Principles and Applications of Soil Microbiology (No. QR111 S674 2005)*. Pearson.

Tang, W.W., Zeng, G.M., Gong, J.L., Liang, J., Xu, P., Zhang, C. and Huang, B.B., 2014. Impact of humic/fulvic acid on the removal of heavy metals from aqueous solutions using nanomaterials: a review. *Science of the Total Environment*, *468*, 1014–1027.

Tewari, L., Saini, J and Arti, 2012. Bioremediation of pesticides by microorganisms: general aspects and recent advances. In: Maheshwary, D.K. and Dubey, R.C. (Eds) *Bioremediation of Pollutants*. I.K. International Publishing House Pvt. Ltd., India. pp. 24–49.

Thongprakaisang, S., Thiantanawat, A., Rangkadilok, N., Suriyo, T. and Satayavivad, J., 2013. Glyphosate induces human breast cancer cells growth via estrogen receptors. *Food and Chemical Toxicology*, *59*, 129–136.

Todd, S.J., Cain, R.B. and Schmidt, S., 2002. Biotransformation of naphthalene and diaryl ethers by green microalgae. *Biodegradation*, *13*(4), 229–238.

Topp, E., Vallayes, T and Soulas, G., 1997. Pesticides: Microbial degradation and effects on microorganisms. In: Van Elsas, J.D., Trevors, J.T. and Wellington, E.M.H. (Eds) *Modern Soil Microbiology*. Mercel Dekker, Inc. New york, USA, 547–575.

Udeigwe, T.K., Eze, P.N., Teboh, J.M. and Stietiya, M.H., 2011. Application, chemistry, and environmental implications of contaminant-immobilization amendments on agricultural soil and water quality. *Environment International*, *37*(1), 258–267.

Van Veen, J.A., van Overbeek, L.S. and van Elsas, J.D., 1997. Fate and activity of microorganisms introduced into soil. *Microbiology and Molecular Biology Reviews*, *61*(2), 121–35.

Verma, S. and Kuila, A., 2019. Bioremediation of heavy metals by microbial process. *Environmental Technology & Innovation*, *14*, 100369.

Wang, L., Chi, X.Q., Zhang, J.J., Sun, D.L. and Zhou, N.Y., 2014. Bioaugmentation of a methyl parathion contaminated soil with *Pseudomonas* sp. strain WBC 3. *International Biodeterioration & Biodegradation*, *87*, 116–121.

World Health Organization, 1990. *Public Health Impact of Pesticides Used Inagriculture*. World Health Organization.

Wu, H., Shen, J., Jiang, X., Liu, X., Sun, X., Li, J., Han, W. and Wang, L., 2018. Bioaugmentation strategy for the treatment of fungicide wastewater by two triazole-degrading strains. *Chemical Engineering Journal*, *349*, 17–24.

Xu, P., Zeng, G.M., Huang, D.L., Feng, C.L., Hu, S., Zhao, M.H., Lai, C., Wei, Z., Huang, C., Xie, G.X. and Liu, Z.F., 2012. Use of iron oxide nanomaterials in wastewater treatment: a review. *Science of the Total Environment*, *424*, 1–10.

Xu, X.H., Liu, X.M., Zhang, L., Mu, Y., Zhu, X.Y., Fang, J.Y., Li, S.P. and Jiang, J.D., 2018. Bioaugmentation of chlorothalonil-contaminated soil with hydrolytically or reductively dehalogenating strain and its effect on soil microbial community. *Journal of Hazardous Materials*, *351*, 240–249.

Yang, C., Liu, N., Guo, X. and Qiao, C., 2006. Cloning of mpd gene from a chlorpyrifos-degrading bacterium and use of this strain in bioremediation of contaminated soil. *FEMS Microbiology Letters*, *265*(1), 118–125.

Yuanfan, H., Jin, Z., Qing, H., Qian, W., Jiandong, J. and Shunpeng, L., 2010. Characterization of a fen-propathrin-degrading strain and construction of a geneticallyengineered microorganism for simultaneous degradation of methyl parathion and enpropathrin. *Journal of Environmental Management*, *91*(11), 2295–2300.

Zaller, J.G., Heigl, F., Ruess, L. and Grabmaier, A., 2014. Glyphosate herbicide affects belowground interactions between earthworms and symbiotic mycorrhizal fungi in a model ecosystem. *Scientific Reports*, *4*(1), 1–8.

Zhang, H., Zhang, Y., Hou, Z., Wang, X., Wang, J., Lu, Z., Zhao, X., Sun, F. and Pan, H., 2016. Biodegradation potential of deltamethrin by the *Bacillus cereus* strain Y1 in both culture and contaminated soil. *International Biodeterioration & Biodegradation*, *106*, 53–59.

Zhang, Q., Wang, B., Cao, Z., Yu, Y., 2012. Plasmid-mediated bioaugmentation for the degradation ofchlorpyrifos in soil. *Journal of Hazardous Materials*, *221–222*, 178–184.

4 Bioaugmentation in the Bioremediation of Petroleum Products

Greeshma Odukkathil and Namasivayam Vasudevan
Anna University

CONTENTS

4.1 INTRODUCTION

Complex mixtures of organic compounds formed during the processing of crude oil in oil refineries are generally called petroleum products. This processing step includes extraction of crude oil from the mine and separation of different petroleum products like gasoline, jet fuel, diesel and heating oil, petrochemical feedstocks, waxes, lubricating oils, and asphalt. They comprise hundreds of hydrocarbon compounds in different proportions and a variety of additives. Gasoline also comprises organic compounds such as benzene and the controversial additive methyl tertiary butyl ether.

The hydrocarbons present in the petroleum products are aliphatic hydrocarbons, aromatic hydrocarbons, resins, and asphaltenes (Steliga 2012). Aliphatic hydrocarbon can be biodegraded by microbes than aliphatic and aromatic hydrocarbon (Hasanuzzaman et al. 2007). Among the various hydrocarbon compounds present in petroleum products, polyaromatic hydrocarbons in the petroleum products like naphthalene, acenaphthene, fluorene, phenanthrene, and pyrene persist long in the environment (Soclo et al. 2000; Chen et al. 2010). The persistent nature of these compounds is mainly due to their less water-soluble nature leading to their grouping with particulate and sedimentary material and less bioavailability (Johnsen et al. 2005) in a petroleum product-contaminated environment.

DOI: 10.1201/9781003187622-4

4.2 PETROLEUM PRODUCTS IN THE ENVIRONMENT

Hydrocarbons present in petroleum products are one of the persistent pollutants present in abundance globally in both terrestrial and aquatic environments. Oil refineries, gas stations, petrochemical industries, oil tanks, vehicles are some of the major sources of this contamination. Aquatic pollution of petroleum products occurs by leakage of petroleum tank and accidental spills of petroleum products during their exploration, processing, refining, storage, and transportation. According to Kvenvolden and Cooper (2003), around 600,000 metric tons of crude oil seepage occurs yearly. This will lead to environmental impact which accounts for 0.1% of oil production. This contamination of the environment with crude oil will cause an adverse effect on humans, plants, and animals.

Marine oil spills are one of the catastrophic incidents which affect the marine environment. A primary cause for this is the oil leakage from a large oil tanker. Exxon Valdez incident, one of the most well-known marine oil spills is caused due to such discharge. The toxic petroleum hydrocarbons in oil disrupt natural habitats. Exploitation of petroleum products also results in deterioration of aquatic and terrestrial environment. Drilling, cleaning of storage tanks for crude oil results in the accumulation of petroleum waste in open space leading to disruption of soil quality. According to Li et al., the oil spill due to transportation is 140 which accounts for 7 million tons of oil. This can even up to as high as 1,000 tons by 2020–2021 (ITOPE 2020). Petroleum products also enter the ocean by natural seepage.

4.3 MICROORGANISMS DEGRADING PETROLEUM PRODUCTS

Microbial degradation of petroleum hydrocarbons is a complex biological process dependent on the type and concentration of the hydrocarbons present. Degradation of petroleum hydrocarbons by microbes is the basic process in the bioremediation of petroleum products. Biodegradation of petroleum products is affected mainly by non/less bioavailability to microorganisms. Petroleum hydrocarbons which are bound to the soil matrix are not easily available for microbial degradation. Hydrocarbons differ in their susceptibility to degradation by microbes. The linear alkanes are more susceptible than branched alkanes which are followed by small aromatics and cyclic alkanes (Ulrici 2000). Even though some organic compounds like high molecular weight polycyclic aromatic hydrocarbons (PAHs) in petroleum products are not easily degradable, certain microbes are capable to utilize them as a sole source of carbon and energy for growth. Microorganisms responsible for degrading petroleum products belong to eukaryotic or prokaryotic organisms (Balba et al. 1998). Microorganism degrading petroleum products are abundantly distributed in soil and water but their population size is less at petroleum product-contaminated sites. Hence, amending the polluted area with highly efficient petroleum products degrading microbes by the process called bioaugmentation (Supaphol et al. 2006) is one cost-effective treatment method for removal of petroleum products from the environment.

Substrate microbe contact is necessary for microbial uptake of hydrocarbons (van Hamme et al. 2003), and the less water-soluble nature of petroleum hydrocarbon molecules reduces this interaction (Burd and Owen 1996) and hence, adherence of microbes to large droplets of oil and their interaction with emulsified oil (Bruheim et al. 1999). Several researches on the aspect of treatment of hydrocarbons by microorganisms have been conducted and are still being explored (Polyak et al. 2018; Jiang et al. 2016; Ramadass et al. 2018). Some of the microorganisms that are effective in petroleum hydrocarbons degradation are from genera *Pseudomonas* sp., *Flavobacterium* sp., *Achromobacter*, *Rhodococcus*, *Acinetobacter* sp., *Bacillus* sp., *Mycobacterium* sp., *Alcaligenes* sp., *Aspergillus* sp., *Alcanivorax* sp., *Mucor*, *Penicillium* sp., and *Candida* sp. (Wu et al. 2017; Nwankwegu and Onwosi 2017; Yaman 2020).

Mixed microbial cultures are widely used for petroleum products degradation because of the presence of microorganisms with different hydrocarbon-degrading capabilities. The effective bioremediation relies on a consortium of organisms instead of the action of a single microorganism

(Grotenhuis et al. 2002). Biodegradation of petroleum products by microbes relies on the type of microbes in the community and how it adapts to the petroleum hydrocarbons (Raghavan and Vivekanandan 1999). Surface active biosurfactants are employed for enhanced oil recovery and as flocculating agents (Rahman et al. 1999). Das and Chandran (2011) isolated *Pseudomonas fluorescens*, *Pseudomonas aeruginosa*, *Bacillus subtilis*, *Alcaligenes* sp., *Acinetobacter woffi*, *Flavobacterium* sp., *Micrococcus roseus*, and *Corynebacterium* sp. from oil-contaminated streams. Some of the petroleum products degrading microbes are listed in Table 4.1.

Marine environment is one of the hotspots of hydrocarbon-degrading bacteria and fungi. Around 25 genera of bacteria and 31 genera of fungi are reported as petroleum hydrocarbon bio-degrading bacteria (Floodgate 1984; Bartha and Bossert 1984; Singh 2006). Many of these bacteria solely consume petroleum hydrocarbons. Bacteria degrading crude petroleum were isolated from soil polluted with petroleum products (Das and Mukherjee 2007). Some bacteria are more specific on the target organic compound they consume. For example, *Acinetobacter* sp. utilizes *n*-alkanes comprising of C10–C40 carbons as the only carbon compound (Throne Holst et al. 2007). For example, bacterial genera such as *Gordonia* sp., *Brevibacterium* sp., *Aeromicrobium* sp., *Dietzia* sp., *Burkholderia* sp., and *Mycobacterium* sp. are microbes consuming long-chain hydrocarbon compounds (Chaillan et al. 2004), and *Sphingomonas* sp. and *Ochromobacter* sp. degrade polyaromatic hydrocarbon (Daugulis and McCracken 2003, Arulazhagan and Vasudevan 2009, 2011).

TABLE 4.1
Microbes Degrading Different Petroleum Hydrocarbons

Petroleum Hydrocarbon Components	Bacterial Species	Main Degradation Profile	References
Aliphatics	*Dietzia* sp.	*n*-Alkanes (C6–C40)	Wang et al. (2011)
	Pseudomonas sp.	*n*-Alkanes (C14–C30)	Sugiura et al. (1997)
	Oleispira antarctica	*n*-Alkanes (C10–C18)	Yakimov et al. (2003)
	Rhodococcus ruber	*n*-Alkanes (C13–C17)	Zhukov et al. (2007)
	Geobacillus	*n*-Alkanes (C15–C36)	Abbasian et al. (2015)
	Thermodenitrifican	Cyclohexane	Lee and Cho (2008)
	Rhodococcus sp.	*n*-Alkanes and branched alkanes	Hara et al. (2003)
	Alcanivorax sp.	Branched and normal alkanes	Brown et al. (2016)
	Gordonia sihwensis		
Aromatics	*Achromobacter xylosoxidans*	Mono-/polyaromatics	Ma et al. (2015)
	Aeribacillus pallidus	Mono-/polyaromatics	Mnif et al. (2015)
	Mycobacterium cosmeticum	Monoaromatics	Zhang et al. (2012)
	Pseudomonas aeruginosa	Monoaromatics	Mukherjee
	Cycloclasticus	Polyaromatics	et al. (2010)
	Neptunomonas naphthovoran	Polyaromatics	Kasai et al. (2002)
	Bacillus Licheniformis Bacillus	Polyaromatics	Hedlund et al. (1999)
	Mojavensis	Polyaromatics	Eskandari
	Sphingomonas, Sphingobium		et al. (2017)
	Novosphingobium		Ghosal et al. (2016)
Resins and asphaltenes	*Pseudomonas* sp.	Resins	Venkateswaran
	Pseudomonas spp., *Bacillus* sp.	Asphaltenes	et al. (1995)
	Citrobacter sp., *Enterobacter* sp., *Staphylococcus* sp., *Lysinibacillus* sp. *Bacillus* sp., *Pseudomonas* sp.	Asphaltenes	Tavassoli et al. (2012) Jahromi et al. (2014)

Recent research identified not less than 79 genera of bacteria that are capable of degrading petroleum hydrocarbons (Tremblay et al. 2017). Several of these bacteria play a crucial role in petroleum hydrocarbon degradation (Sarkar et al. 2017; Varjani 2017; Xu et al. 2017). Some bacterial taxa, for example, *Alkanindiges* sp., exhibit bacterial shift to contaminants (Fuentes et al. 2015). Some obligate hydrocarbonoclastic bacteria like *Alcanivorax* sp., *Marinobacter* sp., *Thallassolituus* sp., *Cycloclasticus* sp., and *Oleispira* sp. are found to show an increase in their population size when exposed to petroleum products (Yakimov et al. 2007). There are also some species which utilize both long-chain and aromatic compounds, and they are *Dietzia* sp. DQ12–45-1b and *Achromobacter xylosoxidans* DN002, respectively (Wang et al. 2011).

Besides bacteria, other microbes belonging to fungi and yeast also have been reported to degrade petroleum hydrocarbons. Among the Fungi, *Amorphoteca*, *Neosartorya*, *Talaromyces*, and *Graphium* are potential degraders and from yeast include *Candida*, *Yarrowia*, and *Pichia* (Chailian et al. 2004) *Candida lipolytica*, *Rhodotorula* mucilaginosa, *Geotrichum* sp, and *Trichosporon mucoides* (Bogusławska-Was and Da Browski 2001). There are approximately 70 genera of known oil-degrading microbes, including bacteria, fungi, and yeast. (Joo et al. 2008). Even though all these microbes are capable of degrading 80%–90% of pure hydrocarbons, their efficiency decreases when they are bioaugmented in diesel or crude oil. The main reasons which reduce the efficiency are temperature variation, humidity, toxicity of other compounds, pH, solubility, viability of the microbes, etc. (Khashayar and Mahsa 2010; Bes̆koski et al. 2012). Bioaugmentation of petroleum product-contaminated environment with hydrocarbon-degrading bacteria enhances the bioremediation and thereby reduces the time of treatment (Mulligan et al. 2001; Bento et al. 2003; Kataoka 2001; Rahman et al. 2003).

The presence of different hydrocarbons of varying chain lengths in petroleum products also limits the degradation of petroleum products. Generally, long-chain hydrocarbons such as n-alkanes undergo rapid biodegradation (Mohanty and Mukherji 2008; Seklemova et al. 2001; Jonge et al. 1997). The structure of hydrocarbon in petroleum products also affects the rate of degradation. Gielnik et al. (2019) have shown that strains of *Rhodococcus* and *Achromobacter* were enhanced the removal of hydrocarbons up to 78% in diesel during the enrichment with municipal solid waste digestate. Autochthonous actinomycetes could decompose hydrocarbons in soil polluted with petroleum waste in the presence of sawdust supplemented with minerals and water (Burghal et al. 2015). Bacterial and fungal isolates such as *Pseudomonas stutzeri* BP10 and *Aspergillus niger* P89 were shown to degrade total petroleum hydrocarbons (TPH) up to 82.3% in soil spiked with 2% crude oil under controlled conditions along with rice husk, sugarcane, vermicompost, or coconut coir as bulking agents (Kumari et al. 2016). *Nocardia* sp. H17-1 was found to effectively degrade n-alkanes with 12–25 carbon than long-chain hydrocarbons with 26 carbon and above.

One of the main step in bioremediation of petroleum products is bioaugmentation, by which petroleum products degrading microbes are amended in polluted soil and water. The microbes usually bioaugmented are single isolates or consortia or a mixed population. Bioaugmentation of bacterial consortium in crude oil-contaminated sites is one successful treatment option. Many researches have explored the capability of using bacterial consortium which is formulated with highly efficient hydrocarbon degraders. Bioaugmentation of a bacterial consortium consisting of *Aeromonas hydrophila*, *Alcaligenes xylosoxidans*, *Gordonia* sp., *Pseudomonas fluorescens*, *Pseudomonas putida*, *Rhodococcus equi*, *S. maltophilia*, and *Xanthomonas* sp. showed increased biodegradation efficiency (89%) remediation of diesel oil-contaminated soil (Suja et al. 2014). Wang et al. (2018) also reported that bacterial consortium enhanced oil degradation than single bacterial isolates. Varjani et al. (2015) reported biodegradation of crude oil by a halotolerant bacterial consortium with *Ochrobactrum* sp., *Stenotrophomonas maltophilia*, and *Pseudomonas aeruginosa*. Besides consortium, bacterial cocultures of exogenous and indigenous bacterial consortium are also found effective in enhancing the biodegradation of crude oil (Tao et al. 2017).

4.4 BIOAUGMENTATION OF MICROBES FOR BIOREMEDIATION

Enhancing the biological processes by adding or augmenting microbes in the water or soil environment is one of the age-old processes named bioaugmentation. Bioaugmentation of microbes capable of degrading toxic pollutants is one of the basic tools for bioremediation of soil/water. Microbes biodegrading petroleum hydrocarbons are usually isolated from the petroleum product-contaminated environment by enrichment technique and grown under laboratory conditions with an organic compound of concern as the sole carbon source. Furthermore, these microbes are scaled up in a bioreactor in specific growth media, and scaled-up biomass is separated by centrifugation. These harvested microbes suspended in buffer are then introduced to the contaminated environment.

Bioaugmentation can be carried out either by enriching a petroleum product-contaminated environment with indigenous microorganisms or with nonindigenous microorganisms. Indigenous bioaugmentation is the process of re-amending an environment with native microbes which are already adapted to petroleum products which are isolated from their site of origin. It can also be carried out by adding exogenous microorganisms if the existing microbial flora lacks petroleum product degrading microbes (Poi et al. 2017). In many of the studies on bioaugmentation technology, microorganisms that biodegrade a specific petroleum hydrocarbon are directly added into the contaminated environment. This requires one or more highly efficient microbial strains which can solely use petroleum hydrocarbon as the main carbon and energy source. The microorganism is usually added directly as a single isolate, consortium, or mixed culture to a contaminated environment either as a free state or in the form of attached microbes on specific biofilm carriers. The performance of these amended microbes in the petroleum product-contaminated environment depends on different parameters like bioavailability (Harms and Zehnder 1995; Rothmel et al. 1994), concentration (Aamand et al. 1995), microbial toxicity (Baud-Grasset and Vogel 1995), physicochemical characteristics of soil/water (Godbout 1995), presence of other competitive microbes, microbiology, and methodology. In the case of petroleum products, the main parameters affecting the performance of bioaugmented microbes are bioavailability of the petroleum hydrocarbons, physiochemical parameters, toxicity of the compounds, and presence of nutrients.

4.4.1 BIOAVAILABILITY OF PETROLEUM HYDROCARBONS

Most of the petroleum hydrocarbons are generally less water soluble which reduces the bioavailable fraction for the biodegradation process. Bioavailability refers to the portion of hydrocarbon in soil/water, which can be utilized by microbes. The extent of degradation of the hydrocarbon can be significantly affected by the constraints in the bioavailability of hydrocarbons. The bioavailability of petroleum hydrocarbon is also affected by the partitioning of hydrocarbons within the soil, contaminant source and matrix, geochemical and physical properties of soil, and aging. All these factors together make the petroleum hydrocarbon less bioavailable to the bioaugmented microbes. The less bioavailable fraction of the petroleum hydrocarbons can be made available by adding emulsifiers like synthetic surfactants or biosurfactants. Use of biosurfactant-producing microbes as enhancers is one of the vowing bioaugmentation techniques for bioremediation of petroleum hydrocarbon-contaminated sites. The other successful technique is the surfactant-aided bioremediation technique. Several studies are carried out in this area for enhanced bioremediation of petroleum hydrocarbons.

4.4.2 PHYSIOCHEMICAL PARAMETERS

Environmental conditions play a pivotal role in the degradation of petroleum products, whether its indigenous microbes or bioaugmented microbes. These conditions are: (i) those that reduce the activity of microbes, such as humidity, temperature, and ionic strength and (ii) clay/organic matter that restrict the mass transfer of the hydrocarbon to the microorganism by diffusion. In addition, advective transport, such as permeability, also affects several aspects of the bioaugmentation process. This affects the adjunct of microorganisms and the addition of nutrient and electron acceptors. Temperature is one of the important factors that affect the degradation of petroleum hydrocarbon as it influences the viscosity, water solubility, and chemical composition of petroleum products. Moreover, the biodegradation rate of hydrocarbon and the structure of the microbial community are also influenced by the change in temperature (Atlas 1981). Hence, optimization of physicochemical parameters is one of the first approaches in the bioaugmentation of microbes in a natural environment.

4.4.3 Toxicity of Organic Compounds

Generally, the toxicity level of petroleum product-contaminated environment will be high as it contains several hydrocarbons groups. Hence, it is one of the biggest tasks during the bioaugmentation of the microbes. In such cases, bioaugmentation of microbial consortium or mixed bacterial culture is a better choice for bioremediation. Developing a consortium capable of growing on mixed organic pollutants by enrichment technique is one of the best options for bioremediation and bioaugmentation of such a bacterial culture in petroleum product-contaminated environment will enhance the degradation processes (Rahman et al. 2002; Nyer et al. 2002). Solubility of the hydrocarbons also results in poor growth of microbes. In such cases, bioaugmentation of biosurfactant-producing microbes will enhance the biodegradation process. In the case of wastewater, the toxicity can be reduced by dilution with other less toxic wastewater or secondary treated wastewater before bioaugmentation of microbes.

4.4.4 Presence of Nutrients

To upgrade the bioremediation process, besides an efficient hydrocarbon-degrading microbes other parameters such as water, oxygen, and available nitrogen and phosphorous sources are also important (Rosenberg et al. 1992). The absence of these nutrients in petroleum product-contaminated sites affects the efficiency of the bioremediation process under real conditions. Hence, under real contaminated sites, amending hydrocarbon biodegrading microbes by bioaugmentation or adding nutrient amendments by biostimulation fosters degradation rates (Jimenez et al. 2006). Delille et al. (2004) also reported that the biodegradation of petroleum products by an indigenous microorganism can be enhanced by adding nutrient amendments in the petroleum products polluted site. Addition of nitrogen and phosphorus to increase the nutrients in the contaminated environment for enhancing biodegradation has been studied by several researchers. Biosolids from sewage treatment plant and nitrogen–phosphorus inorganic fertilizers are efficient biostimulators for biodegradation of petroleum products (Sarkar et al. 2005). Other amendments like commercial fertilizers (Delille et al. 2009), N and P fertilizer together with biosurfactants (Nikolopoulou and Kalogerakis 2008) are also reported to enhance the bioremediation of petroleum products. These approaches propose that the bioaugmentation with the addition of nutrients and biosurfactants based on the nature of a contaminated site will be the best approach to enhance bioremediation (Baek et al. 2007). Loss of viable microbes during inoculation and predation of other microbes are some other issues which affect the bioaugmentation.

4.5 BIOAUGMENTATION TECHNIQUES

Microorganisms can be introduced to petroleum product-contaminated environment by different means (Figure 4.1) either a single isolate biomass, mixed culture biomass, a consortium of bacteria or fungi, biosurfactant-producing bacteria, halophilic bacteria, halothermophilic culture, and immobilized culture. In a soil environment microbes are introduced directly for the bioremediation process. Recently, they are also introduced by amending biochar, clay biochar composites,

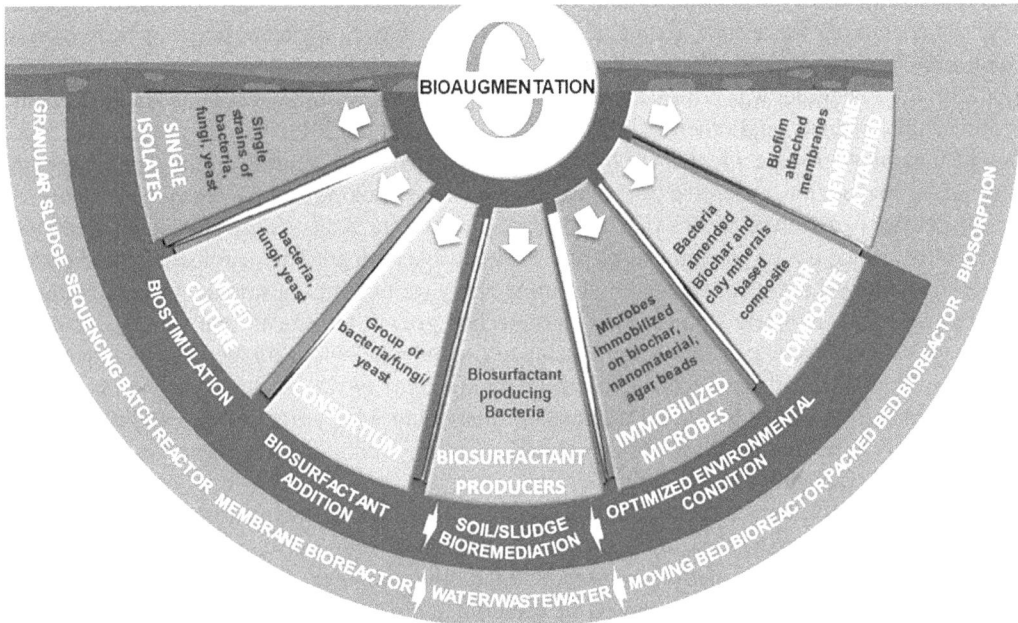

FIGURE 4.1 Bioaugmentation techniques for bioremediation of petroleum products in soil and water/wastewater.

or other composite materials into the soil. Immobilization or encapsulation of microbes is mostly used for bioaugmenting microbes for wastewater treatment. Several biochars made from cellulose and lignin-based materials can also be used as a support material for the growth and reproduction of microorganisms. The specific surface area, porosity, and rich superficial functional group of biochar enhance the adsorption capacity of bacteria and the consequence of contaminants. Hence, amending biochar with bacteria may be a useful approach for the removal of TPH and improving the polluted soil condition. Zhang and Zhang (2019) reported that biochar-immobilized bacteria reduced the TPH and n-alkanes (C_{12-18}), and thus increased biodegradation efficiency with reduced half-life. The study also revealed that the biochar immobilization greatly enhanced the soil properties and activity of bacteria. Chen and Yuan (2012) reported dissipation of PAHs in a contaminated soil amended with immobilized bacteria using biochar and also degradation PAH adsorbed on bacteria amended biochar. Bacteria amended clay biochar composites in the soil are found effective for the removal of pesticides from agriculture runoff (Zhang and Zhang 2019; Greeshma and Vasudevan 2020). According to Mishra et al. (2001), biochar provides physical support for biomass, increased nutrient availability, moisture, improved aeration, and increased viability. It also increases the survival rate of microbes by preventing the extreme environmental condition (Moslemy et al. 2002). It also controls the movement of nutrients, decreases the concentration of toxic compounds in the cells, and prevents predation and competition (McLoughlin 1994). Materials well used and tested for immobilization include agar, agarose, alginate, gelatin, kappa-carrageenan, acrylate copolymers, polyurethane, and polyvinyl alcohol (Cassidy et al. 1996). Gellan gum microbeads encapsulated bacterial consortium showed higher gasoline degradation with a shorter lag phase than free cells (Moslemy et al. 2002). *Pseudomonas monteilii* P26 and *Gordonia* sp. H19 immobilized on polyurethane foam was found to be an effective tool for petroleum bioremediation (Alessandrello et al. 2017).

4.6 BIOREMEDIATION OF PETROLEUM PRODUCT-CONTAMINATED WATER/PETROLEUM WASTEWATER

The petroleum industry discharges a large quantity of pollutants into the environment. Petroleum products containing wastewater generated from refineries have a different range of hydrocarbon compounds which are toxic (Pajoumshariati et al. 2017). At the global level, around 3.35×10^7–5.06×10^7 t/d wastewater was produced (Saber et al. 2017). This wastewater contains hydrocarbons, heavy metals, phenols, and other toxic chemicals (Thakur et al. 2018; Varjani and Sudha 2018). Different technologies are adopted for treating wastewater from the petroleum industry. They are adsorption, coagulation, anaerobic treatment, reverse osmosis, ultrafiltration, chemical destabilization, flocculation, dissolved air flotation, membrane process, etc. Among these, biological treatment is generally carried out by bioaugmenting microbes capable of degrading petroleum hydrocarbons. Biological treatment is widely adopted in the oil industry for treating wastewater. Generally implemented biological treatment is suspended and attached growth biological treatment (Chavan and Mukherji 2008; Srikanth et al. 2018). Aerated lagoon, membrane bioreactor technology, sequencing batch reactor (SBR), and activated sludge treatment are some of the suspended growth biological treatment processes. In the biological treatment of petroleum wastewater nitrogen and phosphorus concentration in the treatment system is reported as a very important parameter for treating oily wastewater (Chavan and Mukherji 2008).

Musa et al. (2015) reported enhanced degradation of petroleum hydrocarbons on augmenting four dominant microbial strains isolated from the petroleum refinery wastewater were identified to best treatment method for petroleum product-contaminated water. According to Arulazhaghan et al. (2017), bioaugmenting halothermophilic bacterial consortium in petroleum wastewater under extreme conditions degrades both the low molecular weight polyaromatic hydrocarbons like naphthalene, anthracene, phenanthrene, and fluorene and the high molecular weight polyaromatic hydrocarbons like pyrene, benzo(k)fluoranthene, and benzo(e)pyrene, respectively. Kumar et al. (2007), also reported that halothermotolerant *Bacillus* strain degraded hydrocarbons mixtures of diesel, gas oil, alkanes, crude oil, and kerosene. Bioaugmentation of attached mixed cultures on HDPE (High-Density Polyethylene) membrane in a moving bed membrane reactor also showed high efficiency of 97% removal of petroleum pollutants (Qaderi et al. 2018). Another technique employed for petroleum wastewater treatment is aerobic granular sludge (AGS) treatment technology. In this treatment, microorganisms that are self-immobilized into aggregates called granular sludge are employed for treating wastewater. This granular sludge has higher sludge settling and biomass retention. The AGS is less influenced by variation in loading impacts and influent quality. The higher sludge settleability reduces sludge loss and improves sludge quantity and the organic loading rate. Chen et al. (2019) achieved a Chemical oxygen demand (COD) removal of 95% and petroleum compounds removal of 90% during the treatment of petroleum wastewater by using AGS technology.

Application of biosurfactants is one of the age-old technologies that is employed for removal of petroleum products from water. For decades, many microbes have been screened for biosurfactant production on petroleum hydrocarbons and petroleum products as such. They are diverse, selective, eco-friendly can be operated under harsh environmental conditions, can be easily produced at large scale, and have high potential application in environmental protection. Biosurfactant enhances the aqueous solubility of petroleum products by emulsifying hydrophobic compounds and thereby bioaccessibility of petroleum hydrocarbon compounds also. Bioavailability of petroleum products to microbes is also enhanced by cell surface hydrophobicity alteration and through membrane permeability (Liang et al. 2017). Hence, biosurfactants/bioaugmenting biosurfactant-producing microbes (Figure 4.2) is one of the efficient technologies for bioremediation and widely utilized for oil recovery, removing oil spills, biodegradation of metal and oil-contaminated soils (Chen et al. 2019). Arulazhaghan et al. (2017) also reported the production of biosurfactants and enhancement of petroleum hydrocarbon degradation upon employing halothermophilic bacterial strains in petroleum wastewater. Bacterial strains of *Pseudomonas* sp were well explored for

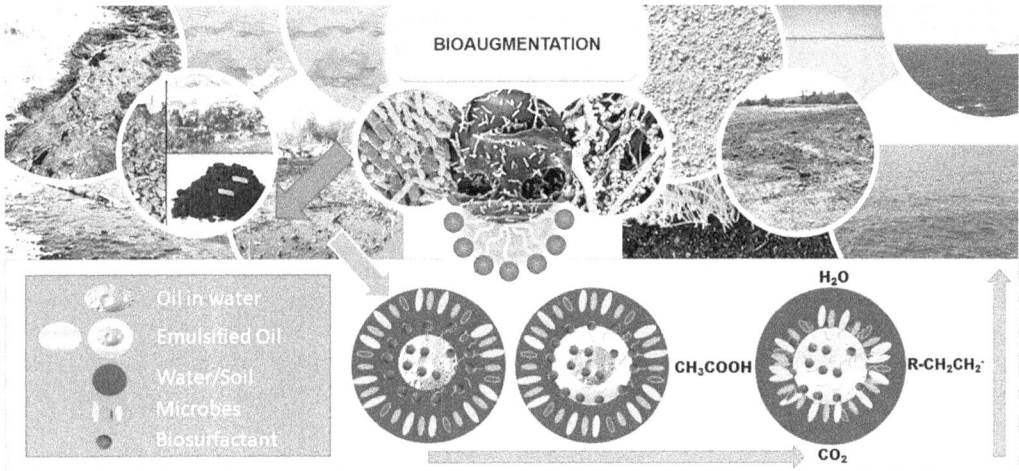

FIGURE 4.2 Bioremediation by bioaugmentation of biosurfactant-producing microbes.

producing biosurfactant production during biodegradation of petroleum products (Vasudevan et al. 2007; Okoro et al. 2012).

Packed bed reactors have also been used in the petroleum wastewater treatment. Generally, microbial cells immobilized on various materials like alginate, chitosan, cellulose, polyurethane foam (PUF), nylon sponge, and synthetic foams are used for microbial cell immobilization by cell entrapment and cell attachment process. Such immobilized microbes are bioaugmented in packed bed reactors for wastewater treatment. Several studies were being carried out to explore the possibilities of treating petroleum wastewater or petroleum product-contaminated water by bioaugmenting immobilized microbes in packed bed reactors. Li et al. (2005) reported that immobilized *Bacillus* sp. used in continuous wastewater treatment systems considerably removed COD from oil-containing wastewater. Banerjee and Ghoshal (2017) also proposed bioaugmentation of immobilized *Bacillus cereus on* calcium alginate packed bed bioreactor as an effective treatment technology for treating petroleum wastewater. Biodegradation of petroleum oil refinery wastewater carried out in a fluidized bed bioreactor with Ca-alginate immobilized phenol-degrading strains of *Bacillus cereus* achieved a removal efficiency of 95% (Banerjee and Ghoshal 2016).

Another wastewater treatment process where microbes are bioaugmented to remove petroleum products is the biosorption technique. Biosorption technique can be adopted to remove hydrocarbons from wastewater in an easy and economical manner (Rajan et al. 2019). Another benefit of biosorption techniques is that biofuels can be produced from the hydrothermal liquefaction of the hydrocarbon biosorbed biomass (Arun et al. 2018; Biswas et al. 2018). Rajan et al. (2019) reported that sorption of 92% hydrocarbon by *Scenedesmus abundans* from petroleum wastewater and their hydrothermal liquefaction for production of biofuel.

SBRs are one of the promising wastewater treatment technologies for treating industrial and nonindustrial wastewater. SBR technology is one of the demonstrated pilot-scale wastewater treatment technology for petroleum industry wastewater treatment (Ghorbanian et al. 2014; Frank 2016; Pajoumshariati et al. 2017). This treatment system was adopted in some refineries, but it's rarely used for treating petroleum industry wastewater (IPIECA 2010; Jafarinejad 2017). In all these treatment processes microorganisms that are naturally occurring, commercial, specific groups, enriched microbes for specific organic pollutants, and acclimatized sewage sludge are bioaugmented for treating the wastewater. A bench-scale sequencing batch reactor was used for treating refinery effluent wastewater (Kutty et al. 2011). Several treatment processes are available for treating petroleum product-contaminated water. Bioaugmenting microbes that can tolerate the toxicity of the hydrocarbons present in the petroleum products, capable of enhancing the bioavailability and degrading the hydrocarbons present in the petroleum products along with the most promising biological treatment technology will be the best option for bioremediation of petroleum product-contaminated water/ petroleum wastewater.

4.7 BIOREMEDIATION OF OIL

One of the most discussed concerns of bioremediation of oil is bioaugmentation. Amending oil-degrading microorganisms for remediation of oil has been suggested as the main approach for bioremediation. The main reason for amending oil biodegrading microorganisms is that natural microbes present in the contaminated site are not an efficient degrader of a wide range of hydrocarbons present in crude oil.

Aldrett et al. (1997) developed 12 bioaugmentation agents which enhanced biodegradation of petroleum hydrocarbon than naturally occurring microbes in 28 days. The developed bioaugmentation agent degraded 60%–65% of TPH in oil. Another bioaugmentation agent TerraZyme showed a high capability for biodegrading oil and also degraded asphaltene fraction (Hozumi et al. 2000).

Vinas et al. (2002) developed three microbial consortia using three different oil products degrading microorganisms which on bioaugmenting in oil degraded saturated, aliphatic, and polyaromatic carbon fraction. The developed consortium also degraded n-alkanes and branched alkanes completely Zrafi-Nouira et al. (2008) bioaugmented an indigenous microbiota of seawater on crude oil and showed a 92.6% degradation of nonaromatic and 68.7% degradation of aromatic hydrocarbon. Gertler et al. (2009) reported the degradation of 95% aliphatic compounds in the oil on bioaugmenting a series of microcosms. Zhang et al. (2012) reported enhanced degradation of oil on simultaneous addition of biosurfactant and bioaugmenting of *Pseudomonas aeruginosa*. Abalos et al. (2004) reported accelerated biodegradation of petroleum hydrocarbons by adding rhamnolipids along with bioaugmented microbes. Autochthonous bioaugmentation is another reported bioaugmentation process where seawater microcosm in combination with inorganic nutrients like uric acid and lecithin in the presence or absence of biosurfactants revealed that the bioaugmentation combined with the addition of biosurfactant and nutrients enhanced the bioremediation of oil.

4.8 BIOREMEDIATION OF PETROLEUM PRODUCT-CONTAMINATED SOIL

In general, the ability of the microorganisms to biodegrade petroleum products is determined by their structure and diversity in the soil environment (Rodríguez-Blanco et al. 2010). Length of the petroleum hydrocarbon chain and structure of the hydrocarbons in the oil also will determine the decomposition rate and efficiency besides the basic characteristics of the soil system that will include water, pH, temperature, mineral nutrients, nitrogen, phosphorus, and organic compounds present. Even though petroleum is an abundant source of carbon for microbes, less availability of nitrogen and phosphate reduces its degradation.

Remediation of petroleum hydrocarbons was enhanced by natural attenuation. Ebuehi et al. (2005) observed a reduction of total petroleum hydrocarbon 1.1004×10^4 mg/kg of the sandy soil to 282 mg/kg after bioaugmentation and tilling in 10 weeks. In some polluted sites, the natural reduction is not a recommended remediation method with less indigenous hydrocarbon-degrading microorganisms. In such cases, microbes capable of degrading hydrocarbons are bioaugmented. This involves the addition of bacterial isolates or consortia or mixed microbial culture with high petroleum hydrocarbon degradation efficiency. Bioaugmentation will accelerate the biodegradation. Generally, these microbes are enriched from petroleum product-contaminated sites (Sarkar et al. 2005).

Bioremediation of petroleum product polluted soil depends upon the ability of the microorganism to biodegrade petroleum present in the soil (Franzmann et al. 2002) or on the population of petroleum-degrading microorganisms present at the polluted site. Under these circumstances, isolated organisms are being used for degrading petroleum hydrocarbons. Some nitrogen-fixing bacteria are also reported for their ability to release the nitrogenase enzyme which enhances bioremediation of crude oil-polluted soil (Odokuma and Ibor 2002). Girigiri et al. (2019) also reported stimulated bioremediation of crude oil-polluted soil by nitrogen-fixing bacteria and phosphate solubilizing bacteria.

Bioaugmentation of biosurfactant-producing bacteria is one of the promising techniques for bioremediation of petroleum hydrocarbon-contaminated soil. Rhamnolipid, a glycolipid biosurfactant from *Pseudomonas aeruginosa* and surfactin from *Bacillus subtilis* were well studied whereas other biosurfactants known in recent years are arthrofactin from *Pseudomonas* species, iturin and lichenysin produced by *Bacillus* species, mannosylerythritol lipids (MEL) from *Candida*, and emulsan from *Acinetobacter* species (Das et al. 2008).

Bioaugmentation of biosurfactant producers in hydrocarbon polluted soils is one of the best options especially for enhancing the biodegradation of soil-bound petroleum hydrocarbons. According to Mnif et al. (2015), biosurfactants enhance the rate of biodegradation. Multifunctional biosurfactant producers were also isolated by Nimrat et al. (2019) which reduced the total phenolic content in oil-polluted soils. Ali et al. (2020) used biosurfactant-producing bacterium such as *Bacillus cereus* WR146 along with compost and nutrients to improve the reduction of crude oil content from 4.20% to 3.73% to 2.42%. Trejos-Delgado et al. (2020) reported 52% of TPH degradation in Tween 80 amended soil within 80 days and 76% of TPH degradation in compost amended soil within 60 days.

The three fungal strains namely *Aspergillus niger*, *Penicillium ochrochloron*, and *Trichodema viride* could degrade 71.9% of petroleum hydrocarbons within the first 30–40 days (Esssabri et al. 2019). Al Disi et al. (2017) have studied the weathering process of harsh oil-polluted soil with 5% and 10% of 39 cultures where *Pseudomonas* and *Citrobacter* sp. have been shown to have the ability to degrade short-chain *n*-alkanes than long-chain *n*-alkanes under at different carbon to nitrogen ratio. Bioaugmentation of yeast *Candida catenulata* CM1 in diesel-contaminated soil degraded 77% diesel in soil with 2% (w/w) diesel (Joo et al. 2008).

Cow dung in combination with agricultural waste materials like palm kernel, the husk is also used as an amendment for enhancing bioremediation of oil-contaminated soil. The study revealed that this combination enhances the biodegradation of oil (Ofegbu et al. 2014). Use of cow dung could improve the bioremediation efficiency. Neethu et al. (2019) have isolated microorganisms capable of degrading hydrocarbon pollutants from cow dung such as *Micrococcus* sp., *Bacillus* sp., *Pseudomonas* sp., *Enterobacter* sp., *Proteus kleibsilla*, *Aspergillus* sp., *Rhizopus*, and *Penicillium*. Olawale et al. (2020) observed that the application of cow dung and poultry dung and cow dung/poultry in petroleum-contaminated soil could bring about 84%–96% of petrol. The main microorganisms involved in this process were *Bacillus*, *Aspergillus*, and *Escherichia coli*.

Agricultural waste like wheat bran and swine wastewater is also been used for bioremediation of soil polluted with oil (Zhang et al. 2020). Burghal et al. (2015) reported that gram-positive bacteria from the actinomycetes group and autochthonous microorganisms could decompose hydrocarbons levels from 52 to 10.6 g/kg in the presence of 1.5% sawdust and necessary minerals and water. Zhang and Zhang (2019) reported TPH removal efficiencies of 58.08%, 45.31%, 29.85%, 38.63% and 8.69% with (i) immobilized microorganisms at 5% (w/w), (ii) sterilized biochar at 5% (w/w) with 10% bacteria (v/w), (iii) sterilized biochar at 5% (w/w), (iv) bacteria at 10% (v/w), and (v) control, respectively.

Bioaugmentation of Petrophilic microorganisms is also a better option for petroleum product-contaminated soil the microbes show diverse metabolic ability to alter products of hydrocarbon oxidation into necessary substrates (Van Eyk 1997). The use of petrophilic microbes consisting of petrophilic fungi, petrophilic bacteria, and *Azotobacter* sp. degraded hydrocarbons at the rate of 0.22 ppm/day (Fauzi and Suryatmana 2016). Helmy et al. (2016) reported that the addition of petrophilic consortia and biosurfactants to petroleum oil-contaminated soil could show a removal of efficiency up to 46% and 85%, respectively.

4.9   BIOREMEDIATION OF OILY SLUDGE

A large quantity of oily sludge is produced during the water–oil separation process in the oil fields. Oily sludge contains hundreds of hydrocarbons which are toxic and induce mutation and cancer-causing compounds. This is a serious environmental problem as there is economic disposal option. TPH contained in oily sludge is detrimental to human health and the environment. They are therefore classified as priority environmental pollutants by the US Environmental Protection Agency and their release into the environment is strictly controlled. Bioremediation is one of the promising technologies for treating oily sludge. Bioaugmentation of microbes in oily sludge is a well-proposed technology for the treatment of oily sludge. Among the microbes, bioaugmenting bacterial consortium is one of the common options in practice. Generally, mixed bacterial consortium is more advantageous to pure cultures because of synergistic interactions among the microbes (Cerqueira et al. 2011; Tahhan et al. 2011; Mukred et al. 2008). Among the petroleum hydrocarbon-degrading bacteria, *Pseudomonas* is the most abundant bacterium found in oil-contaminated soil. *Acinetobacter* and *Rhodococcus* are the other two common bacterial strains in petroleum-contaminated habitats. These bacteria are commonly used for bioaugmentation studies. Some yeast species are also capable of degrading crude oil compounds. Bioaugmentation with five exogenous yeast coculture supplementation removed 80% of high molecular weight PAHs in oilfield-produced water (Hesham et al. 2019). Bioaugmentation of *Pseudomonas aeruginosa* and biostimulation enhanced the biodegradation of hydrocarbons compared with individual application of bioaugmentation, biostimulation, natural attenuation, and abiotic factors. According to Varjani et al. (2020), simultaneous application of *P. aeruginosa* NCIM 5514 and nutrients removed 92.97% oil from the petroleum sludge decrease in oily sludge. The studies on sludge degradation revealed that nutrient supplementation is an important factor to enhance the biodegradation of petroleum hydrocarbons (Jasmine and Mukherji 2015). Mishra et al. (2001) proposed the bioaugmentation of a carrier-based consortium for the treatment of oily sludge. Ke et al. (2021) demonstrated that mixed strains together with bran prepared in form of SCBA (solid complex bacterial agents) enhanced the bioremediation of oily sludge. Vasudevan and Rajaram (2001) also reported bioremediation of oil sludge-contaminated soil by bioaugmenting a bacterial consortium and biostimulation with inorganic nutrients, compost and a bulking agent such as wheat bran enhanced the removal of hydrocarbon.

Several bacteria are reported to utilize petroleum hydrocarbons as the only carbon source and energy, such as *Pseudomonas*, *Achromobacter*, *Candida digboiensis*, *Micromonospora*, *Bacillus subtilis*, and *Leuconostoc mesenteroides* (Cerqueira et al. 2012). Bioaugmentation of microbial culture comprising of different bacteria has shown better performance than the single strain in the treatment of oily sludge (Guerra et al. 2018). It has demonstrated that bioaugmenting microbes capable of degrading petroleum products have strong ability to remove total petroleum hydrocarbon and organic matters in oily sludge with complete mineralization (Silva-Castro et al. 2013; Mandal et al. 2012). A combination of composting and bioaugmentation (Dzondo-Gadet et al. 2005) is also being explored. Cerqueira et al. (2014) demonstrated that the application of bioaugmentation and biostimulation enhanced the biodegradation of TPH present in the petroleum sludge, which indicated their potential applications for the treatment of the soils contaminated with petroleum products.

Earthworm-assisted bioremediation is another option for bioremediation of oily sludge. In such a case to enhance the bioremediation process and to increase the survivability and reproduction rate of earthworms, microbes are bioaugmented in the oily sludge (Chachina et al. 2015) and this will improve the process. Koolivand et al. (2020) proposed a treatment option for the removal of TPH using bioaugmented composting by hydrocarbon-degrading bacteria and vermicomposting by *Eisenia fetida*, individually and in combination with bacterial composting and vermi composting. A synergistic effect of the hydrocarbon-degrading bacteria and earthworms in TPH removal was observed and a higher removal efficiency was observed. Generally, due to the high oil content in the sludge and also due to high organic content, bioremediation is usually carried out by adding amendments and essential nutrients before bioaugmenting the hydrocarbon-degrading microbes.

The above studies also conclude that bioaugmentation of hydrocarbon-degrading bacteria in oily sludge for bioremediation will enhance the process.

4.10 CONCLUSION

Bioremediation is an inventive technique, which provides mitigation of petroleum products using microorganisms. It converts hazardous petroleum products to less toxic compounds without affecting the environment. The main advantages of bioremediation are its economic, eco-friendly, and ease of use of the process. Bioaugmentation is one of the main strategies for bioremediation. Bioaugmentation is one of the options which could bring about a solution to petroleum pollutants in soil, water, and other environments. Even though the indigenous microbes or heterotrophic microorganisms are available in the given situation, other factors such as the availability of nutrients such as N and P and other soil characteristics such as pH, the temperature may be a deterrent in the progress of bioremediation. Under such conditions, allochthonous species or petrophilic species are said to play a vital role in the remediation process. The availability of biosurfactant-producing organisms and versatile microorganisms which can sustain adverse environmental conditions such as pH, temperature, and toxic compounds may play an important role in the process. In general, the bioremediation process may differ from the given situation to the other. Hence, it may be pertinent to meticulously plan a remediation process depending on the type of contaminant group present ex. alkanes or aromatic as well as the toxic or more recalcitrant groups such as asphaltenes and heavy metals under the given situation. Besides the indigenous microbial population, bioaugmentation of allochthonous population along with biosurfactant-producing strains with versatile capability may have a desired effect in the remediation process.

REFERENCES

Aamand, J., G. Bruntse, M. Jepsen, C. Jorgensen, and B.K. Jensen. 1995. Degradation of PAHs in soil by indigenous end inoculated bacteria. In *Bioaugmentation for Site Remediation*, R.E. Hinchee, J. Fredrickson, and B.C. Alleman (Eds.), Columbus, OH: Battelle Press, 121–127.

Abalos, A., M.Viñas, J. Sabaté, M.A. Manresa, and A.M. Solanas. 2004. Enhanced biodegradation of casablanca crude oil by a microbial consortium in presence of a rhamnolipid produced by *Pseudomonas aeruginosa* AT10. *Biodegradation* 15(4):249–260. doi:10.1023/b:biod.0000042915. 28757.fb.

Al Disi, Z., S. Jaoua, D. Al Thani, S. Al Meer, and N. Zouari. 2016. Considering the specific impact of harsh conditions and oil weathering on diversity, adaptation and activity of hydrocarbon degrading bacteria in strategies of bioremediation of harsh oily polluted soils. *BioMed Research International* 2017:1–12.

Aldrett, S., J.S. Bonner, T.J. McDonald, M.A. Mills, and R.L. Autenrieth. 1997. Degradation of crude oil enhanced by commercial microbial cultures. *International Oil Spill Conference*, (1):995–996.

Alessandrello, M.J., M. S. J. Tomás, E.E. Raimondo, D.L. Vulloc, and M.A. Ferreroa. 2017. Petroleum oil removal by immobilized bacterial cells on polyurethane foam under different temperature conditions *Marine Pollution Bulletin* 122 (1–2):156–160.

Arulazhagan, P. and N. Vasudevan. 2009. Role of moderately halophilic bacterial consortium in the biodegradation of polycyclic aromatic hydrocarbons. *Marine Pollution Bulletin* 58(2):256. doi:10.1016/j.marpolbul.2008.09.017. ISSN: 0025-326X.

Arulazhagan, P. and N. Vasudevan. 2011. Biodegradation of polycyclic aromatic hydrocarbons by a halotolerant bacterial strain *Ochrobactrum* sp. VA1. *Marine Pollution Bulletin* 62:388. doi:10.1016/j.marpolbul. 2010.09.020. ISSN: 0025-326X.

Arulazhagan, P., Q. Huda, B. J.M. Al-Badry, J. J. Godon, and D. Jeyakumar. 2017. Role of a halothermophilic bacterial consortium for the biodegradation of PAHs and the treatment of petroleum wastewater at extreme conditions. *International Biodeterioration and Biodegradation* 121:44–54.

Arun, J., P. Varshini, P.K. Prithvinath, V. Priyadarshini, and K.P. Gopinath. 2018. Enrichment of bio-oil after hydrothermal liquefaction (HTL) of microalgae *C. vulgaris* grown in wastewater: Bio-char and post HTL wastewater utilization studies. *Bioresource Technology* 261:182–187.

Atlas, R.M. 1981. Microbial degradation of petroleum hydrocarbons: An environmental perspective. *Microbiology Reviews* 45(1):180–209.

Baek, K.H., B.D. Yoon, B.H. Kim, D.H. Cho, I.S. Lee, H.M. Oh, and H.S. Kim. 2007. Monitoring of microbial diversity and activity during bioremediation of crude OH-contaminated soil with different treatments. *Journal of Microbiology and Biotechnology* 17:67–73.

Balba, M.T., R. Al-Daher, N. Al-Awadhi, H. Chino, and H. Tsuji. 1998. Bioremediation of oil-contaminated desert soil: The Kuwaiti experience. *Environment International* 24(1–2):163–173.

Banerjee, A. and A.K. Ghoshal. 2016. Biodegradation of real petroleum wastewater by immobilized hyper phenol-tolerant strains of *Bacillus cereus* in a fluidized bed bioreactor. *Journal of Biotechnology* 6:137.

Banerjee, A. and A.K. Ghoshal. 2017. Biodegradation of an actual petroleum wastewater in a packed bed reactor by an immobilized biomass of *Bacillus cereus. Journal of Environmental Chemical Engineering.* doi:10.1016/j.jece.2017.03.008.

Bartha, R. and I. Bossert. 1984. The treatment and disposal of petroleum refinery wastes. In *Petroleum Microbiology*, R. M. Atlas (Ed.), New York: Macmillan, 553–577.

Baud-Grasset, F. and T.M. Vogel 1995. Bioaugmentation: Biotreatment of contaminated soil by adding adapted bacteria. In *Bioaugmentation for Site Remediation*, R.E. Hinchee, J. Fredrickson, and B.C. Alleman (Eds.), Columbus, OH: Battelle Press, 39–48.

Bogusławska-Was, E. and W. Da̧browski. 2001. The seasonal variability of yeasts and yeast-like organisms in water and bottom sediment of the Szczecin Lagoon. *International Journal of Hygiene and Environmental Health* 203(5–6): 451–458.

Bruheim, P., H. Bredholt, and K. Eimhjellen. 1999. Effects of surfactant mixtures, including corexit 9527, on bacterial oxidation of acetate and alkanes in crude oil. *Applied and Environmental Microbiology* 65:1658–1661.

Burd, G. and P. Owen. 1996. Bacterial degradation of polycyclic aromatic hydrocarbons on agar plates: The role of biosurfactants. *Ward Biotechnology Techniques* 10:371–374.

Burghal, A.A., N.A. Al-Mudaffar, and K.H. Mahdi. 2015. Ex situ bioremediation of soil contaminated with crude oil by use of actinomycetes consortia for process bioaugmentation. *European Journal of Experimental Biology* 5:24–50.

Cassidy, M.B., H. Lee, and J.T. Trevors. 1996. Environmental applications of immobilized microbial cells: A review. *Indian Journal of Microbiology* 16:79–101.

Cerqueira, V.S., E.B. Hollenbach, F. Maboni, F.A. Camargo, R.P. Maria do Carmo, and F.M. Bento. 2012. Bioprospection and selection of bacteria isolated from environments contaminated with petrochemical residues for application in bioremediation. *World Journal of Microbiology and Biotechnology* 28:1203–1222.

Cerqueira V.S., E.B. Hollenbach, F. Maboni, M.H. Vainstein, F.A. Camargo, M.D.C.R. Peralba, and F.M. Bento. 2011. Biodegradation potential of oily sludge by pure and mixed bacterial cultures. *Bioresource Technology* 102:11003–11010.

Cerqueira, V.S., R.P. Maria do Carmo, F.A.O. Camargo, and F.M. Bento. 2014. Comparison of bioremediation strategies for soil impacted with petrochemical oily sludge. *International Biodeterioration and Biodegradation Part B* 95:338–345.

Chaillan, F., A. Le Fleche, and E. Bury. 2004. Identification and biodegradation potential of tropical aerobic hydrocarbon degrading microorganisms. *Research in Microbiology* 155(7):587–595.

Chandran, P. and N. Das. 2010. Biosurfactant production and diesel oil degradation by yeast species *Trichosporon asahii* isolated from petroleum hydrocarbon contaminated soil. *International Journal of Engineering Science and Technology* 2:6942–6953.

Chandran, P. and N. Das. 2012. Role of sophorolipid biosurfactant in degradation of diesel oil by Candida tropicalis. *Bioremediation Journal* 16:19–30.

Chavan, A. and S. Mukherji. 2008. Treatment of hydrocarbon-rich wastewater using oil degrading bacteria and phototrophic microorganisms in rotating biological contactor: Effect of N:P ratio. *Journal of Hazardous Materials* 154:63–72.

Chen, B.L. and M.X. Yuan. 2012. Enhanced dissipation of polycyclic aromatic hydrocarbons in the presence of fresh plant residues and their extracts. *Environmental Pollution* 161:199–205.

Chen, B.L., Y.S. Wang, and D.F. Hu. 2010. Biosorption and biodegradation of polycyclic aromatic hydrocarbons in aqueous solutions by a consortium of white-rot fungi. *Journal of Hazardous Materials* 179(1–3):845–851.

Chen, C., J. Ming, B.A. Yoza, J. Liang, Q.X. Li, H. Guo, Z. Liu, J. Deng, and Q. Wang. 2019. Characterization of aerobic granular sludge used for the treatment of petroleum wastewater. *Bioresource Technology* 271:353–359.

Das, K. and A.K. Mukherjee. 2007. Crude petroleum-oil biodegradation efficiency of *Bacillus subtilis* and *Pseudomonas aeruginosa* strains isolated from a petroleumoil contaminated soil from North–East India. *Bioremediation Journal* 98:1339–1341.

Das, N. and P. Chandran. 2011. Microbial degradation of petroleum hydrocarbon contaminants: An overview. *Biotechnology Research International* 2011: Article ID: 941810. https://doi.org/10.4061/2011/941810.

Das, P., S. Mukherjee, and R. Sen. 2008. Genetic regulations of the biosynthesis of microbial surfactants: An overview. *Biotechnology & Genetic Engineering Reviews* 25(1):165–186.

Delille, D., E. Pelletier, A. Rodriguez-Blanco, and J.F. Ghiglione. 2009. Effects of nutrient and temperature on degradation of petroleum hydrocarbons in sub-Antarctic coastal seawater. *Polar Biology* 32:1521–1528. doi:10.1007/s00300-009-0652-z.

Delille D., F. Coulon, and E. Pelletier. 2004. Biostimulation of natural microbial assemblages in oil-amended vegetated and desert sub-Antarctic soils. *Microbial Ecology* 47:407–415.

Ebuehi, O.A.T., I.B. Abibo, P.D. Shekwolo, K.I. Sigismund, A. Adoki, and I.C. Okoro. 2005. Remediation of crude oil contaminated soil by enhanced natural attenuation technique. *Journal of Applied Sciences and Environmental Management* 9:103–106.

Fauzi, M. and P. Suryatmana. 2016. Bioremediation of crude oil waste contaminated soil using petrophilic consortium and *Azotobacter* sp. *Journal of Degraded and Mining Lands Management* 3:521–526.

Floodgate, G. 1984. The fate of petroleum in marine ecosystems. In *Petroleum Microbiology*, R. M. Atlas (Ed.), New York: Macmillan, 355–398.

Frank, V.B. 2016. *Co-Treatment of Domestic and Oil & Gas Wastewater with a Hybrid Sequencing Batch Reactor, Membrane Bioreactor*, Master of Science (Civil and Environmental Engineering) Thesis, Colorado School of Mines.

Franzmann, P.D., W.J. Robertson, L.R. Zappia, and G.B. Davis. 2002. The role of microbial populations in the containment of aromatic hydrocarbons in the subsurface. *Biodegradation* 13:65–78.

Ghorbanian, M., G. Moussavi, and M. Farzadkia. 2014. Investigating the performance of an up-flow anoxic fixed bed bioreactor and a sequencing anoxic batch reactor for the biodegradation of hydrocarbons in petroleum contaminated saline water. *International Biodeterioration and Biodegradation* 90:106–114.

Girigiri, B., C.N. Ariole, and H.O. Stanley. 2019. Biofertilizer from nitrogen fixing and phosphate solubilizing bacteria. *American Journal of Nanosciences*:27–38.

Godbout, J.G. Y. Comeau, and C.W. Greet. 1995. Soil characteristics effects on introduced bacterial survival and activity. In *Bioaugmentation for Site Remediation*, R.E. Hinchee, J. Fredrickson, and B.C. Alleman (Eds.), Columbus, OH: Battelle Press, 115–120.

Grotenhuis, J.T.C., B. Muijs, M. Wagelmans, J. van de Gun, J. Doze, and J. Gieteling. 2002. Monitoring PAH bioavailability during in-situ remediation of contaminated sediment. SedNet workshop, Chemical analysis and risk assessment of emerging contaminants in sediments and dredged material, Barcelona, Spain.

Guerra, A.B., J.S. Oliveira, R.C.B. Silva-Portela, W. Araújo, A.C. Carlos, A.T.R. Vasconcelos, A.T. Freitasc, Y.S. Domingosd, M.F. Fariasd, G.T. Fernandesde, and L.F. Agnez-Limaa. 2018. Metagenome enrichment approach used for selection of oil-degrading bacteria consortia for drill cutting residue bioremediation. *Environmental Pollution* 235:869–880.

Harms, H. and A. Zehnder. 1995. Bioavailabllity of sorbed 3-chlorodibenzofuran. *Applied and Environmental Microbiology* 61:27–33.

Hasanuzzaman, M., A. Ueno, H. Ito, Y. Ito, Y. Yamamoto, I. Yumoto, and H. Okuyama. 2007. Degradation of long-chain n-alkanes (C36 and C40) by Pseudomonas aeruginosa strain WatG. *International Journal of Biodeterioration & Biodegradation* 59(1):40–43.

Hozumi, T., H. Tsutsumi, and M. Kono. 2000. Bioremediation on the shore after an oil spill from the Nakhodka in the Sea of Japan. I. Chemistry and characteristics of the heavy oil loaded on the Nakhodka and biodegradation tests on the oil by a bioremediation agent with microbiological cultures in the laboratory. *Marine Pollution Bulletin* 40:308–314.

IPIECA. 2010. Petroleum refining water/wastewater use and management, IPIECA Operations Best Practice Series, London, United Kingdom.

Jafarinejad, Sh. 2017. *Petroleum Waste Treatment and Pollution Control*, First edition, Elsevier Inc., Butterworth-Heinemann.

Jasmine, J. and S. Mukherji. 2015. Characterization of oily sludge from a refinery and biodegradability assessment using various hydrocarbon degrading strains and reconstitute consortia. *Journal of Environmental Management* 149:118–125.

Jiang, Y., U.J. Yves, H. Sun, X. Hu, H. Zhan, and Y. Wu. 2016. Distribution, compositional pattern and sources of polycyclic aromatic hydrocarbons in urban soils of an industrial city, Lanzhou, China. *Ecotoxicology and Environmental Safety* 126:154–162.

Jimenez, N., M. Vinas, J. Sabate, S. Diez, J.M. Bayona, A.M. Solanas, and J. Albaiges. 2006. The Prestige oil spill. 2. Enhance biodegradation of a heavy fuel oil under field conditions by the use of an oleophilic fertilizer. *Environmental Science and Technology* 40:2578–2585. doi:10.1021/es052370z.

Johnsen, A.R., L.Y. Wickb, and H. Harms. 2005. Principles of microbial PAH-degradation in soil. *Environmental Pollution* 133:71–846.

Joo, H.S., P.M. Ndegwa, M. Shoda, and C.G. Phae. 2008. Bioremediation of oil contaminated soil using Candida catenulata and food waste. *Environmental Pollution* 156:891–896.

Koolivand, A., R. Saeedi, F. Coulon, V. Kumar, J. Villaseñor, F. Asghari, and F. Hesampoor. 2020. Bioremediation of petroleum hydrocarbons by vermicomposting process bioaugmentated with indigenous bacterial consortium isolated from petroleum oily sludge. *Ecotoxicology and Environmental Safety* 198:110645. doi:10.1016/j.ecoenv.2020.110645.

Kumar, M., V. Leon, A.D.S. Materano, and O.A. Ilzins. 2007. A halotolerant and thermotolerant Bacillus sp. degrades hydrocarbons and produces tensio-active emulsifying agent. *World Journal of Microbiology and Biotechnology* 23:211–220.

Kutty, S.R.M., H.A. Gasim, P.F. Khamaruddin, and A. Malakahmad. 2011. Biological treatability study for refinery wastewater using bench scale sequencing batch reactor systems, *Water Resources Management VI, WIT Transactions on Ecology and the Environment* 145:691–699. doi:10.2495/WRM11062.

Li, Q., C. Kang, and C. Zhang. 2005. Wastewater produced from an oilfield and continuous treatment with an oil-degrading bacterium. *Process Biochemistry* 40:873–877.

Liang, X., R. Shi, M. Radosevich, F. Zhao, Y. Zhang, S. Han, and Y. Zhang. 2017. Anaerobic lipopeptide biosurfactant production by an engineered bacterial strain for in situ microbial enhanced oil recovery. *RSC Advances* 7(33):20667–20676.

Mandal, A.K., P.M. Sarma, C.P. Jeyaseelan, V.A. Channashettar, B. Singh, B. Lal, and J. Datta. 2012. Large scale bioremediation of petroleum hydrocarbon contaminated waste at Indian oil refineries-case studies. *International Journal of Life Science and Pharma Research* 2:114–128.

McLoughlin, A.J. 1994. Controlled release of immobilized cells as a strategy to regulate ecological competence of inocula. In *Biotechnics/Wastewater*, T. Scheper (Ed.), Berlin: Springer, 1–45.

Mishra, S., J. Jyot, R.C. Kuhad, and B. Lal. 2001. Evaluation of inoculum addition to stimulate *in situ* bioremediation of oily-sludge-contaminated soil. *Applied and Environmental Microbiology* 67:1675–1681.

Mnif, I., S. Mnif, R. Sahnoun, S. Maktouf, Y. Ayedi, S.E. Chaabouni, and D. Ghribi. 2015. Biodegradation of diesel oil by a novel microbial consortium: Comparison between coinoculation with biosurfactant-producing strain and exogenously added biosurfactants. *Environmental Science and Pollution Research* 22(19):14852–14861.

Moslemy, P., R.J. Neufeld, and S.R. Guiot. 2002 Biodegradation of gasoline by gellan gum-encapsulated bacterial cells. *Biotechnology and Bioengineering* 80:175–184. doi:10.1002/bit.10358.

Mukred, A.M, A.A. Hamid, A. Hamzah, and W.M.W. Yusoff. 2008. Development of three bacteria consortium for the bioremediation of crude petroleum–oil in contaminated. *Journal of Biological Science* 8:73–79.

Musa, N.M., S. Abdulsalam, I.A. Suleiman, and A. Sale. 2015. Bioremediation of petroleum refinery wastewater effluent via augmented native microbes. *Journal of Emerging Trends in Engineering and Applied Sciences* 6(1):1–6.

Neethu, T.M., P.K. Dubey, A.R. Kaswala, and K.G. Patel. 2019. Cow dung as a bioremediation agent to petroleum hydrocarbon contaminated agricultural soils. *Current Journal of Applied Science and Technology* 38:1–9.

Nimrat, S., S. Lookchan, T. Boonthai, and V. Vuthiphandchai. 2019. Bioremediation of petroleum contaminated soils by lipopeptide producing *Bacillus subtilis* SE1. *African Journal of Biotechnology* 18:494–501.

Nyer, E.K, F. Payne, and S. Suthersan. 2002. Environment vs. bacteria or let's play 'name that bacteria'. *Ground Water Monitoring and Remediation* 23:36–45.

Odokuma, L.O. and M.N. Ibor 2002. Nitrogen-fixing bacteria enhanced bioremediation of a crude oil polluted soil. *Global Journal of Pure and Applied Sciences* 8(4):455–468.

Okoro, C.C., A. Agrawal, and C. Callbeck. 2012. Simultaneous biosurfactant production and hydrocarbon biodegradation by the resident aerobic bacterial flora of oil production skimmer pit at elevated temperature and saline conditions. *International Journal of Ecology and Environmenttal Science* 38(2–3):109–118.

Olawale, O., K.S. Obayomi, S.O. Dahunsi, and O. Folarin. 2020. Bioremediation of artificially contaminated soil with petroleum using animal waste: Cow and poultry dung. *Cogent Engineering* 7(1):1721409. doi:10.1080/23311916.2020.1721409.

Pajoumshariati, S., N. Zare, and B. Bonakdarpour. 2017. Considering membrane sequencing batch reactors for the biological treatment of petroleum refinery wastewaters. *Journal of Membrane Science* 523:542–550.

Poi, P., E. Shahsavari, A. Aburto-Medina, and A.S. Ball. 2017. Bioaugmentation: An effective commercial technology for the removal of phenols from wastewater. *Microbiology Australia* 38(2). doi:10.1071/MA17035.

Polyaka, Y.M., L.G. Bakina, M.V. Chugunova, N.V. Mayachkin, A,O. Gerasimov, and V.M. Bureb. 2018. Effect of remediation strategies on biological activity of oil-contaminated soil - A field study. *International Journal of Biodeterioration & Biodegradation* 126:57–68.

Qaderi, F., A.H. Sayahzadeh, and M. Azizic. 2018. Efficiency optimization of petroleum wastewater treatment by using of serial moving bed biofilm reactors. *Journal of Cleaner Production* 192:665–677.

Raghavan, P.U.M. and M. Vivekanandan. 1999. Biodegradation of crude oil by naturally occurring Pseudomonas putida. *Applied Biochemistry and Biotechnology* 44:29–32.

Rahman, K.S.M., J.M. Banat, J. Thahira, T. Thayumanavan, and P. Lakshmanaperumalsamy. 2002. Bioremediation of gasoline contaminated soil by a bacterial consortium amended with poultry litter, coir pith, and rhamnolipid biosurfactant. *Bioresource Technology* 81:25–32.

Ramadass, K., M. Megharaj, and K. Venkateswarlu. 2018. Bioavailability of weathered hydrocarbons in engine oil-contaminated soil: Impact of bioaugmentation mediated by Pseudomonas spp. on bioremediation. *Science of the Total Environment* 636:968–974.

Rodriguez-Blanco, A., V. Antoine, and E. Pelletier. 2010. Effects of temperature and fertilization on total vs. active bacterial communities exposed to crude and diesel oil pollution in NW Mediterranean Sea. *Environmental Pollution* 158:663–673.

Rosenberg, E., R. Legmann, A. Kushmaro, R. Taube, E. Adler, and E.Z. Ron. 1992. Petroleum bioremediation-a multiphase problem. *Biodegradation* 3:337–350.

Rothmel, R.K., J. Gaudet, W.H. School, M.J.R. Shannon, R. Krishnamoorthy, J.R. Smith, and R. Unterman. 1994. Biostimulation versus bioaugmentation: Two strategies for treating PCB-contaminated soils and sediments. In *Abstracts of the 4th General Meeting of the American Society for Microbiology*. Las Vegas, NV: ASM, 1 gg4: #Q–153.

Saber, D., D. Mauro, T. Sirivedhin, D. Saber, D. Mauro, and T. Sirivedhin. 2017. Environmental forensics investigation in sediments near a former manufactured gas plant site. *Environmental Forensics Investigation in Sediments near a Former Manufactured Gas Plant Site*. doi:10.1080/15275920500506881.

Sarkar, D., M. Ferguson, R. Datta, and S. Birnbaum. 2005. Bioremediation of petroleum hydrocarbons in contaminated soils: Comparison of biosolids addition, carbon supplementation, and monitored natural attenuation. *Environmental Pollution* 136:187–195.

Sarkar, P., A. Roy, S. Pal, B. Mohapatra, S.K. Kazy, M.K. Maiti, and P. Sar. 2017. Enrichment and characterization of hydrocarbon-degrading bacteria from petroleum refinery waste as potent bioaugmentation agent for in situ bioremediation. *Bioresource Technology* 242, 15–27. doi:10.1016/j.biortech.2017.05.010.

Singh, H. 2006. *Mycoremediation: Fungal Bioremediation*, New York, NY: Wiley Interscience.

Soclo, H.H., P. Garrigues, and M. Ewald. 2000. Origin of polycyclic aromatic hydrocarbons (PAHs) in coastal marine sediments: Case studies in Cotonou (Benin) and Aquitaine (France) areas. *Marine Pollution Bulletin* 40:387–396.

Srikanth, S., M. Kumar, D. Singh, M.P. Singh, S.K. Puri, and S.S.V. Ramakumar. 2018. Long-term operation of electro biocatalytic reactor for carbon dioxide transformation into organic molecules. *Bioresource Technology* 265:66–74.

Steliga, T. 2012. Role of fungi in biodegradation of petroleum hydrocarbons in drill waste. *Polish Journal of Environmental Studies* 21(2):471–479.

Suja, F., F. Rahim, M.R. Taha, N. Hambali, M.R. Razali, A. Khalid, and A. Hamzah. 2014. Effects of lacal microbial bioaugmentation and biostimulation on the bioremediation of total petroleum hydrocarbons (TPH) in crude oil contaminated soil based on laboratory and field observations. *International Biodeterioration & Biodegradation* 90:115–122.

Supaphol, S., S. Panichsakpatana, S. Trakulnaleamsai, N. Tungkananuruk, P. Roughjanajirapa, and A.G. O'Donnel. 2006. The selection of mixed microbial inocula in environmental biotechnology: Example using petroleum contaminated tropical soils. *Journal of Microbiological Methods* 65:432–441.

Tahhan, R.A., T.G. Ammari, S.J. Goussous, and H.I. Al-Shdaifat. 2011. Enhancing the biodegradation of total petroleum hydrocarbons in oily sludge by a modified bioaugmentation strategy. *International Biodeterioration and Biodegradation* 65:130–134.

Thakur, C., V.C. Srivastava, J.D. Mall, and A.D. Hiwarkar. 2018. Mechanistic study and multi response optimization of the electrochemical treatment of petroleum. *Clean: Soil, Air, Water* 46(3):1700624. doi:10.1002/clen.201700624.

Throne-Holst, M., A. Wentzel, T.E. Ellingsen, H. Kotlar, and S.B. Zotche. 2007. Identification of novel genes involved in long-chain n-alkane degradation by Acinetobacter sp. strain DSM 17874. *Applied and Environmental Microbiology* 73(10):3327–3332.

Ulrici, W. 2000. Contaminant soil areas, different countries and contaminant monitoring of contaminants. In *Environmental Process II. Soil Decontamination Biotechnology*, H.J. Rehm and G. Reed (Eds.), vol. 11, 5–42.

Van Eyk, J. 1997. Petroleum Bioventing. A.A. Balkema, Ritternam, The Netherlands. Ayotamuno, M.J., R.B. Kogbara and J.C. Agunwamba. 2006. Bioremediation of a petroleum hydrocarbon polluted agricultural soil at various levels of soil tillage in Portharcourt, Nigeria. *Nigerian Journal of Technology* 25:44–51.

van Hamme, J.D, A. Singh, and O.P. Ward. 2003. Recent advances in petroleum microbiology. *Microbiology and Molecular Biology Reviews* 67:503–549.

Varjani, S.J. 2017. Microbial degradation of petroleum hydrocarbons. *Bioresource Technology* 223:277–286.

Varjani, S.J., E. Gnansounou, and A. Pandey. 2017. Comprehensive review on toxicity of persistent organic pollutants from petroleum refinery waste and their degradation by microorganisms. *Chemosphere* 188:280–291.

Varjani, S.J., R. Joshi, V.K. Srivastava, H.H. Ngo, and W. Guo. 2020. Treatment of wastewater from petroleum industry: Current practices and perspectives. *Environmental Science and Pollution* 27:27172–27180.

Varjani, S.J., D.P. Rana, A.K. Jain, S. Bateja, and V.N. Upasani. 2015. Synergistic ex-situ biodegradation of crude oil by halotolerant bacterial consortium of indigenous strains isolated from on shore sites of Gujarat, India. *International Biodeterioration & Biodegradation* 103:116–124.

Varjani, S.J. and M.C. Sudha. 2018. Treatment technologies for emerging organic contaminants removal from wastewater. In *Water Remediation*, S. Bhattacharya, A.B. Gupta, A. Gupta, and A. Pandey (Eds.), Singapore: Springer Nature, vol. 452, 91–115.

Vasudevan, N. and P. Rajaram. 2001. Bioremediation of oil sludge-contaminated soil. *Environment International* 26(5). doi:10.1016/S0160-4120(01)00020-4, ISSN: 0160-4120.

Vasudevan, N., S. Bharathi, and P. Arulazhagan. 2007. Role of plasmid in the degradation of petroleum hydrocarbon by Pseudomonas fluorescens NS1. *Journal of Environmental Science Health Part A* 42:1141–1146.

Vinas M, M. Grifoll, J. Sabate, and A.M. Solanas. 2002. Biodegradation of a crude oil by three microbial consortia of different origins and metabolic capabilities. *Journal Industrial Microbiology and Biotechnology* 28:252–260.

Yaman, C. 2020. Performance and kinetics of bioaugmentation, biostimulation, and natural attenuation processes for bioremediation of crude oil-contaminated soils. *Processes* 8(883):5–14.

Zhang, C., D. Wu, and H. Ren. 2020. Bioremediation of oil contaminated soil using agricultural wastes via microbial consortium. *Scientific Reports* 10:9188.

Zhang, L. and X. Zhang. 2019. Bioremediation of petroleum hydrocarbon contaminated soil by petroleum degrading bacteria immobilized on biochar. *RSC Advances* 9:5304–35311.

Zhang, X., D. Xu, C. Zhu, T. Lundaa, and K.E. Scherr. 2012. Isolation and identification of biosurfactant producing and crude oil degrading Pseudomonas aeruginosa strains. *Chemical Engineering Journal* 209: 138–146.

Zrafi-Nouira, I., S. Guermazi, R. Chouari, N.M.D. Safi, E. Pelletier, A. Bakhrouf, and A. Sghir. 2008. Molecular diversity analysis and bacterial population dynamics of an adapted seawater microbiota during the degradation of Tunisian zarzatine oil. *Biodegradation* 20(4):467–486. doi:10.1007/s10532-008-9235-x.

5 Bioaugmentation for Removal of Cyanides

S.Z.Z. Cobongela
Nanotechnology Innovation Centre

CONTENT

5.1 INTRODUCTION

Cyanides are potentially deadly compounds that exist in various chemical forms. They can be found as solids, aqueous (solids dissolved in water), and gaseous forms. Cyanides are generally distributed in the environment especially water and soil. They are carbon–nitrogen radicals characterized by the presence of C≡N cyano-group with high toxicity. Cyanide toxicity potency is determined by the presence of free cyanide such as hydrogen cyanide (HCN) and cyanide ion (CN^-). Metal and nitrile cyanides are less potent compared to free cyanide. Industrial effluent is a huge contributor to the cyanide found in the environment. The industries with high cyanide effluents include metal finishing, iron and steel, electronics, plastics, mining, and others (Patil and Paknikar 2014). Cyanide compounds also occur naturally in the environment. They are produced by some bacteria, fungi, arthropods, and plants (Jones 1998). Plants particularly produce cyanide as a protection against herbivores. This is similar to the defense secretion phenomenon seen in arthropods. Fungi and bacteria produce and secrete cyanide as an antibiotic to inhibit the growth of the competitive organism (Parker et al. 1988). Cyanides can also be generated by chemicals known as cyanogen

DOI: 10.1201/9781003187622-5

5.2   CYANIDE CLASSIFICATION

Cyanide radicals are fast-acting and quickly combine with metal to form simple or complex salts. There are two cyanide gases known as hydrogen cyanide (HCN) and cyanogen chloride (CNCl). In room temperatures, HCN exists as a pale blue or colorless liquid and colorless gas at high temperatures. Sodium cyanide (NaCN), potassium cyanide (KCN), and $Zn(CN)_2$ are examples of simple crystal forms of cyanide and are classified as weak acids dissociable (WAD) complexes. Cyanide also exists in strong metal cyanides such as potassium ferricyanide ($K_3[Fe(CN)_6]$) and other cyanides conjugated to gold or silver (Ebbs 2004; Akcil 2003). Cyanide does not adsorb in the soil, it remains in an aqueous solution thereby prone to react with other compounds. Metallic contaminants from industrial effluents such as iron, copper, zinc, and nickel also react with cyanide to form metal cyanide (Patil and Paknikar 2014; Gonçalves, Pinto and Granato 1998). These metallo-cyanide complexes have high chemical and biological stability compared to other forms of complexes such as sodium cyanide and KCN which easily dissociate at neutral pH (Young and Jordan; Botz).

Cyanides are typically white powder with an almond-like smell (Dash, Gaur and Balomajumder 2009). The almond-like smell might not be strange as cyanide is also found in almonds and other kinds of nuts and legumes such as cashew and beans, respectively. Though cyanide is known to be toxic to living organisms in the ecosystem especially aquatic life, it is however naturally produced by some organisms including fungi, algae, photosynthetic bacteria, plants, and others (Dubey and Holmes 1995). A concentration of about 0.2 and 0.02–0.076 mg/L has been deemed toxic to most fish species and microorganisms. Unpolluted water streams contain between 0.001 and 0.05 mg/L of total cyanide (Wild, Rudd and Neller 1994). The concentration of total cyanide on polluted river water is usually between 0.01 and 10 mg/L (Ikebukuro et al. 1996). The acceptable effluent concentration is between 0.01 and 10 mg/L (Wild, Rudd and Neller 1994). However, some industries collect the waste over time and increase cyanide concentration to about 10,000–30,000 mg/L (Sabatini et al. 2012).

5.3   DANGERS OF CYANIDE IN HUMANS

Exposure to cyanide often occurs through inhalation, ingestion, and absorption through skin and eyes. It is therefore distributed to all body organs and tissues via the bloodstream (Zheng et al. 2004). In cells, cyanide interferes with the respiratory system in the mitochondria, particularly cytochrome C oxidase. Cyanide irreversibly binds to the ferric ions on cytochrome C oxidase and inhibits the final electron transfer to oxygen, which in turn prevents adenosine triphosphate (ATP) production. The toxicity of cyanide is fundamentally the inhibition of the aerobic cellular respiratory system. It shifts the cellular metabolism from aerobic to anaerobic even in the presence of sufficient oxygen. Anaerobic metabolism increases the production of lactic acid. The inhibition of the cytochrome C complex subsequently affects vital organs such as the brain and heart.

5.4   CYANIDE REMEDIATION

There are several known physiochemical and biological processes effective in the degradation and removal of cyanide from waste effluent. The physicochemical processes include physical, adsorption, complexation, and oxidation (Young and Jordan). Dilution, distillation, electrowinning, and membranes are examples of physical methods (Rosehart 1973; Kuhn 1971; Roberts and Jackson 1971). Adsorption, coagulation, and complexation are used as separation methods for the removal of cyanide while oxidation is used as the degradation method. Oxidation process aids in breaking the triple bond between carbon and nitrogen thus completely destructing cyanide into less and/ or nontoxic by-products. Chemical methods that degrade cyanide comprise electrolytic decomposition, incineration, ozonation, alkaline chlorination, etc. (Hillemanns 1989). Other treatment methods (ion exchange, copper-catalyzed hydrogen peroxide, Caro's acid, reverse osmosis, thermal hydrolysis, INCO process, etc.) are incapable of completely degrading cyanide and are expensive (Patil and Paknikar 2000). In addition, physiochemical processes often require a number of pre- and/or post-treatments to completely remove cyanide and release discharge that meets environmental criteria (Demopoulos and Cheng 2004).

5.5 ADVANTAGES OF BIOAUGMENTATION PROCESSES

Biological processes mostly employ oxidation pathways. Generally, microorganisms use biosorption for concentration/recovery, and biodegradation is usually accelerated by enzymes. Bioremediation has gained interest in wastewater treatment industries due to its economic and environmental-friendly advantages. The two most important advantages of bioremoval of cyanide are: less expensive and efficient compared to chemical/physical treatments and natural oxidation, respectively. Biological processes degrade cyanide while using the by-products as an energy source. For example, the degradation of cyanide results in increased ammonia which is in turn used by the microorganisms as an energy source. Also, bioremediation eliminates the acidification step which is often required in physiochemical processes prior to degradation to limit volatilization due to cyanhydric acid. Physiochemical processes also result in further toxic residues as degradation metabolites (Kao et al. 2003). Therefore, biological processes subsequently reduce degradation steps while decreasing the toxicity of the by-product in the resultant discharge (Dumestre et al. 1997). Cyanide-degrading microorganisms are commonly found and isolated from industrial and coke wastewater (Kao et al. 2003; Park et al. 2008; Akcil et al. 2003).

5.6 BIOAUGMENTATION OF CYANIDES

The very first biodegradation process of cyanide was commercially established in the mid-1980s with the work done by Mudder and Whitlock published in 1984 (Mudder and Whitlock 1984). However, it has been observed for the first time in the early 20th century (Akcil and Mudder 2003). Since then, several microorganisms have demonstrated effective biological degradation of cyanide and cyanide-containing compounds. Both aerobic and anaerobic microorganisms have been used for application in various cyanide treatment facilities. They are incorporated in *in situ* full-scale active and passive biological treatment facilities.

Donghee Park and colleagues conducted a study where they used *Cryptococcus humicolus*, a yeast that degrades cyanide, and other cyanide-degrading microorganisms to degrade ferric cyanide in a full-scale cokes wastewater (Park et al. 2008). This consortium of microorganisms is inoculated in a fluid-bed reactor containing the wastewater, glucose, KCN, and other necessary nutrients for 2 months. However, the removal of the cyanide is very poor due to poor settlement of microbial congregates, lack of carbon sources, and slow biodegradability of ferric cyanide. *Cryptococcus humicolus* was first isolated in 2002 from coke wastewater similar to the one used in the above study (Kwon et al. 2002). This yeast strain degraded free and metallo-cyanides.

5.6.1 BIOAUGMENTATION OF THIOCYANATE

Thiocyanate (SCN^-) is one of the predominant cyanide species present in industrial effluent. It is usually found in concentrations ranging from 17 to 1500 mg/L (Luthy and Tallon 1980). Several heterotrophic and chemolithotrophic bacteria are capable of degrading thiocyanate by using it as a sole energy source. A bacterium, belonging to *Thiobacillus* genus, is isolated from sewage effluent and well water containing thiocyanate (Happold et al. 1954). This genus is known to utilize carbon dioxide (CO_2) as the sole carbon source while utilizing inorganic sulfur compounds for energy. This particular bacterium was no different from the genus as it oxidized thiocyanate present in the growth media. The growth conditions for optimum thiocyanate oxidation were at 30°C temperature with a pH ranging from 6.8 to 7.6. It was later named *Thiobacillus thiocyanoxidans* after the study conducted by Happold and colleagues. Over six decades later, researchers are still reporting on the thiocyanate degrading ability of *Thiobacillus* genus. However, the developed and new analytical technologies have improved the understanding of the degradation pathways in a biological reaction. Currently, there are two pathways associated with bacterial degradation of thiocyanate (Berben, Balkema, et al. 2017). These are the cyanate pathway and carbonyl sulfide pathway, commonly known as the COS pathway. Recent studies have reported COS pathway is one of the observed degradation mechanisms for thiocyanate by *Thiobacillus* genus (Li et al. 2020). This is confirmed by instruments such as Raman spectroscopy.

Biological degradation of thiocyanate often requires a two-stage activated sludge (Li et al. 2020).

A different bacterium that degrades thiocyanate is *Thioalkalivibrio* genus. It is a haloalkaliphile that metabolizes a number of thiolated compounds including thiocyanate at pH 10 and above (Sorokin et al. 2001). *Thioalkalivibrio* genus acts by oxidizing thiocyanate using it as a sole electron donor and nitrogen source. Examples of *Thioalkalivibrio* genus species are *thiocyanoxidans*, *paradoxus, sibericum, aerophilum, denitrificans*, etc. (Sorokin et al. 2001; Sorokin et al. 2002). *Thioalkalivibrio* genus bacteria in the studies conducted by Sorokin and colleagues were isolated from soda lakes. Soda lakes are known to harbor vast alkaline and anaerobic/aerobic microbial communities (Imhoff et al. 1979). This diverse microbial community also contains a huge population of sulfur-oxidizing bacteria which are good candidates in degrading thiocyanate.

5.7 TECHNIQUES IN BIOAUGMENTATION OF CYANIDES

The biological degradation is conducted in bioreactors with activated sludge. The activated sludge process is often aerated with oxygen and contains selected microorganisms. The aerobic microorganism's culture is cultivated and activated to grow and degrade compounds of interest. A sludge process usually contains a clarifier to separate wastewater solid/liquid material, a reactor containing microorganisms, a liquid/solid biomass separator, and a sludge recycling system. In cyanide biodegradation, sequencing batch reactor (SBR) system and packed-bed (fixed-bed) reactor are commonly used activated sludge processes (Campos et al. 2006; Sirianuntapiboon et al. 2008; Kumar et al. 2015).

Occasionally, a combination of aerobic and anaerobic microorganisms yields better degradation results. This is observed in a study done by Novak and colleagues on an up-flow anaerobic sludge blanket (UASB) reactor (Novak et al. 2013). Packed-bed bioreactors are tubular shaped and immobilize microorganisms on the walls/bed of the reactor. Simultaneous adsorption and biodegradation (SAB) technique is often accomplished on packed-bed bioreactors. This technique allows the reuse of the microorganisms in a continuous mode while eliminating the biomass separation step. Alternatively, the SBR process has the biomass in the activated sludge and removes the biomass after a desirable treatment of the waste.

5.8 CYANIDE BIOAUGMENTATION CONDITIONS

Cyanide is generally toxic to most microorganisms, however, several species from bacteria and fungi kingdom have developed resistance (Raybuck 1992). Commonly, cyanide-degrading microorganisms are alkaliphiles (Dumestre et al. 1997; Luque-Almagro et al. 2005; Adjei and Ohta 2000). However, some microorganisms can degrade cyanide in acidic conditions (Barclay et al. 1998). Biodegradation is mostly dependent on external conditions such as pH, oxygen, temperature, and cyanide concentration and solubility. In the presence or absence of oxygen, microorganisms degrade cyanide through aerobic or anaerobic processes, respectively. Microorganisms detoxify and utilize cyanide as their nitrogen and carbon source. They degrade cyanide in five different pathways namely: oxidation, reduction, hydrolytic, substitution, and/or transfer reactions (Ebbs 2004). An example of an aerobic process in the degradation of cyanide and thiocyanate involves two oxidation steps shown in reactions 1 and 2 below, respectively. *Pseudomonas fluorescens* is an example of cyanide degradation under aerobic conditions (Harris and Knowles).

5.9 CYANIDE AND THIOCYANATE DEGRADATION PROCESSES

The first step involves oxidative breakdown on total cyanide (including metal cyanides) and or thio-cyanate. Precipitated free metals from metallo-cyanides are subsequently adsorbed in the biofilm. For example, *Rhizopus oryzae* is a fungus that degrades and simultaneously adsorbs and biode-grades FeCN and ZnCN (Dash, Gaur and Balomajumder 2009). Degradation efficiency of SAB is higher than degradation alone. The immobilized *Rhizopus oryzae*, immobilized using activated granular carbon, also adsorb the free iron and zinc ions from FeCN and ZnCN degradation.

Concurrently, ammonia and bicarbonate are produced as the main metabolites (Equations 5.1 and 5.2). The second step of biodegradation comprises two nitrification processes for conversion of ammonia to nitrite as an intermediate metabolite leading to the production of nitrate as the final by-product (Equations 5.3 and 5.4) (Akcil and Mudder 2003). Simultaneously, bicarbonate serves as a food source for the nitrifying of the microorganisms while releasing carbon dioxide. Furthermore, the alkalinity of the bicarbonate favors the growth of alkaliphiles. *Klebsiella oxytoca* is an example of bacteria capable of degrading cyanide under anaerobic conditions (Chen et al. 2009; Kao et al. 2004). Anaerobic conditions usually yield methane and ammonia as by-products (Equation 5.5). Anaerobic microorganisms degrade cyanide mostly via a reduction reaction. In some cases, cyanide hydrolytic pathways have been observed in anaerobic microorganisms (Luque-Almagro et al. 2005). However, bioaugmentation of cyanide is mostly catalyzed by enzymes.

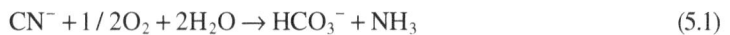

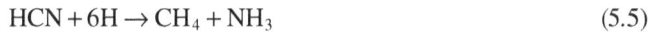

$$CN^- + 1/2O_2 + 2H_2O \rightarrow HCO_3^- + NH_3 \tag{5.1}$$

$$SCN^- + 2H_2O + 5/2O_2 \rightarrow SO_4^{2-} + HCO_3^- + NH_3 \tag{5.2}$$

$$NH_3 + 3/2O_2 \rightarrow NO_2^- + H^+ + H_2O \tag{5.3}$$

$$NO_2^- + 1/2O_2 \rightarrow NO_3^- \tag{5.4}$$

$$HCN + 6H \rightarrow CH_4 + NH_3 \tag{5.5}$$

5.10 CYANIDE-DEGRADING ENZYMES

A species of *Fusarium oxysporum* also degrades cyanide while yielding formamide and formate as intermediates. *Fusarium oxysporum* is a fungus that degrades cyanide while producing formamide hydrolyase to further degrade formamide to formate (Pereira, Pires and Roseiro 1999). Formamide is also toxic to the environment and needs to be bioremediated. A study done by Campos and colleagues is an example of bioremediation of formamide in cyanide biodegradation (Campos et al. 2006). Two microorganisms, *Fusarium oxysporum* together with a formamide degrading bacterium of *Methylobacterium* species, are merged to achieve complete degradation of cyanide and bioremediation of formamide. Other studies make use of microorganism consortium incorporating more than two microbial species to achieve better degradation outcomes (Valentina et al. 2018; Mekuto et al. 2013). The mixed culture degradation also assists in increasing the biodegradation rate in higher concentrations of cyanide that would otherwise be toxic to a single culture (Valentina, Hadisoebroto and Rinanti 2018). Changing other biomass variants such as food, temperature, and pH also leads to better degradation activity of the microorganisms. For instance, *Bacillus pumilus* intracellularly degrades free cyanide (KCN). Cyanide degradation by *Bacillus pumilus* whole-cell and/or free-cell extract is linked to enzyme activity (Meyers et al. 1991). Most of these organisms act by producing specific enzymes that are capable of breaking down cyanide. They do this as a way to circumvent cyanide toxicity (Wang et al. 1992). Several cyanide-degrading enzymes have been identified and isolated to date. Some enzymes are directly involved in the degradation of cyanide or cyanide-containing and cyanide-related compounds.

5.10.1 Cyanide Oxidation (Oxygenase, Monooxygenase, Dioxygenase)

Pseudomonas species have been associated with the oxidative pathway using cyanide oxygenase activity (Luque-Almagro et al. 2005; Harris and Knowles). *Pseudomonas putida* degrades free cyanide such as KCN and cyanide complexes $K_2[Ni(CN)_4]$; however, it produces $Ni(CN)_2$ as a major metabolite (Silva-Avalos et al. 1990). Conversely, some *Pseudomonas* species also catalyze the degradation of $Ni(CN)_4{}^{2-}$ (Dorr and Knowles 1989).

Enzymatic oxidation may occur in different reactions catalyzed by oxygenase enzymes. Ammonia and carbon dioxide are the final by-products of enzymatic degradation of cyanide. Cyanide dioxygenase catalyzes the oxidation of cyanide in one single reaction (Equation 5.6). Reactions (5.7) and (5.8) show a two-step reaction catalyzed by monooxygenase and cyanase, respectively. It has been reported in the literature that some cyanide dioxygenase enzymes convert cyanide directly to ammonia and carbon dioxide without yielding cyanate as an intermediate. An example of this dioxygenase is isolated from an *Escherichia coli* strain (Figueira et al. 1996).

$$HCN + O_2 + 2e^- + 2H^+ \rightarrow NH_3 + CO_2 \tag{5.6}$$

$$HCN + O_2 + 2e^- + 2H \rightarrow OCN^- + H_2O \tag{5.7}$$

$$OCN^- + HCO_3{}^- + 2H^+ \rightarrow NH_3 + 2CO_2 \tag{5.8}$$

The formation of glucose-cyanide complex is essential prior to degradation. Monooxygenase yields cyanate (OCN^-) as an intermediate while cyanase converts cyanate to ammonia and carbon dioxide. The cyanate biodegradation catalysis by cyanase is bicarbonate-dependent (refer to reaction C). *Pseudomonas fluorescens* produces both oxygenase and cyanase for complete bioaugmentation of cyanides (Dorr and Knowles 1989). Both these enzymes are physiologically distinct and their activities are independent. *Pseudomonas pseudoalcaligenes* strain assimilates sodium cyanide using cyanase enzyme regulated by cyanate, cyanide, and cyanmetallic complexes. Optimum activity was established at alkaline conditions, pH 11.5, and 30 mM concentration of free cyanide. This strain also utilizes the by-products such as ammonium, nitrate, cyanate, cyanoacetamide, nitroferricyanide, and a variety of cyanide-metal complexes. Cyanase is activated during the growth phase in the presence of cyanide and cyanate while ammonia and nitrates inhibited cyanase production (Luque-Almagro et al. 2005). The cyanase with the highest catalytic efficiency rate reported to date is produced by the fungus *Thermomyces lanuginosus* (Ranjan et al. 2021).

5.10.1.1 Pterin-Dependent Oxygenase

The cytosolic enzyme cyanide oxygenase requires pterin and NADH as cofactors (Kunz et al. 2001; Mahendran et al. 2020). Therefore, they can be classified as pterin-dependent hydroxylase. The degradation pathway is different from the aforementioned oxygenases. This pterin-dependent oxygenase catalysis produces formamide as an intermediate (Equation 5.9) and eventually yields formate and ammonia (Equation 5.10). The final step produces carbon dioxide and biopterin as by-products (Equation 5.11). Therefore, the final by-products of cyanide enzymatic degradation are ammonia, carbon dioxide, and biopterin (Equation 5.12). Biopterin also diminishes cyanide toxicity in cyanide-degrading bacteria such as *Bacillus pumilus* (Mahendran et al. 2018). Since formate is also toxic to microorganisms, it is also oxidized further to carbon dioxide and ammonia. Microorganisms such as *Pseudomonas fluorescens* do this by simultaneously producing formate oxidizing enzyme, formate dehydrogenase (Kunz et al. 2001; Fernandez et al. 2004).

$$HCN + O_2 + 2NADH + 2H^+ \rightarrow HCONH_2 + H_2O + 2NAD^+ \tag{5.9}$$

$$HCONH_2 + H_2O \rightarrow HCO_2H + NH_3 \tag{5.10}$$

$$HCO_2H + NAD^+ \rightarrow CO_2 + NADH + H^+ \tag{5.11}$$

$$HCN + O_2 + NADH + H^+ \rightarrow CO_2 + NH_3 + NAD^+ \tag{5.12}$$

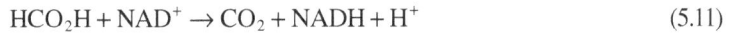

5.10.2 Cyanide Reduction (Nitrogenase)

Cyanide reduction is accelerated by nitrogenase enzyme naturally produced by microorganisms (Chen et al. 2009; Seefeldt et al. 2013). The nitrogenase enzyme is inactivated by oxygen. It is often used by bacteria to fix atmospheric nitrogen gas. Nitrogenase reduces HCN to generate ammonia and methane (Equation 5.13). *Klebsiella oxytoca* is an example of bacteria that degrades cyanide by producing nitrogenase enzyme (Kao et al. 2003; Liu et al. 1997). Later studies still confirm the activity of the nitrogenase produced by *Klebsiella oxytoca* to degrade free cyanide and cyanide complex $K_2[Ni(CN)_4]$ (etracyanonickelate) to ammonia and methane (Chen et al. 2009).

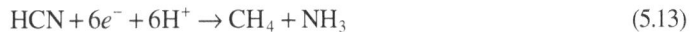

$$HCN + 6e^- + 6H^+ \rightarrow CH_4 + NH_3 \tag{5.13}$$

5.10.3 Cyanide Hydrolysis (Nitrilase, Nitrile Hydratase, Cyanide Hydratase, Cyanide Dihydratase, Formamidase)

There are several enzyme-catalyzed cyanide hydrolysis pathways. The enzymes include nitrilase, nitrile hydratase, cyanide hydratase, and cyanide dihydratase. Nitrilase directly converts nitriles or organocyanides (R-CN) in the presence of water to carboxylic acid and ammonia (Equation 5.14) (Benedik and Sewell 2018). *Pseudomonas pseudoalcaligenes* is one of the bacteria that produce nitrilase that hydrolyzes KCN (Acera et al. 2017). Nitrilase is robust and does not require the addition of cofactors for catalysis. It is a good candidate for large-scale cyanide biodegradation. *Aspergillus niger* produces both nitrilase and nitrile hydratase with optimum activity at pH 8–9 and 45°C (Rinágelová et al. 2014).

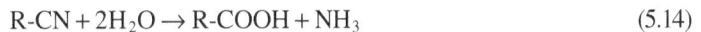

$$R\text{-}CN + 2H_2O \rightarrow R\text{-}COOH + NH_3 \tag{5.14}$$

Nitrilase catalysis is a one-step hydrolysis pathway while nitrile hydratase requires two steps. Nitrile hydratase firstly hydrolyzes organocyanides to amides (Equation 5.15) while amidase converts the amides to carboxylic acid and ammonia (Equation 5.16) as final biodegradation metabolites. Nitrilase also contributes to amide formation as an intermediate through enantioselective process (Fernandes et al. 2006; Benedik and Sewell 2018).

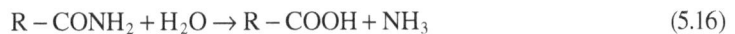

$$R - CN + H_2O \rightarrow R - CONH_2 \tag{5.15}$$

$$R - CONH_2 + H_2O \rightarrow R - COOH + NH_3 \tag{5.16}$$

Cyanide hydratase and cyanide dihydratase (cyanidase) are from nitrilase superfamily and both degrade cyanide (Pace and Brenner 2001). Cyanide hydratase is usually a fungal enzyme while cyanide dihydratase is often produced by bacteria. Like cyanide oxygenase, cyanide hydratase hydrolyzes cyanide to formamide (Equation 5.17). *Gloeocercospora sorghi* is a phytogenic fungus, and other fungal species such as *Stemphylium loti* and *Leptosphaeria maculans* produce cyanide hydratase to convert HCN to formamide (Wang et al. 1992; Basile et al. 2008; O'Reilly and Turner 2003; Nolan et al. 2003). *Fusarium solani* is another well-studied fungus that degrades cyanide by firstly hydrolyzing KCN to formamide using cyanide hydratase under alkaline conditions while assimilating formamide to a less toxic by-product, formate (Dumestre et al. 1997).

Cyanide dihydratase degrades cyanide to formate and ammonia (Equation 5.18). The most studied cyanide dihydratase is produced by *Bacillus pumilus* (Jandhyala et al. 2003; Crum et al. 2016;

Meyers et al. 1991). These two enzymes both depend on other enzymes to completely degrade cyanide and resultant toxic metabolites. The formamide produced by cyanide hydratase is further degraded to formate and ammonia by enzyme formamidase (Equation 5.19). *Pseudomonas putida* and *Fusarium solani* show a catalytic activity of formamidade (Babu et al. 1996; Dumestre et al. 1997).

Formate dehydrogenase further degrades formate to carbon dioxide (Equation 5.20). *Arabidopsis thaliana* plant species produces both cyanide dihydratase and formate dehydrogenase for complete degradation of cyanide to carbon dioxide and ammonia (Kebeish et al. 2015). However, *Alcaligenes xylosoxidans* subspecies *denitrificans* hydrolyzes HCN with high efficacy in the presence of water to give out ammonia and formate using cyanide dihydratase; usually, this pathway yields formamide as an intermediate, (Ingvorsen et al. 1991). *Pseudomonas stutzeri* also uses cyanide dehydratase to convert cyanide to ammonia and formate (Watanabe et al. 1998).

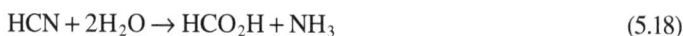

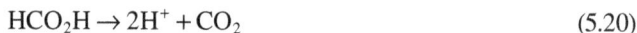

$$HCN + H_2O \rightarrow HCONH_2 \tag{5.17}$$

$$HCN + 2H_2O \rightarrow HCO_2H + NH_3 \tag{5.18}$$

$$HCONH_2 + H_2O \rightarrow HCO_2H + NH_3 \tag{5.19}$$

$$HCO_2H \rightarrow 2H^+ + CO_2 \tag{5.20}$$

5.10.3.1 Thiocyanate Hydrolysis (Rhodanese, Thiocyanate Hydrolase, Thiocyanate Dehydrogenase, Carbonyl Sulfide Hydrolase)

Other hydrolysis degradation processes involve a substitution reaction from cyanide to thiocyanate and sulfite in the presence of thiosulfate using rhodanese also known as thiosulfate or cyanide sulfurtranferase (Equation 5.21) (Cipollone et al. 2007). This enzyme is conventionally used by plants and microorganisms to detoxify cyanide. It is produced by leaves of *Manihot utilissima* plant and bacteria such as *Bacillus subtilis* and *Pseudomonas aeruginosa* (Cipollone et al. 2008). The resultant thiocyanate undergoes two degradation processes using a series of enzymes. The first thiocyanate degradation process is through carbonyl sulfide or COS pathway which involves the use of thiocyanate hydrolase.

Thiocyanate hydrolase catalyzes the hydrolysis of thiocyanate nitrile bonds in the presence of water to carbonyl sulfide and ammonia (Equation 5.22). The first thiocyanate degradation by thiocyanate hydrolyzing enzymes produced by microorganisms was first observed in *Thiobacillus thioparus* bacteria (Katayama et al. 1992; Bezsudnova et al. 2007). The COS is then hydrolyzed further to hydrogen sulfide and carbon dioxide catalyzed by carbonyl sulfide hydrolase (Equation 5.23). The second thiocyanate degradation process is known as cyanate pathway, it involves the use of thiocyanate dehydrogenase. Thiocyanate dehydrogenase catalyzes the conversion of thiocyanate to cyanate as the major metabolite (Equation 5.24) (Tikhonova et al. 2020). One of the thiocyanate dehydrogenase-producing species is a haloalkaliphilic sulfur-oxidizing bacterium of *Thioalkalivibrio* genus (Berben, Overmars, et al. 2017). The cyanate is further oxidized to ammonia and carbon dioxide using cyanate (Equation 5.8).

$$HCN + S_2O_3^- \rightarrow SCN^- + SO_3^{2-} \tag{5.21}$$

$$SCN^- + H_2O \rightarrow COS + NH_3 \tag{5.22}$$

$$COS + H_2O \rightarrow H_2S + CO_2 \tag{5.23}$$

$$SCN^- + H_2O \rightarrow OCN^- + S^0 + 2H \tag{5.24}$$

5.10.4 CYANIDE SUBSTITUTION/TRANSFER (β-CYANOALANINE SYNTHASE, β-CYANOALANINE NITRILASE, β-CYANOALANINE HYDRATASE, ASPARAGINASE)

Substitution/transfer reaction involves the use of enzymes and amino acids. β-cyanoalanine (βCA) synthase detoxifies cyanide to hydrogen sulfide and βCA in the presence of cysteine (Equation 5.25) (Hatzfeld et al. 2000; Piotrowski et al. 2001). An example of βCA synthase enzyme produced by *Oryza sativa* also degrades other forms of cyanides such as KCN and $K_3Fe(CN)_6$ (Yu et al. 2012). βCA nitrilase coverts βCA to ammonia and aspartic acid in one-way step (Equation 5.26) (Piotrowski et al. 2001). The βCA is also converted in a two-way step using two enzymes, βCA hydratase, and asparaginase. The βCA is converted by βCA hydratase enzyme to asparagine (Equation 5.27) while asparaginase converts asparagine to ammonia and aspartic acid (Equation 5.28).

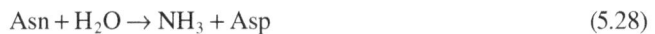

$$HCN + Cys \rightarrow H_2S + \beta CA \tag{5.25}$$

$$\beta CA + H_2O \rightarrow NH_3 + Asp \tag{5.26}$$

$$\beta CA + H_2O \rightarrow Asn \tag{5.27}$$

$$Asn + H_2O \rightarrow NH_3 + Asp \tag{5.28}$$

5.11 CONCLUSION

Cyanide is a highly toxic compound that is produced naturally by plants and other microorganisms. A massive percentage of cyanide in the environment is from industrial wastewater. Cyanide exists as free cyanide, WAD, and metal cyanide. There is a number of physical and chemical remediation processes to date. These have been presented with vast disadvantages ranging from high costs to producing other hazardous by-products. Remediation of cyanide using biological processes has proven to be the most environmentally friendly process in the degradation and removal of cyanide and cyanide complexes. Although cyanide is toxic to the environment, it can also be destroyed by naturally occurring microorganisms. Several fungi, bacteria, plants, and others have been shown to completely degrade cyanide under controlled conditions. Most of the cyanide-degrading microorganisms produce enzymes to catalyze and accelerate the biodegradation of cyanides. Cyanide-degrading enzymes are of high recommendation for cyanide bioremediation in cyanide-contaminated industrial wastewater. They catalyze the degradation in different pathways such as oxidation, hydrolysis, reduction, etc.

REFERENCES

Acera, F., M.I. Carmona, F. Castillo, A. Quesada and R. Blasco. 2017. A Cyanide-Induced 3-Cyanoalanine Nitrilase in the Cyanide-Assimilating Bacterium Pseudomonas pseudoalcaligenes Strain CECT 5344. *Applied and Environmental Microbiology* 83 9. doi:10.1128/AEM.00089-17, https://www.ncbi.nlm.nih.gov/pmc/articles/PMC5394316/.

Adjei, M.D. and Y. Ohta 2000. Factors affecting the biodegradation of cyanide by Burkholderia cepacia strain C-3. *Journal of Bioscience and Bioengineering* 89 3: 274–277. doi:10.1016/S1389-1723(00)88833-7.

Akcil, A. 2003. Destruction of cyanide in gold mill effluents: biological versus chemical treatments. *Biotechnology Advances* 21 6: 501–511. doi:10.1016/s0734-9750(03)00099-5.

Akcil, A, A.G. Karahan, H. Ciftci and O. Sagdic. 2003. Biological treatment of cyanide by natural isolated bacteria (*Pseudomonas* sp.). *Minerals Engineering* 16, 7: 643–649. doi:10.1016/S0892-6875(03)00101-8.

Akcil, A. and T. Mudder. 2003. Microbial destruction of cyanide wastes in gold mining: process review. *Biotechnology Letters* 25 6: 445–450. doi:10.1023/A:1022608213814.

Babu, G.R.V., O.K. Vijaya, V.L. Ross, J.H. Wolfram and K.D. Chapatwala. 1996. Cell-free extract(s) of Pseudomonas putida catalyzes the conversion of cyanides, cyanates, thiocyanates, formamide, and cyanide-containing mine waters into ammonia. *Applied Microbiology and Biotechnology* 45 1–2: 273–277. doi:10.1007/s002530050683.

Barclay, M., A. Hart, C.J. Knowles, J.C. L. Meeussen and V.A. Tett. 1998. Biodegradation of metal cyanides by mixed and pure cultures of fungi. *Enzyme and Microbial Technology* 22 4: 223–231. doi:10.1016/S0141-0229(97)00171-3.

Basile, L.J., R.C. Willson, B. Trevor Sewell and M.J. Benedik. 2008. Genome mining of cyanide-degrading nitrilases from filamentous fungi. *Applied Microbiology and Biotechnology* 80 3: 427. doi:10.1007/s00253-008-1559-2.

Benedik, M.J. and B. Trevor Sewell. 2018. Cyanide-degrading nitrilases in nature. *The Journal of General and Applied Microbiology* 64 2: 90–93. doi:10.2323/jgam.2017.06.002.

Berben, T., C. Balkema, D.Y. Sorokin and G. Muyzer. 2017. Analysis of the Genes Involved in Thiocyanate Oxidation during Growth in Continuous Culture of the Haloalkaliphilic Sulfur-Oxidizing Bacterium Thioalkalivibrio thiocyanoxidans ARh 2T Using Transcriptomics. *mSystems* 2 6. doi:10.1128/mSystems.00102-17. https://www.ncbi.nlm.nih.gov/pmc/articles/PMC5744179/ (zugegriffen: 16. März 2021).

Berben, T., L. Overmars, D.Y. Sorokin and G. Muyzer. 2017. Comparative Genome Analysis of Three Thiocyanate Oxidizing Thioalkalivibrio Species Isolated from Soda Lakes. *Frontiers in Microbiology* 8. doi:10.3389/fmicb.2017.00254. https://www.frontiersin.org/articles/10.3389/fmicb.2017.00254/full (zugegriffen: 29. April 2021).

Bezsudnova, E.Y., D.Y. Sorokin, T.V. Tikhonova and V.O. Popov. 2007. Thiocyanate hydrolase, the primary enzyme initiating thiocyanate degradation in the novel obligately chemolithoautotrophic halophilic sulfur-oxidizing bacterium Thiohalophilus thiocyanoxidans. *Biochimica et Biophysica Acta (BBA) - Proteins and Proteomics* 1774, 12: 1563–1570. doi:10.1016/j.bbapap.2007.09.003.

Botz, M.M. *Overview of Cyanide Treatment Methods*. 10.

Campos, M.G., P. Pereira and J. Carlos Roseiro. 2006. Packed-bed reactor for the integrated biodegradation of cyanide and formamide by immobilised *Fusarium oxysporum* CCMI 876 and *Methylobacterium* sp. RXM CCMI 908. *Enzyme and Microbial Technology* 38, 6: 848–854. doi:10.1016/j.enzmictec.2005.08.008.

Chen, C.Y., C.M. Kao, S.C. Chen and T.Y. Chen. 2009. Biodegradation of tetracyanonickelate by Klebsiella oxytoca under anaerobic conditions. *Desalination* 249, 3: 1212–1216. doi:10.1016/j.desal.2009.06.036.

Cipollone, R., P. Ascenzi, P. Tomao, F. Imperi and P. Visca. 2008. Enzymatic Detoxification of Cyanide: Clues from Pseudomonas aeruginosa Rhodanese. *Microbial Physiology* 15 2–3: 199–211. doi:10.1159/000121331.

Cipollone, R., P. Ascenzi and P. Visca. 2007. Common themes and variations in the rhodanese superfamily. *IUBMB Life* 59 2: 51–59. doi:10.1080/15216540701206859.

Crum, M.A., B. Trevor Sewell and M.J. Benedik. 2016. *Bacillus pumilus* Cyanide Dihydratase Mutants with Higher Catalytic Activity. *Frontiers in Microbiology* 7. doi:10.3389/fmicb.2016.01264. https://www.frontiersin.org/articles/10.3389/fmicb.2016.01264/full.

Dash, R.R., A. Gaur and C. Balomajumder. 2009. Cyanide in industrial wastewaters and its removal: A review on biotreatment. *Journal of Hazardous Materials* 163, 1: 1–11. doi:10.1016/j.jhazmat.2008.06.051.

Demopoulos, G.P. and T.C. Cheng. 2004. A case study of CIP tails slurry treatment: comparison of cyanide recovery to cyanide destruction. *The European Journal of Mineral Processing and Environmental Protection* 4 1: 9.

Dorr, P.K. and C.J. Knowles. 1989. Cyanide oxygenase and cyanase activities of Pseudomonas fluorescens NCIMB 11764. *FEMS Microbiology Letters* 60 3: 289–294. doi:10.1111/j.1574-6968.1989.tb03488.x.

Dubey, S. K. and D. S. Holmes. 1995. Biological cyanide destruction mediated by microorganisms. *World Journal of Microbiology and Biotechnology* 11 3: 257–265. doi:10.1007/BF00367095.

Dumestre, A., T. Chone, J. Portal, M. Gerard and J. Berthelin. 1997. Cyanide Degradation under Alkaline Conditions by a Strain of Fusarium solani Isolated from Contaminated Soils. *Applied and Environmental Microbiology* 63 7: 2729–2734.

Ebbs, S. 2004. Biological degradation of cyanide compounds. *Current Opinion in Biotechnology* 15 3: 231–236. doi:10.1016/j.copbio.2004.03.006.

Fernandes, B.C., M. Cesar Mateo, C. Kiziak, A. Chmura, J. Wacker, F. van Rantwijk, A. Stolz and R.A. Sheldon. 2006. Nitrile Hydratase Activity of a Recombinant Nitrilase. *Advanced Synthesis & Catalysis* 348 18: 2597–2603. doi: 10.1002/adsc.200600269.

Fernandez, R.F., E. Dolghih and D.A. Kunz. 2004. Enzymatic Assimilation of Cyanide via Pterin-Dependent Oxygenolytic Cleavage to Ammonia and Formate in Pseudomonas fluorescens NCIMB 11764. *Applied and Environmental Microbiology* 70 1: 121–128. doi:10.1128/AEM.70.1.121-128.2004.

Figueira, M.M., V.S. Ciminelli, M.C. de Andrade and V.R. Linardi. 1996. Cyanide degradation by an Escherichia coli strain. *Canadian Journal of Microbiology* 42 5: 519–523. doi:10.1139/m96-070.

Gonçalves, M.M.M., A.F. Pinto and M. Granato. 1998. Biodegradation of Free Cyanide, Thiocyanate and Metal Complexed Cyanides in Solutions with Different Compositions. *Environmental Technology* 19 2: 133–142. doi:10.1080/09593331908616665.

Happold, F.C., K.I. Johnstone, H.J. Rogers and J.B. Youatt. 1954. The Isolation and Characteristics of an Organism. *The Journal of General Microbiology* 10:261–266.

Harris, R. and J.Y.R. Christopher. 1983. Knowles. Isolation and Growth of a Pseudomonas Species that Utilizes Cyanide as a Source of Nitrogen. *Microbiology* 129 4: 1005–1011. doi:10.1099/00221287-129-4-1005.

Hatzfeld, Y., A. Maruyama, A. Schmidt, M. Noji, K. Ishizawa and K. Saito. 2000. β-Cyanoalanine Synthase Is a Mitochondrial Cysteine Synthase-Like Protein in Spinach and Arabidopsis. *Plant Physiology* 123 3: 1163–1172.

Hillemanns, R. 1989. Metal Cyanide Containing Wastes – Treatment Technologies. S.A.K. Von Palmer, M.A. Breton, T.J. Nunno, D.M. Sullivan and N.F. Surprenant. Noyes Data Corporation, Park Ridge 1988. XIV, 721 S., zahlr. Abb. u. Tab., geb., US-$ 74. *Chemie Ingenieur Technik* 61 9: 766–766. doi:10.1002/cite.330610930.

Ikebukuro, K., A. Miyata, S. Jin Cho, Y. Nomura, S. Mok chang, Y. Yamauchi, Y. Hasebe, S. Uchiyama and I. Karube. 1996. Microbial cyanide sensor for monitoring river water. *Journal of Biotechnology* 48 1: 73–80. doi:10.1016/0168-1656(96)01399-5.

Imhoff, J.F., H.G. Sahl, Gaber S.H. Soliman and H.G. Trüper. 1979. The Wadi Natrun: Chemical composition and microbial mass developments in alkaline brines of Eutrophic Desert Lakes. *Geomicrobiology Journal* 1 3: 219–234. doi:10.1080/01490457909377733.

Ingvorsen, K., B. Højer-Pedersen and S.E. Godtfredsen. 1991. Novel cyanide-hydrolyzing enzyme from Alcaligenes xylosoxidans subsp. denitrificans. *Applied and Environmental Microbiology* 57 6: 1783–1789.

Jandhyala, D., M. Berman, P.R. Meyers, B. Trevor Sewell, R.C. Willson and M.J. Benedik. 2003. CynD, the Cyanide Dihydratase from *Bacillus pumilus*: Gene Cloning and Structural Studies. *Applied and Environmental Microbiology* 69 8: 4794–4805. doi:10.1128/AEM.69.8.4794-4805.2003.

Jones, D.A. 1998. Why Are So Many Food Plants Cyanogenic? *Phytochemistry* 47 2: 155–162. doi:10.1016/S0031-9422(97)00425-1.

Kao, C.M., C.C. Lin, J.K. Liu, Y.L. Chen, L.T. Wu and S.C. Chen. 2004. Biodegradation of the Metal–Cyano Complex Tetracyanonickelate (II) by Klebsiella oxytoca. *Enzyme and Microbial Technology* 35 5: 405–410. doi:10.1016/j.enzmictec.2004.05.010.

Kao, C.M., J.K. Liu, H.R. Lou, C.S. Lin and S.C. Chen. 2003. Biotransformation of Cyanide to Methane and Ammonia by *Klebsiella oxytoca*. *Chemosphere* 50 8: 1055–1061. doi:10.1016/S0045-6535(02)00624-0.

Katayama, Y., Y. Narahara, Y. Inoue, F. Amano, T. Kanagawa and H. Kuraishi. 1992. A Thiocyanate Hydrolase of *Thiobacillus thioparus*. A Novel Enzyme Catalyzing the Formation of Carbonyl Sulfide from Thiocyanate. *The Journal of Biological Chemistry* 267 13: 9170–9175.

Kebeish, R., M. Aboelmy, A. El-Naggar, Y. El-Ayouty and C. Peterhansel. 2015. Simultaneous Overexpression of Cyanidase and Formate Dehydrogenase in Arabidopsis Thaliana Chloroplasts Enhanced Cyanide Metabolism and Cyanide Tolerance. *Environmental and Experimental Botany* 110: 19–26. doi:10.1016/j.envexpbot.2014.09.004.

Kuhn, A.T. 1971. Electrolytic Decomposition of Cyanides, Phenols and Thiocyanates in Effluent Streams-A Literature Review. *Journal of Applied Chemistry and Biotechnology* 21 2: 29–34. doi:10.1002/jctb.5020210201.

Kumar, V., V. Kumar and T. Chand Bhalla. 2015. Packed Bed Reactor for Degradation of Simulated Cyanide-Containing Wastewater. *3 Biotech* 5 5: 641–646. doi:10.1007/s13205-014-0261-6.

Kunz, D.A., R.F. Fernandez and P. Parab. 2001. Evidence That Bacterial Cyanide Oxygenase Is a Pterin-Dependent Hydroxylase. *Biochemical and Biophysical Research Communications* 287 2: 514–518. doi:10.1006/bbrc.2001.5611.

Kwon, H.K., S.H. Woo and J.M. Park. 2002. Degradation of tetracyanonickelate (II) by *Cryptococcus humicolus* MCN2. *FEMS Microbiology Letters* 214 2: 211–216. doi:10.1111/j.1574-6968.2002.tb11349.x.

Li, L., F. Yue, Y. Li, A. Yang, J. Li, Y. Lv and X. Zhong. 2020. Degradation Pathway and Microbial Mechanism of High-Concentration Thiocyanate in Gold Mine Tailings Wastewater. *RSC Advances* 10 43: 25679–25684. doi:10.1039/D0RA03330H.

Liu, J.K., C.H. Liu and C.S. Lin. 1997. The Role of Nitrogenase in a Cyanide-Degrading Klebsiella oxytoca Strain. *Proceedings of the National Science Council, Republic of China. Part B, Life Sciences* 21 2: 37–42.

Luque-Almagro, V.M., M.-J. Huertas, M. Martínez-Luque, C. Moreno-Vivián, M. Dolores Roldán, L. Jesús García-Gil, F. Castillo and R. Blasco. 2005. Bacterial Degradation of Cyanide and Its Metal Complexes under Alkaline Conditions. *Applied and Environmental Microbiology* 71 2: 940–947. doi:10.1128/AEM.71.2.940-947.2005.

Luthy, R.G. and J.T. Tallon. 1980. Biological Treatment of a Coal Gasification Process Wastewater. *Water Research* 14 9: 1269–1282. doi:10.1016/0043-1354(80)90186-4.

Mahendran, R., B., Sabna, M. Thandeeswaran, k.G. Kiran, M. Vijayasarathy, J. Angayarkanni and G. Muthusamy. 2020. Microbial (Enzymatic) Degradation of Cyanide to Produce Pterins as Cofactors. *Current Microbiology* 77 4: 578–587. doi:10.1007/s00284-019-01694-9.

Mahendran, R., M. Thandeeswaran, G. Kiran, M. Arulkumar, K.A. Ayub Nawaz, J. Jabastin, B. Janani, T.A. Thomas and J. Angayarkanni. 2018. Evaluation of Pterin, a Promising Drug Candidate from Cyanide Degrading Bacteria. *Current Microbiology* 75 6: 684–693. doi:10.1007/s00284-018-1433-0.

Mekuto, L., V.A. Jackson and S. Karabo Obed Ntwampe. 2013. *Biodegradation of Free Cyanide Using Bacillus sp. Consortium Dominated by Bacillus safensis, Lichenformis and Tequilensis Strains: A Bioprocess Supported Solely with Whey.* http://digitalknowledge.cput.ac.za/handle/11189/2038.

Meyers, P.R., P. Gokool, D.E. Rawlings and D.R. Woods. 1991. An Efficient Cyanide-Degrading *Bacillus pumilus* strain. *Journal of General Microbiology* 137 6: 1397–1400. doi:10.1099/00221287-137-6-1397.

Mudder, T.I. and J.L. Whitlock. 1984. Biological Treatment of Cyanidation Waste Waters. *Mining, Metallurgy & Exploration* 1 2: 161–165. doi:10.1007/BF03402571.

Nolan, L.M., P.A. Harnedy, P. Turner, A.B. Hearne and C. O'Reilly. 2003. The Cyanide Hydratase Enzyme of *Fusarium lateritium* also Has Nitrilase Activity. *FEMS Microbiology Letters* 221 2: 161–165. doi:10.1016/S0378-1097(03)00170-8.

Novak, D., I.H. Franke-Whittle, E. Tratar Pirc, V. Jerman, H. Insam, R. Marinšek Logar and B. Stres. 2013. Biotic and Abiotic Processes Contribute to Successful Anaerobic Degradation of Cyanide by UASB Reactor Biomass Treating Brewery Waste Water. *Water Research* 47 11: 3644–3653. doi:10.1016/j.watres.2013.04.027.

O'Reilly, C. and P.D. Turner. 2003. The Nitrilase Family of CN Hydrolysing Enzymes – A Comparative Study. *Journal of Applied Microbiology* 95 6: 1161–1174. doi:10.1046/j.1365-2672.2003.02123.x.

Pace, H.C. and C. Brenner. 2001. The Nitrilase Superfamily: Classification, Structure and Function. *Genome Biology* 2 1: reviews0001.1. doi:10.1186/gb-2001-2-1-reviews0001.

Park, D., D. Sung Lee, Y. Mo Kim and J.M. Park. 2008. Bioaugmentation of Cyanide-Degrading Microorganisms in a Full-Scale Cokes Wastewater Treatment Facility. *Bioresource Technology* 99 6: 2092–2096. doi:10.1016/j.biortech.2007.03.027.

Parker, W. L., M. L. Rathnum, J. H. Johnson, J. S. Wells, P. A. Principe and R. B. Sykes. 1988. Aerocyanidin, a New Antibiotic Produced by *Chromobacterium violaceum*. *The Journal of Antibiotics* 41 4: 454–460. doi:10.7164/antibiotics.41.454.

Patil, Y. and K. Paknikar. 2014. Development of a Process for Biodetoxification of Metal Cyanides from Wastewaters. SSRN Scholarly Paper. Rochester, NY: Social Science Research Network. https://papers.ssrn.com/abstract=2523788.

Patil, Y.B. and K.M. Paknikar. 2000. Biodetoxification of Silver–Cyanide from Electroplating Industry Wastewater. *Letters in Applied Microbiology* 30 1: 33–37. doi:10.1046/j.1472-765x.2000.00648.x.

Pereira, P., A. Sofia Pires and J. Carlos Roseiro. 1999. The Effect of Culture Aging, Cyanide Concentration and Induction Time on Formamide Hydro-Lyase Activity of *Fusarium oxysporum* CCMI 876. *Enzyme and Microbial Technology* 25 8: 736–744. doi:10.1016/S0141-0229(99)00105-2.

Piotrowski, M., S. Schönfelder and E.W. Weiler. 2001. The *Arabidopsis thaliana* Isogene NIT4 and Its Orthologs in Tobacco Encode Beta-Cyano-L-Alanine Hydratase/Nitrilase. *The Journal of Biological Chemistry* 276 4: 2616–2621. doi:10.1074/jbc.M007890200.

Ranjan, B., P.H. Choi, S. Pillai, K. Permaul, L. Tong and S. Singh. 2021. Crystal Structure of a Thermophilic Fungal Cyanase and Its Implications on the Catalytic Mechanism for Bioremediation. *Scientific Reports* 11 1. doi:10.1038/s41598-020-79489-3. https://www.osti.gov/pages/biblio/1765222.

Raybuck, S.A. 1992. Microbes and Microbial Enzymes for Cyanide Degradation. *Biodegradation* 3 1. doi:10.1007/BF00189632, http://link.springer.com/10.1007/BF00189632.

Rinágelová, A., O. Kaplan, A.B. Veselá, M. Chmátal, A. Křenková, O. Plíhal, F. Pasquarelli, M. Cantarella and L. Martínková. 2014. Cyanide Hydratase from *Aspergillus niger* K10: Overproduction in *Escherichia coli*, Purification, Characterization and Use in Continuous Cyanide Degradation. *Process Biochemistry* 49 3: 445–450. doi:10.1016/j.procbio.2013.12.008.

Roberts, R.F. and B. Jackson. 1971. The Determination of Small Amounts of Cyanide in the Presence of Ferrocyanide by Distillation under Reduced Pressure. *Analyst* 96 1140: 209–212. doi:10.1039/AN9719600209.

Rosehart, R.G. 1973. Mine Water Purification by Reverse Osmosis. *The Canadian Journal of Chemical Engineering* 51 6: 788–789. doi:10.1002/cjce.5450510629.

Sabatini, L., C. Ferrini, M. Micheloni, A. Pianetti, B. Citterio, C. Parlani and F. Bruscolini. 2012. Isolation of a Strain of *Aspergillus fumigatus* Able to Grow in Minimal Medium Added with an Industrial Cyanide Waste. *World Journal of Microbiology and Biotechnology* 28 1: 165–173. doi:10.1007/s11274-011-0805-4.

Seefeldt, L.C., Z.-Y. Yang, S. Duval and D.R. Dean. 2013. Nitrogenase Reduction of Carbon-Containing Compounds. *Biochimica et Biophysica Acta (BBA) - Bioenergetics* 1827 8: 1102–1111. doi:10.1016/j.bbabio.2013.04.003.

Silva-Avalos, J., M.G. Richmond, O. Nagappan and D.A. Kunz. 1990. Degradation of the Metal-Cyano Complex Tetracyanonickelate(II) by Cyanide-Utilizing Bacterial Isolates. *Applied and Environmental Microbiology* 56 12: 3664–3670.

Sirianuntapiboon, S., K. Chairattanawan and M. Rarunroeng. 2008. Biological Removal of Cyanide Compounds from Electroplating Wastewater (EPWW) by Sequencing Batch Reactor (SBR) System. *Journal of Hazardous Materials* 154 1: 526–534. doi:10.1016/j.jhazmat.2007.10.056.

Sorokin, D. Y., A. M. Lysenko, L. L. Mityushina, T. P. Tourova, B. E. Jones, F. A. Rainey, L. A. Robertson and G. J. Kuenen. 2001. Thioalkalimicrobium Aerophilum gen. nov., sp. nov. and Thioalkalimicrobium sibericum sp. nov., and Thioalkalivibrio versutus gen. nov., sp. nov., Thioalkalivibrio nitratis sp.nov., Novel and Thioalkalivibrio denitrificancs sp. nov., Novel Obligately Alkaliphilic and Obligately Chemolithoautotrophic Sulfur-Oxidizing Bacteria from Soda Lakes. *International Journal of Systematic and Evolutionary Microbiology* 51 Pt 2: 565–580. doi:10.1099/00207713-51-2-565.

Sorokin, D.Y., T.P. Tourova, A.M. Lysenko, L.L. Mityushina and J. Gijs Kuenen. 2002. Thioalkalivibrio thiocyanoxidans sp. nov. and Thioalkalivibrio paradoxus sp. nov., Novel Alkaliphilic, Obligately Autotrophic, Sulfur-Oxidizing Bacteria Capable of Growth on Thiocyanate, from Soda Lakes. *International Journal of Systematic and Evolutionary Microbiology* 52 Pt 2: 657–664. doi:10.1099/00207713-52-2-657.

Tikhonova, T.V., D.Y. Sorokin, W.R. Hagen, M.G. Khrenova, G. Muyzer, T.V. Rakitina, I.G. Shabalin, A.A. Trofimov, S.I. Tsallagov and V.O. Popov. 2020. Trinuclear Copper Biocatalytic Center Forms an Active Site of Thiocyanate Dehydrogenase. *Proceedings of the National Academy of Sciences* 117 10: 5280–5290. doi:10.1073/pnas.1922133117.

Valentina, C., R. Hadisoebroto and A. Rinanti. 2018. Removing Cyanide by Mixed Culture at Liquid Media with Variation in pH and Cyanide Concentration. Hg. von Ade Gafar Abdullah and Asep Bayu Dani Nandiyanto. *MATEC Web of Conferences* 197: 13016. doi:10.1051/matecconf/201819713016.

Wang, P., D.E. Matthews and H.D. VanEtten. 1992. Purification and characterization of cyanide hydratase from the phytopathogenic fungus *Gloeocercospora sorghi*. *Archives of Biochemistry and Biophysics* 298 2: 569–575. doi:10.1016/0003-9861(92)90451-2.

Watanabe, A., K. Yano, K. Ikebukuro and I. Karube. 1998. Cyanide Hydrolysis in a Cyanide-Degrading Bacterium, *Pseudomonas stutzeri* AK61, by Cyanidase. *Microbiology (Reading, England)* 144 Pt 6: 1677–1682. doi:10.1099/00221287-144-6-1677.

Wild, S.R., T. Rudd and A. Neller. 1994. Fate and Effects of Cyanide during Wastewater Treatment Processes. *Science of the Total Environment* 156 2: 93–107. doi:10.1016/0048-9697(94)90346-8.

Young, C.A. and T.S. Jordan. *Cyanide Remediation: Current and Past Technologies.* 26.

Yu, X.-Z., P.-C. Lu and Z. Yu. 2012. On the Role of β-Cyanoalanine Synthase (CAS) in Metabolism of Free Cyanide and Ferri-Cyanide by Rice Seedlings. *Ecotoxicology (London, England)* 21 2: 548–556. doi:10.1007/s10646-011-0815-x.

Zheng, A., D.A. Dzombak and R.G. Luthy. 2004. Formation of Free Cyanide and Cyanogen Chloride from Chloramination of Publicly Owned Treatment Works Secondary Effluent: Laboratory Study with Model Compounds. *Water Environment Research* 76 2: 113–120.

6 Bioaugmentation to Remove Recalcitrant Pollutants in Industrial Wastewater

L.P. Ananthalekshmi
Mahatma Gandhi University

Indu C. Nair
SAS SNDP Yogam College

K. Jayachandran
Mahatma Gandhi University

CONTENTS

6.1 GENERAL INTRODUCTION

Recalcitrant compounds are produced by anthropogenic activities and are a large class of pollutants persisting in the environment. These are usually organic, consisting of long and complex hydrocarbon chains most of which can cause permanent damage to the ecosystem. Many of these compounds are highly mutagenic and are potent carcinogens. Recalcitrant compounds are less degradable and include heavy metals, organic compounds, and synthetic polymers and are present in soil and water resources. Organic compounds like benzene, toluene, phenyl compounds, xylene, aromatic hydrocarbons, and chlorine derivatives are present in a very high concentration in industrial effluents. Environmentally friendly methods, involving the application of microorganisms or enzymes are universally accepted for their remediation. These methods are widely popular because they have no side effects, are eco-friendly, and cause no harm to living beings. Although there are many methods of bioremediation, recalcitrant degradation is still challenging and is of serious concern. Exploring this field of research is important for the appropriate selection and establishment of suitable bioaugmentation methods to address the harmful effects of recalcitrant pollutants in industrial wastewater.

The recalcitrant pollutants emerge from a wide range of sources. They are industrial, petroleum wastes, oil waste, pesticides, effluent, and even pharmaceutical by-products in origin. Most of the

DOI: 10.1201/9781003187622-6

recalcitrant pollutants are water insoluble and can adversely affect either humans or any life forms in the nature. Persistent organic pollutants (POPs) already exist in the body of humans, marine animals, and other animals. They accumulate through direct exposure and also through step by step phenomenon of biomagnifications via different trophic levels of food chains. Bioaccumulation of many pesticides like dichlorodiphenyltrichloroethane (DDT) in the human body has already been extensively reported and studied. All these recalcitrant generally, are reported as a class of POPs. When POPs are present in a very minimal amount it gets stored in the living body, when the concentration increases it starts affecting the physiological systems by disrupting the signaling pathways or inducing gene alteration that leads to many health hazards.

POPs are almost omnipresent and are resistant to biodegradation. They are present in all the resources – soil, air, water and are accumulated in the human body. Accumulation of POPs in a very small amount itself can cause damage to the reproductive system, immune system, nervous system and also affects other physiological conditions. Many compounds act as endocrine disrupters causing serious damage to the human physiological system. Parabens are esters of p-hydrobenzoic acid capable of causing hormonal imbalance and breast cancer (Argenta et al. 2021). Pharmaceuticals like diclofenac and ibuprofen are bioaccumulative and affect other living organisms (Koumaki et al. 2021). Diethyl phthalate causes reproductive problems, eye irritation, and respiratory problems (Maszenan, Liu, and Ng 2011).

Even though many chemical augmenting processes exist they have an alternative effect that they may also damage living beings. On the contrary, the biological process gives a complete result with less harm to the nature. Presently green methods of bioremediation, using any of the life forms are preferred for the removal of recalcitrant pollutants of which bioaugmentation is the most promising option. Mostly Bacteria, fungi, algae, and plants are used for bioaugmentation.

Bioaugmentation is the application of any microorganism to enhance the biodegradation of a pollutant. Many bacteria are exploited for the process of bioaugmentation. A combination of bioaugmentation with a suitable mechanism may increase the rate of biodegradation. Bioaugmentation coupled with phytoremediation, technique of genetic engineering, suitable bioreactors with bacterial consortium for degradation, may speed up the process.

Currently, biological methods offer a promising platform for the selection of sustainable remedies for recalcitrant abolition. Each type of POPs can be effectively removed by the selection of an approachable bioremediation method. Bioaugmenting with suitable microorganisms, coupled with biosorption, biostimulation, phytoremediation, and acceptable chemical methods are to be formulated for the removal of recalcitrant industrial pollutants for re-establishing sustainable life on mother earth.

6.2 BIOAUGMENTATION

Bioaugmentation is a key method of bioremediation to exploit the maximum chances of the capacity of a particular microorganism for degrading recalcitrant compounds to maintain a pollution-free environment. This mechanism can be used for the remediation of contaminants in soil and water. Different methods of bioaugmentation are being developed according to the need, type of pollutant, and required advantages. Biostimulation and bioattenuation are the two other green methods that can be combined with bioaugmentation for an effective bioremediation of the entire class of pollutants. Biostimulation is the process of stimulating the organism to grow and to perform metabolism in extreme conditions by adding nutrients and compounds acting as electron donors and acceptors. A combination of physical, chemical, and biological treatment to reduce toxicity is termed bioattenuation. There can be a great possibility of utilizing the combined methods of bioaugmentation with biostimulation and attenuation for the effective degradation of recalcitrant pollutants. The entire process of biodegradation by the above green methods depends on the environmental factors and other conditions comprising biotic and abiotic factors. As these are green methods they are effective and long lasting, they offer promising outcomes with no side effects.

6.3 MAJOR RECALCITRANT POLLUTANTS IN INDUSTRIAL WASTEWATER

Recalcitrant pollutants are of large classes and arise due to anthropogenic activities. They are present almost everywhere. Major recalcitrant pollutants in industrial wastewater include polycyclic aromatic hydrocarbons (PAHs) such as pyrenes, naphthalene, phenanthrene, benzopyrene, organochlorides including benzyl chloride, phthalates consisting of butyl octyl phthalate, dibutyl phthalate, benzyl butyl phthalate, phthalic acid, diisobutyl phthalate, di-(-2ethy hexyl) phthalate, 1,2-dibenzyl phthalate, phenolic compounds, aromatic hydrocarbons, and nitroaromatic compounds (Table 6.1). All these cause major health issues in humans and adverse effects on other life forms. The method of recalcitrant removal depends on the type of compound, its toxicity, its removal efficiency, and the factors affecting the removal.

6.4 BIOAUGMENTATION OF RECALCITRANT POLLUTANTS IN INDUSTRIAL WASTEWATER

Industries generally require many raw materials and pure water supply as an important input for their production and generate contaminated wastewater with unknown constituents of recalcitrant compounds. Among the pollutants, the aromatic compounds are more recalcitrant. Benzene, toluene ethyl benzene, and xylene are the major aromatic pollutants in industrial wastewater. Phenol derivatives, phthalate esters, chlorinated compounds, heavy metals, azodyes, pesticides, and PAHs are other major contaminants in industrial wastewater necessitating the implementation of the methods of bioaugmentation.

Biodegradation is a natural process of breaking down of compounds into less toxic constituents by using microorganisms. The process may involve direct metabolism or co-metabolism. In co-metabolism, the bacteria use any pollutant or recalcitrant as a carbon source along with another carbon source, mostly carbohydrate. The microorganisms metabolize and covert the compound into other substitutes, which under further reactions may make it a nontoxic compound. The type of biodegradation is based on the compound selected for the degradation. For aromatic compounds their benzene ring undergoes either ortho- or meta-cleavage. Meta-cleavage pathway is mediated by 2,3-dioxygenase activity and 1,2 dioxygenase results in ortho-cleavage pathway, resulting in the formation of catechol finally resulting in complete mineralization of the pollutant.

Many attempts to explore the methods of bioaugmentation have been reported recently for the remediation of specific recalcitrant pollutants in industrial wastewater. Heavy metal augmentation was performed using the technique of cell immobilization in an activated sludge reactor. Heavy metals inhibit natural aerobic denitrification. *Comamonas* sp. *ZF-3* isolated from seed sludge was reported as a highly efficient strain in degrading organic compounds present in cooking wastewater. *Comamonas* sp. *ZF-3* was coupled with biofilm-based anoxic filter fluidized bed reactor system (Yuan, Li, and Zhong 2020). *Rhodococcus pyrindinivorans F5* was reported to augment oil from oily wastewater (Mazumder et al. 2020). Ethinyl estradiol, an endocrine-disrupting compound was found to be degraded by *Rhodopsuedomonas palustris*. A photo-assisted microbial fuel cell combined with the process of attenuation with *Rhodopsuedomonas palustris* was explored for the remediation of POPs including polychlorinated biphenyls. Natural attenuation, biostimulation with fungal cultures immobilized on sugarcane bagasse were also being attempted (Sadañoski et al. 2020). The major POPs can be grouped into six major classes – hydrocarbons, pesticides, industrial chemicals, phthalate, and its esters, chlorine derivatives, PAHs, and finally pharmaceutical by-products.

6.4.1 BIOAUGMENTATION OF HYDROCARBONS

Bioremediation methods involve a green approach and as its open ways for addressing the otherwise challenging issues in environmental pollution, they have been already implemented for field applications. Biostimulation with bioaugmentation could be coupled for bioremediation of diesel polluted

TABLE 6.1

Methods of Bioaugmentation Evaluated for the Removal Recalcitrant Pollutants in Industrial Wastewater

S. No.	Pollutants	Bioaugmentation Method	Application	References
1.	Heavy metals and petroleum hydrocarbons	Natural attenuation, phytoremediation	Petroleum hydrocarbons are removed by natural attenuation process including biodegradation, dispersion, sorption, volatilization, (bio)chemical stabilization	Agnello et al. (2016)
		Bioaugmentation-assisted phytoremediation using Alfa Alfa and *Pseudomonas aeruginosa*	Heavy metals like Cu, Zn, and Pb are absorbed	Ahmadi et al. (2019)
2.	Saline and recalcitrant petrochemical wastewater	Bacterial consortium *Pseudomonas pseudoalcaligenes* strain R1, *Bacillus subtilis* subsp. *inaquosorum* R2, and *Shewanellachilikensis* strain AM1	The bacterial consortium effectively removed the recalcitrant from the saline water	
3.	Parabens	Aerobic granular sludge system	Methyl paraben, ethyl parabens are removed by aerobic granular system	Argenta et al. (2021)
4.	Antimicrobial triclocarbon and PAHs	Electro biostimulation	Actively removes the polycyclic aromatic hydrocarbons depending on the electroactivity of the microorganism	Bai et al. (2021)
5.	Recalcitrant from Land fill leachate	Electro-coagulation coupled with biological treatment	Effective method for removing recalcitrant from landfill leachate	Djelal, Lelievre, and Ricordel (2014)
6.	Various organic compounds	Plasmid-mediated bioaugmentation	Biodegradation of organic compounds by genetic engineering	Garbisu et al. (2017)
7.	Various recalcitrant	Nanohybrids	Diverse of nanomaterials can be used for bioaugmentation	Guo et al. (2020)
8.	Heavy metals	Biosorption	Marine algae serve as a good biosorption agent and help in removing heavy metals in aqueous forms	He and Chen (2014)
9.	Pyridine	Continuous-flow self-forming dynamic membrane bioreactor	Bioaugmented continuous-flow self-forming dynamic membrane bioreactor degrades pyrimidine and other nitrogen containing aromatic compounds	Hou et al. (2016)
10.	1,2,3-Trichloropropane (TCP)	Genetically engineered microbes	TCP is degraded by engineering TCP degrading enzymes	Janssen and Stucki (2020)
11.	Petrochemical wastewater	UV-assisted advanced oxidation	Industrial wastewater treatment	Kakavandi and Ahmadi (2019)

seawater and petroleum hydrocarbons (Xue et al. 2021). Bioattenuation with *Pseudomonas aeruginosa*, isolated from hydrocarbon polluted regions of petroleum sites was also being reported (Varjani 2017). After the bioattenuation, the organism was acclimatized to perform extended biodegradation particularly of hydrocarbons from oily wastes (Varjani and Upasani 2021). The protection and preservation of oil-degrading bacteria play an important role in its bioremediation. A good protective agent increases the stability of the organism and often includes a specific percentage concentration of sucrose, trehalose, glycerin, and beta cyclodextrin (Li et al. 2021a). Aromatic compounds including benzene, toluene, ethyl benzene, and xylene (BTEX) were remediated using a combined mechanism of coupled biostimulation and bioaugmentation using a central composite design (Li et al. 2021b). Parabens, another recalcitrant compounds, were proved to be augmented by aerobic granular sludge systems (Argenta et al. 2021). Triclocarban and PAHs were degraded by a mechanism of combined bioaugmentation with electrobiostimulation (Bai et al. 2021). Bioaugmentation with *Comamonas testosteroni* was reported to accelerate pyridine degradation by activating monooxygenase enzyme (Zhu et al. 2021). Another important perspective of bioaugmentation was by exploring the chances of microbial genetic engineering for sake of degradation (Janssen and Stucki 2020). Mostly, PAHs undergo biotransformation by specific bacterial isolate, viz. *Sphingomonas yanoikuyae* JAR02 degrading benzopyrene (Rentz, Alvarez, and Schnoor 2008) and *Roseobacter* degrading pyrene (Zhou et al. 2020).

Microbial community can be effectively used for recalcitrant pollutant removal from the environment. The different methods are selected according to the nature of the recalcitrant compound. Pyridine degrading strain *Comamonas testosteroni* was identified with potential degrading capacity for other compounds toluene, quinolone, phenol, etc. *Comamonas testosteroni* was found to accelerate the mono oxygenation of pyridine (Zhu et al. 2021). Another approach to bioaugmentation is the use of genetically modified organisms. Methods were developed to augment chlorinated compounds using genetic manipulation of plasmids of selected bacteria – *Agrobacterium radiobacter* and *Psuedomonas putida* (Janssen and Stucki 2020). Petroleum biodegradation by-products consist of a diverse class of polycyclic aromatic compounds. A combined approach of composting with microbial consortium (*Bacillus*, *Citrobacter*, *Klebsiella*, *Pseudomonas* and *Rhodobacter*) and bioaugmentation was done for removing these PAHs (Wu et al. 2020).

PAHs including pyrene, benzopyrene, and phenanthrene are highly persistent due to their complicated ring structure and a long chain of hydrocarbons. All of them were reported to be carcinogenic and found to be causing hemolytic anemia (Nzila, Razzak, and Zhu 2016). *Rhodotorula mucilaginosa,* a basidiomycete fungus was being recently reported to degrade PAHs (Smith, Lethbridge and Burns 1999; Martínez-Ávila et al. 2021). *Pseudomonas aeruginosa*, *Pseudomonas* sp., and *Ralstonia* sp. are detected with genes capable of degrading PAHs (Sangkharak et al. 2020). Benzopyrene consists of five aromatic rings fused together which makes it more recalcitrant. It is known to be a potential carcinogen, besides it is neurotoxic and immunotoxic and induces reproductive damages (Nzila and Musa 2020). Pyrene could be enzymatically degraded by fungus *Lasiodiplodia theobromae* (Cao et al. 2020). Phenanthrene has three fused aromatic rings and contains nitrogen. It also has similar toxic levels in the human body (Anyanwu and Semple 2016). Phenanthrene was also being degraded by an Ascomycete fungus isolated from soil. *Aspergillus oryzae* MF13 and *Aspergillus flavipes* QCS12 were reported to degrade phenanthrene and other PAHs in a more effective way (Cruz-Izquierdo et al. 2021).

6.4.2 BIOAUGMENTATION OF PESTICIDES

Pesticides undergo enzymatic transformation and microbial degradation. Biodegradation involves the removal of functional groups of pesticides by decarboxylation, demethylation, dehalogenation, and ring cleavage (Liu et al. 2021). Aldrin, endrin, DDT, heptachlor, chlordane, and toxaphene are the major persistent pesticides that accumulate in living beings. Aldrin is a chlorinated cyclodiene and is a broad-spectrum insecticide that causes adverse effects in humans including dizziness, headaches, and gastrointestinal problems. It was also found to be responsible for inhalation problems in other animals (Otani and Sakata 2007). It was reported to be degraded by white-rot fungi (Xiao et al. 2011). Endrin is also an organochloride, similar to aldrin, and is highly toxic. It can cause muscle damage, bone damage and may even cause death (Singh et al. 2008). Phytoremediation of endrin with cucurbits was being reported (Matsumoto et al. 2009). DDT is an insecticide and has been identified as a persistent pollutant. It is a potent carcinogen and is ubiquitous in all life forms. Many bacterial strains were being reported to degrade DDT and some bacteria were reported to transform DDT into its less toxic forms (Pan et al. 2016). Heptachlor is another class of organochloride which is an insecticide. It is toxic to humans and aquatic life, causes skin irritation, nausea, vomiting, and diarrhea. There were biological as well as chemical degradation of heptachlor being reported (Pokethitiyook and Poolpak 2012). Chlordane and toxaphene are also organochlorides and inhibit many receptors in physiological system. They were also identified as toxic to aquatic ecosystems and animals (Sanborn et al. 1976). All these pesticides are identified as POPs by the United Nations Environment Programme.

Methanogens are genetically engineered to enhance the degradation of pesticides. *Methanotroph* and also *Methylomonas* sp. LW13 was being reported to remove bensulfo methane, an herbicide. A gene encoding bensulfo methane hydrosylase was transferred into E.coli strain for bioaugmentation. (Liu et al. 2021). A combined approach of biostimulation with bioaugmentation was done for remedying 2,4 dichlorophenoxyacetic acid (Barba et al. 2021). Different microbes streamlined into various methodologies could be used for degrading pesticides (Kumar and Sachan 2021). *Solanum torvum* and *Withania somnifera* were used for phytoremediation of certain organochlorides. These plants were reported to absorb some of the pesticides (Singh and Singh 2017). Earthworms were also being used to remove DDT (Xu et al. 2021).

6.4.3 BIOAUGMENTATION OF POLYCHLORINATED BIPHENYLS AND HEXACHLOROBENZENE

Polychlorinated biphenyls and hexachlorobenzene are the major POPs present in industrial effluent. Polychlorinated biphenyls are highly toxic, cause cancer, and suppress the immune system. Bioaugmentation is the best remedial step that can be adopted for the removal of polychlorinated biphenyl (Jing, Fusi, and Kjellerup 2018). Hexachlorobenzene forms dioxin which is more recalcitrant (van Birgelen 1998) and damages liver kidney and thyroid. Some of the basidiomycetes were reported for the bioremediation of hexachlorobenzene (Matheus, Bononi, and Machado 2000).

Microbes generally used for transforming hexachlorobenzene and polychlorinated biphenyls are very dynamic and versatile (Field and Sierra-Alvarez 2008b). Poly chlorinated bi phenyl and hexachloro benzene are the major industrial pollutants which are present in industrial sewage (Quinlivan, Ghassemi, and Leshendok 1975). Combined biostimulation and bioaugmentation cause reductive dechlorination of chlorinated biphenyls (Sudjarid et al. 2012). Combined microbial degradation as well as phytoremediation could also be done to remove the pollutants from industrial effluent (Rylott, Johnston, and Bruce 2015). Plants have the ability to absorb and translocate the chlorinated biphenyls and a combination of bacteria and suitable plants is a promising step for enhanced absorption. The bacterial and plant interaction mediated the easy uptake of pesticides (Arslan et al. 2015).

Hexachlorobenzene disturbs the immune system and other systems of the human body. Hexachlorobenzene is the most important constituent of industrial effluent. Polychlorinated bisphenyls could undergo degradation by polychlorinated biphenyl-degrading enzymes present or genetically engineered into certain bacteria (Fava and Piccolo 2002).

6.4.4 BIOAUGMENTATION OF PHTHALATE AND ITS ESTERS

Phthalates and phthalic acid esters are released from plastic components and are best-known endocrine receptor disruptors. Phthalates are also carcinogens and cause reproductive toxicity (Bradley et al. 2013). These persistent compounds have emerged as by-products during the manufacturing of plastic and other synthetic polymers. Phthalates are a large class of chemicals including, dibutyl phthalate, benzyl butyl phthalate, phthalic acid, diisobutyl phthalate, di-(-2ethy hexyl) phthalate, 1,2-dibenzyl phthalate, and many more. Phthalates are transformed by certain bacteria into its esters and corresponding acids. Some of the genetically engineered microbes were also used as a tool in bioaugmentation methods for phthalates (Kong et al. 2019). Phthalates can be degraded aerobically and anaerobically according to the nature of the microorganism selected. Various species of *Pseudomonas*, *Rhodoccous*, and *Acinetobacter* were reported to degrade phthalates and phthalic acid esters (Nahurira et al. 2017). Dibutyl phthalate could be degraded by *Bacillus subtilis* and proper bioremediation of the pollutant was done using this method (Huang et al. 2018). Di 2 ethyl hexyl phthalate could be degraded by Rhodococcus sp (Zhao et al. 2019). Di-n-octyl phthalate was reported to be degraded using a sequencing batch reactor (SBR) (Zhang et al. 2018). Phthalate hydrolases encoding genes were isolated from soil metagenomic library as an effective step towards phthalate degradation (Qiu et al. 2020). Benzyl butyl phthalate and dimethyl phthalate were reported to be degraded by *Bacillus marisflavi* RR014 (Kaur et al. 2021).

6.4.5 BIOAUGMENTATION OF CHLORINE DERIVATIVES – DIOXINS AND FURANS

Dioxins and furans are produced during many industrial processes. Both interact with female hormones and causes imbalances in reproductive systems (Lambertino et al. 2021). Tetrachlorodibenzo-p-dioxin is the most toxic dioxin known. Furan ethers are present in cigarette smoke, agrochemicals, etc. Both dioxins and furans are chlorine-derived by-products and are highly toxic to humans. These are neurotoxic and immunotoxin. They cause reproductive damage and skin irritation (Loganathan and Masunaga 2009). Landfilling, vitrification, photolytic destruction are the remedial ways for dioxin and furan accumulation (Rathna, Varjani, and Nakkeeran 2018). Some species of fungi are reported to degrade furans and dioxins.

Dioxins were reported to be co metabolically used as the sole carbon source by white-rot fungi and *Pseudomonas veronii* (Field and Sierra-Alvarez 2008a). Furans undergoes transesterification and polycondensation to break down into its components (Papageorgiou et al. 2016). Biostimulation and bioaugmentation induce dechlorination of dioxins and furnas (Ahn et al. 2008). Catabolic megaplasmids in *Rhodococcus* could also co-transform dioxins (Sun et al. 2017). Biostimulation and bioaugmentation are combined in many cases for furan degradation. *Penicillium brasilianum* and *Fusarium solani* were used in the bioremediation of dioxin-contaminated soil. These fungi are saprophytic and are reported to reduce the toxic level of dioxin in the soil (Delsarte et al. 2021). *Agromyces* and *Arthrobacter* were reported to degrade dibenzofuran, dibenzo-*p*-dioxin, and 2-monochlorodibenzo-*p*-dioxin (Almnehlawi et al. 2021)

6.4.6 BIOAUGMENTATION OF PHARMACEUTICAL BY-PRODUCTS

Bioaugmentation of endocrine-disrupting contaminants (EDCs) like nonyl phenol, bisphenol A and pharmaceutical contaminants ibuprofen, diclofenac, naproxen, and ketoprofen present in the municipal sewage were augmented using special high-rate activated sludge systems (Koumaki et al. 2021). Ethinyloestradiol is an important recalcitrant and is a pharmaceutical pollutant that causes physiological changes in the human body. These pollutants were reported to be removed by chemical augmentation methods such as coagulation, flocculation, chemical precipitation, reverse osmosis, and activated carbon adsorption (Kurniawan, Lo, and Chan 2006). Many pharmaceutical by-products are removed from wastewater by an activated carbon system. All these methods focus only on separating or partial removal of recalcitrant from the contaminated phase and never address the complete removal or mineralization of the recalcitrant pollutants (Delgado et al. 2019). Advanced oxidation and electrochemical methods are applicable for augmenting pharmaceutical by-products. The application of the new methods of bioaugmentation for the remediation of pharmaceutical pollutants is yet to be explored in an extended way.

REFERENCES

Agnello, A. C., M. Bagard, E. D. van Hullebusch, G. Esposito, and D. Huguenot. 2016. "Comparative biore-mediation of heavy metals and petroleum hydrocarbons co-contaminated soil by natural attenuation, phytoremediation, bioaugmentation and bioaugmentation-assisted phytoremediation." Science of the Total Environment 563–564: 693–703. doi: 10.1016/j.scitotenv.2015.10.061.

Ahmadi, M., M. Ahmadmoazzam, R. Saeedi, M. Abtahi, S. Ghafari, and S. Jorfi. 2019. "Biological treatment of a saline and recalcitrant petrochemical wastewater by using a newly isolated halo-tolerant bacterial consortium in MBBR." Desalination and Water Treatment 167: 84–95. doi: 10.5004/dwt.2019.24627.

Ahn, Y.-B., L. Fang, D.E. Fennell, and M.M. HÄggblom. 2008. "Biostimulation and bioaugmentation to enhance dechlorination of polychlorinated dibenzo-p-dioxins in contaminated sediments." FEMS Microbiology Ecology 66 (2): 271–281. doi: 10.1111/j.1574-6941.2008.00557.x.

Almnehlawi, H.S., R.K. Dean, S.L. Capozzi, L.A. Rodenburg, G.J. Zylstra, and D.E. Fennell. 2021. "Agromyces and Arthrobacter isolates from surficial sediments of the Passaic River degrade dibenzofuran, dibenzo-p-dioxin and 2-monochlorodibenzo-p-dioxin." Bioremediation Journal: 1–25. doi: 10.1080/10889868.2021.1892027.

Anyanwu, I.N., and K.T. Semple. 2016. "Assessment of the effects of phenanthrene and its nitrogen heterocy-clic analogues on microbial activity in soil." SpringerPlus 5 (1). doi: 10.1186/s40064-016-1918-x.

Argenta, T.S., A.R. Mendes Barros, C.d.A. de Carvalho, A.B. dos Santos, and P.I. Milen Firmino. 2021. "Parabens in aerobic granular sludge systems: Impacts on granulation and insights into removal mechanisms." Science of the Total Environment 753. doi: 10.1016/j.scitotenv.2020.142105.

Arslan, M., A. Imran, Q. Mahmood Khan, and M. Afzal. 2015. "Plant–bacteria partnerships for the remedia-tion of persistent organic pollutants." Environmental Science and Pollution Research 24 (5): 4322–4336. doi: 10.1007/s11356-015-4935-3.

Bai, Y., B. Liang, H. Yun, Y. Zhao, Z. Li, M. Qi, X. Ma, C. Huang, and A. Wang. 2021. "Combined bio-augmentation with electro-biostimulation for improved bioremediation of antimicrobial triclocarban and PAHs complexly contaminated sediments." Journal of Hazardous Materials 403. doi: 10.1016/j.jhazmat.2020.123937.

Barba, S., J. Villaseñor, M.A. Rodrigo, and P. Cañizares. 2021. "Biostimulation versus bioaugmentation for the electro-bioremediation of 2,4-dichlorophenoxyacetic acid polluted soils." Journal of Environmental Management 277. doi: 10.1016/j.jenvman.2020.111424.

Bradley, E.L., R.A. Burden, K. Bentayeb, M. Driffield, N. Harmer, D.N. Mortimer, D.R. Speck, J. Ticha, and L. Castle. 2013. "Exposure to phthalic acid, phthalate diesters and phthalate monoesters from foodstuffs: UK total diet study results." Food Additives & Contaminants: Part A 30 (4): 735–742. doi: 10.1080/19440049.2013.781684.

Cao, H., C. Wang, H. Liu, W. Jia, and H. Sun. 2020. "Enzyme activities during benzo[a]pyrene degradation by the fungus Lasiodiplodia theobromae isolated from a polluted soil." Scientific Reports 10 (1). doi: 10.1038/s41598-020-57692-6.

Cruz-Izquierdo, R.I., A.D. Paz-González, F. Reyes-Espinosa, L.K. Vazquez-Jimenez, M. Salinas-Sandoval, M.I. González-Domínguez, and G. Rivera. 2021. "Analysis of phenanthrene degradation by Ascomycota fungi isolated from contaminated soil from Reynosa, Mexico." Letters in Applied Microbiology 72 (5): 542–555. doi: 10.1111/lam.13451.

Delgado, Nasly, Alberto Capparelli, A. Navarro, and D. Marino. 2019. "Pharmaceutical emerging pollutants removal from water using powdered activated carbon: Study of kinetics and adsorption equilibrium." Journal of Environmental Management 236: 301–308 doi: 10.1016/j.jenvman.2019.01.116.

Delsarte, I., E. Veignie, Y. Landkocz, and C. Rafin. 2021. "Bioremediation Performance of Two Telluric Saprotrophic Fungi, Penicillium brasilianum and Fusarium solani, in Aged Dioxin-contaminated Soil Microcosms." Soil and Sediment Contamination: An International Journal: 1–2. doi: 10.1080/15320383.2021.1890692.

Djelal, H., Y. Lelievre, and C. Ricordel. 2014. "Combination of electro-coagulation and biological treatment by bioaugmentation for landfill leachate." Desalination and Water Treatment 54 (11): 2986–2993. doi: 10.1080/19443994.2014.908146.

Fava, F., and A. Piccolo. 2002. "Effects of humic substances on the bioavailability and aerobic biodegradation of polychlorinated biphenyls in a model soil." Biotechnology and Bioengineering 77 (2): 204–211. doi: 10.1002/bit.10140.

Field, J.A., and R. Sierra-Alvarez. 2008a. "Microbial degradation of chlorinated dioxins." Chemosphere 71 (6): 1005–1018. doi: 10.1016/j.chemosphere.2007.10.039.

Field, J.A., and R. Sierra-Alvarez. 2008b. "Microbial transformation and degradation of polychlorinated biphenyls." *Environmental Pollution* 155 (1): 1–12. doi: 10.1016/j.envpol.2007.10.016.

Garbisu, C., O. Garaiyurrebaso, L. Epelde, E. Grohmann, and I. Alkorta. 2017. "Plasmid-mediated bioaugmentation for the bioremediation of contaminated soils." *Frontiers in Microbiology* 8. doi: 10.3389/fmicb.2017.01966.

Guo, Z., J.J. Richardson, B. Kong, and K. Liang. 2020. "Nanobiohybrids: Materials approaches for bioaugmentation." *Science Advances* 6 (12). doi: 10.1126/sciadv.aaz0330.

He, J., and J. Paul Chen. 2014. "A comprehensive review on biosorption of heavy metals by algal biomass: Materials, performances, chemistry, and modeling simulation tools." *Bioresource Technology* 160: 67–78 doi: 10.1016/j.biortech.2014.01.068.

Hou, C., J. Shen, D. Zhang, Y. Han, D. Ma, X. Sun, J. Li, W. Han, L. Wang, and X. Liu. 2016. "Bioaugmentation of a continuous-flow self-forming dynamic membrane bioreactor for the treatment of wastewater containing high-strength pyridine." *Environmental Science and Pollution Research* 24 (4): 3437–3447. doi: 10.1007/s11356-016-8121-z.

Huang, Y.-H., X.-J. Huang, X.-H. Chen, Q.-Y. Cai, S. Chen, C.-H. Mo, H. Lü, and M.-H. Wong. 2018. "Biodegradation of di-butyl phthalate (DBP) by a novel endophytic bacterium Bacillus subtilis and its bioaugmentation for removing DBP from vegetation slurry." *Journal of Environmental Management* 224: 1–9. doi: 10.1016/j.jenvman.2018.07.023.

Janssen, D.B., and G. Stucki. 2020. "Perspectives of genetically engineered microbes for groundwater bioremediation." *Environmental Science: Processes & Impacts* 22 (3): 487–499. doi: 10.1039/c9em00601j.

Jing, R., S. Fusi, and B.V. Kjellerup. 2018. "Remediation of polychlorinated biphenyls (PCBs) in contaminated soils and sediment: State of knowledge and perspectives." *Frontiers in Environmental Science* 6. doi: 10.3389/fenvs.2018.00079.

Kakavandi, B., and M. Ahmadi. 2019. "Efficient treatment of saline recalcitrant petrochemical wastewater using heterogeneous UV-assisted sono-Fenton process." *Ultrasonics Sonochemistry* 56: 25–36 doi: 10.1016/j.ultsonch.2019.03.005.

Kaur, R., A. Kumari, G. Sharma, D. Singh, and R. Kaur. 2021. "Biodegradation of endocrine disrupting chemicals benzyl butyl phthalate and dimethyl phthalate by Bacillus marisflavi RR014." *Journal of Applied Microbiology*. doi: 10.1111/jam.15045.

Kong, X., D. Jin, X. Tai, H. Yu, G. Duan, X. Yan, J. Pan, J. Song, and Y. Deng. 2019. "Bioremediation of dibutyl phthalate in a simulated agricultural ecosystem by Gordonia sp. strain QH-11 and the microbial ecological effects in soil." *Science of the Total Environment* 667: 691–700 doi: 10.1016/j.scitotenv.2019.02.385.

Koumaki, E., C. Noutsopoulos, D. Mamais, G. Fragkiskatos, and A. Andreadakis. 2021. "Fate of emerging contaminants in high-rate activated sludge systems." *International Journal of Environmental Research and Public Health* 18 (2). doi: 10.3390/ijerph18020400.

Kumar, P., and S.G. Sachan. 2021. "Exploring microbes as bioremediation tools for the degradation of pesticides." In: Shah, M.P. (ed) *Advanced Oxidation Processes for Effluent Treatment Plants*, 51–67.

Kurniawan, T., W. Lo, and G. Chan. 2006. "Physico-chemical treatments for removal of recalcitrant contaminants from landfill leachate." *Journal of Hazardous Materials* 129 (1–3): 80–100. doi: 10.1016/j.jhazmat.2005.08.010.

Lambertino, A., V. Persky, S. Freels, H. Anderson, T. Unterman, S. Awadalla, and M. Turyk. 2021. "Associations of PCBS, dioxins and furans with follicle-stimulating hormone and luteinizing hormone in postmenopausal women: National Health and Nutrition Examination Survey 1999–2002." *Chemosphere* 262. doi: 10.1016/j.chemosphere.2020.128309.

Li, H., Y. Li, M. Bao, and S. Li. 2021a. "Solid inoculants as a practice for bioaugmentation to enhance bioremediation of hydrocarbon contaminated areas." *Chemosphere* 263. doi: 10.1016/j.chemosphere.2020.128175.

Li, J., Q. Lu, E.A. Odey, K.S. Lok, B. Pan, Y. Zhang, and H. Shim. 2021b. "Coupling of biostimulation and bioaugmentation for enhanced bioremoval of chloroethylenes and BTEX from clayey soil." *Ecotoxicology*. doi: 10.1007/s10646-020-02323-z.

Liu, Y., H. Zhang, X. He, and J. Liu. 2021. "Genetically engineered methanotroph as a platform for bioaugmentation of chemical pesticide contaminated soil." *ACS Synthetic Biology* 10 (3): 487–494. doi: 10.1021/acssynbio.0c00532.

Loganathan, B.G., and S. Masunaga. 2009. "PCBs, dioxins, and furans: Human exposure and health effects." In: Gupta, R.C (ed) *Handbook of Toxicology of Chemical Warfare Agents*, 245–253.

Martínez-Ávila, L., H. Peidro-Guzmán, Y. Pérez-Llano, T. Moreno-Perlín, A. Sánchez-Reyes, E. Aranda, G. Ángeles de Paz, A. Fernández-Silva, J.L. Folch-Mallol, H. Cabana, N. Gunde-Cimerman, and R.A. Batista-García. 2021. "Tracking gene expression, metabolic profiles, and biochemical analysis in the halotolerant basidiomycetous yeast Rhodotorula mucilaginosa EXF-1630 during benzo[a]pyrene and phenanthrene biodegradation under hypersaline conditions." *Environmental Pollution* 271. doi: 10.1016/j.envpol.2020.116358.

Maszenan, A. M., Y. Liu, and W.J. Ng. 2011. "Bioremediation of wastewaters with recalcitrant organic compounds and metals by aerobic granules." *Biotechnology Advances* 29 (1): 111–123. doi: 10.1016/j.biotechadv.2010.09.004.

Matheus, D.R., V. Lúcia Ramos Bononi, and K. Maria Gomes Machado. 2000. *World Journal of Microbiology and Biotechnology* 16 (5): 415–421. doi: 10.1023/a:1008910128114.

Matsumoto, E., Y. Kawanaka, S.-J. Yun, and H. Oyaizu. 2009. "Bioremediation of the organochlorine pesticides, dieldrin and endrin, and their occurrence in the environment." *Applied Microbiology and Biotechnology* 84 (2): 205–216. doi: 10.1007/s00253-009-2094-5.

Mazumder, A., S. Das, D. Sen, and C. Bhattacharjee. 2020. "Kinetic analysis and parametric optimization for bioaugmentation of oil from oily wastewater with hydrocarbonoclastic *Rhodococcus pyridinivorans* F5 strain." *Environmental Technology & Innovation* 17. doi: 10.1016/j.eti.2020.100630.

Nahurira, R., L. Ren, J. Song, Y. Jia, J. Wang, S. Fan, H. Wang, and Y. Yan. 2017. "Degradation of di(2-ethylhexyl) phthalate by a novel gordonia alkanivorans strain YC-RL2." *Current Microbiology* 74 (3): 309–319. doi: 10.1007/s00284-016-1159-9.

Nzila, A., and M.M. Musa. 2020. "Current status of and future perspectives in bacterial degradation of benzo[a]pyrene." *International Journal of Environmental Research and Public Health* 18 (1). doi: 10.3390/ijerph18010262.

Nzila, A., S. Razzak, and J. Zhu. 2016. "Bioaugmentation: An emerging strategy of industrial wastewater treatment for reuse and discharge." *International Journal of Environmental Research and Public Health* 13 (9). doi: 10.3390/ijerph13090846.

Otani, T., N. Seike, and Y. Sakata. 2007. "Differential uptake of dieldrin and endrin from soil by several plant families and Cucurbitagenera." *Soil Science and Plant Nutrition* 53 (1):86–94. doi: 10.1111/j.1747-0765.2007.00102.x.

Pan, X., D. Lin, Y. Zheng, Q. Zhang, Y. Yin, L. Cai, H. Fang, and Y. Yu. 2016. "Biodegradation of DDT by Stenotrophomonas sp. DDT-1: Characterization and genome functional analysis." *Scientific Reports* 6 (1). doi: 10.1038/srep21332.

Papageorgiou, G.Z., D.G. Papageorgiou, Z. Terzopoulou, and D.N. Bikiaris. 2016. "Production of bio-based 2,5-furan dicarboxylate polyesters: Recent progress and critical aspects in their synthesis and thermal properties." *European Polymer Journal* 83: 202–229 doi: 10.1016/j.eurpolymj.2016.08.004.

Pokethitiyook, P., and T. Poolpak. 2012. "Heptachlor and its metabolite: Accumulation and degradation in sediment." In Soundarajan, R.P. (ed) *Pesticides - Recent Trends in Pesticide Residue Assay*. InTEch publishers, Croatia, http://dx.doi.org/10.5772/3329.

Qiu, J., Y. Zhang, Y. Shi, J. Jiang, S. Wu, L. Li, Y. Shao, and Z. Xin. 2020. "Identification and characterization of a novel phthalate-degrading hydrolase from a soil metagenomic library." *Ecotoxicology and Environmental Safety* 190. doi: 10.1016/j.ecoenv.2019.110148.

Quinlivan, S.C., M. Ghassemi, and T.V. Leshendok. 1975. "Sources, characteristics and treatment and disposal of industrial wastes containing hexachlorobenzene." *Journal of Hazardous Materials* 1 (4): 343–359. doi: 10.1016/0304-3894(75)80006-9.

Rathna, R., S. Varjani, and E. Nakkeeran. 2018. "Recent developments and prospects of dioxins and furans remediation." *Journal of Environmental Management* 223: 797–806. doi: 10.1016/j.jenvman.2018.06.095.

Rentz, J.A., Pedro J.J. Alvarez, and J.L. Schnoor. 2008. "Benzo[a]pyrene degradation by Sphingomonas yanoikuyae JAR02." *Environmental Pollution* 151 (3): 669–677. doi: 10.1016/j.envpol.2007.02.018.

Rylott, E.L., E.J. Johnston, and N.C. Bruce. 2015. "Harnessing microbial gene pools to remediate persistent organic pollutants using genetically modified plants—a viable technology?" *Journal of Experimental Botany* 66 (21): 6519–6533. doi: 10.1093/jxb/erv384.

Sadañoski, M.A., A.S. Tatarin, M.L. Barchuk, M. Gonzalez, C.N. Pegoraro, M.I. Fonseca, L.N. Levin, and L.L. Villalba. 2020. "Evaluation of bioremediation strategies for treating recalcitrant halo-organic pollutants in soil environments." *Ecotoxicology and Environmental Safety* 202. doi: 10.1016/j.ecoenv.2020.110929.

Sanborn, J.R., R.L. Metcalf, W.N. Bruce, and P.-Y. Lu. 1976. "The fate of chlordane and toxaphene in a terrestrial-aquatic model ecosystem." *Environmental Entomology* 5 (3): 533–538. doi: 10.1093/ee/5.3.533.

Sangkharak, K., A. Choonut, T. Rakkan, and P. Prasertsan. 2020. "The degradation of phenanthrene, pyrene, and fluoranthene and its conversion into medium-chain-length polyhydroxyalkanoate by novel polycyclic aromatic hydrocarbon-degrading bacteria." *Current Microbiology* 77 (6): 897–909. doi: 10.1007/s00284-020-01883-x.

Singh, B.K., R. Chander Kuhad, A. Singh, R. Lal, and K. K. Tripathi. 2008. "Biochemical and molecular basis of pesticide degradation by microorganisms." *Critical Reviews in Biotechnology* 19 (3): 197–225. doi: 10.1080/0738-859991229242.

Singh, T., and D.K. Singh. 2017. "Phytoremediation of organochlorine pesticides: Concept, method, and recent developments." *International Journal of Phytoremediation* 19 (9): 834–843. doi: 10.1080/15226514.2017.1290579.

Smith, M.J., G. Lethbridge, and R.G. Burns. 1999. "Fate of phenanthrene, pyrene and benzo[a]pyrene during biodegradation of crude oil added to two soils." *FEMS Microbiology Letters* 173 (2): 445–452. doi: 10.1111/j.1574-6968.1999.tb13537.x.

Sudjarid, W., I.M. Chen, W. Monkong, and J. Anotai. 2012. "Reductive dechlorination of 2,3,4-chlorobiphenyl by biostimulation and bioaugmentation." *Environmental Engineering Science* 29 (4): 255–261. doi: 10.1089/ees.2011.0228.

Sun, J., Y. Qiu, P. Ding, P. Peng, H. Yang, and L. Li. 2017. "Conjugative transfer of dioxin–catabolic mega-plasmids and bioaugmentation prospects of a Rhodococcus sp." *Environmental Science & Technology* 51 (11): 6298–6307. doi: 10.1021/acs.est.7b00188.

van Birgelen, A. P. 1998. "Hexachlorobenzene as a possible major contributor to the dioxin activity of human milk." *Environmental Health Perspectives* 106 (11): 683–688. doi: 10.1289/ehp.106-1533492.

Varjani, S., and V.N. Upasani. 2021. "Bioaugmentation of Pseudomonas aeruginosa NCIM 5514 – A novel oily waste degrader for treatment of petroleum hydrocarbons." *Bioresource Technology* 319. doi: 10.1016/j.biortech.2020.124240.

Varjani, S.J. 2017. "Microbial degradation of petroleum hydrocarbons." *Bioresource Technology* 223: 277–286. doi: 10.1016/j.biortech.2016.10.037.

Wu, M., X. Guo, J. Wu, and K. Chen. 2020. "Effect of compost amendment and bioaugmentation on PAH degradation and microbial community shifting in petroleum-contaminated soil." *Chemosphere* 256. doi: 10.1016/j.chemosphere.2020.126998.

Xiao, P., T. Mori, I. Kamei, H. Kiyota, K. Takagi, and R. Kondo. 2011. "Novel metabolic pathways of organochlorine pesticides dieldrin and aldrin by the white rot fungi of the genus Phlebia." *Chemosphere* 85 (2): 218–224. doi: 10.1016/j.chemosphere.2011.06.028.

Xu, H.-J., J. Bai, W. Li, J. Colin Murrell, Y. Zhang, J. Wang, C. Luo, and Y. Li. 2021. "Mechanisms of the enhanced DDT removal from soils by earthworms: Identification of DDT degraders in drilosphere and non-drilosphere matrices." *Journal of Hazardous Materials* 404. doi: 10.1016/j.jhazmat.2020.124006.

Xue, J., K. Shi, C. Chen, Y. Bai, Q. Cui, N. Li, X. Fu, and Y. Qiao. 2021. "Evaluation of response of dynamics change in bioaugmentation process in diesel-polluted seawater via high-throughput sequencing: Degradation characteristic, community structure, functional genes." *Journal of Hazardous Materials* 403. doi: 10.1016/j.jhazmat.2020.123569.

Yuan, K., S. Li, and F. Zhong. 2020. "Treatment of coking wastewater in biofilm-based bioaugmentation process: Biofilm formation and microbial community analysis." *Journal of Hazardous Materials* 400. doi: 10.1016/j.jhazmat.2020.123117.

Zhang, K., H.-b. Luo, W. Chen, and J. Chen. 2018. "Application of bioaugmentation strategy in biodegradation of di-n-octyl phthalate (DOP) in sequencing batch reactor (SBR)." *DEStech Transactions on Computer Science and Engineering*. doi: 10.12783/dtcse/pcmm2018/23710.

Zhao, H.-M., H. Du, C.-Q. Huang, S. Li, X.-H. Zeng, X.-J. Huang, L. Xiang, H. Li, Y.-W. Li, Q.-Y. Cai, C.-H. Mo, and Z. He. 2019. "Bioaugmentation of exogenous strain Rhodococcus sp. 2G can efficiently mitigate di(2-ethylhexyl) phthalate contamination to vegetable cultivation." *Journal of Agricultural and Food Chemistry* 67 (25): 6940–6949. doi: 10.1021/acs.jafc.9b01875.

Zhou, H., S. Zhang, J. Xie, H. Wei, Z. Hu, and H. Wang. 2020. "Pyrene biodegradation and its potential pathway involving Roseobacter clade bacteria." *International Biodeterioration & Biodegradation* 150. doi: 10.1016/j.ibiod.2020.104961.

Zhu, G., Y. Zhang, S. Chen, L. Wang, Z. Zhang, and B.E. Rittmann. 2021. "How bioaugmentation with Comamonas testosteroni accelerates pyridine mono-oxygenation and mineralization." *Environmental Research* 193. doi: 10.1016/j.envres.2020.110553.

7 Application of Nanomaterials in the Bioaugmentation of Heavily Polluted Environment

V.C. Akubude
Federal University of Technology

T.F Oyewusi
Adeleke University

V.C. Okafor and P.C. Obumseli
Federal University of Technology

A.O. Igwe
Michael Okpara University of Agriculture

CONTENTS

DOI: 10.1201/9781003187622-7

7.1 INTRODUCTION

Environmental cleanup from pollutants is essential to ensure the safety of living things within it. Application of nanomaterials in cleanup of heavily polluted sites has been investigated by several researchers. They have specific characteristics that make them unique and therefore offer a potential solution in effective environmental cleanup. Nanomaterials exist at nanoscale level which implies wider surface area and better reactivity. Integrating nanomaterial with bioremediation which is presenting gaining the attention of the research community and industries will offer a better approach to environmental sanitation. This is because combining the unique benefits of nanomaterial with the advantages of bioremediation (such as environmental-friendly and cost-effective) offers a promising future to pollution control. Bioremediation provides a biological solution to pollution control. This process utilizes microorganisms like bacteria, fungi, protists, or their enzyme in degrading the environmental pollutants into a less harmful state. There are diverse approaches to bioremediation as shown in Figure 7.1. Bioaugmentation is one out of many of them that involve the use of selected/well-engineered microbes with high biodegradation capacities to treat the polluted environment. This approach has gained research interest but it is less efficient when dealing with a high amount of contaminants and xenobiotics or refractory compounds, resulting in unstable treatment efficiencies and recovery time. Research has shown that combining nanomaterials in the bioremediation process offers an effective remediation process. They have a high reactivity capacity as they penetrate the pollutant increase the removal efficiency. Several unique nanomaterials are been developed for wastewater treatment and soil remediation. And these materials are relatively cheap, appreciable removal efficiency and easy regeneration (Chen et al. 2017; Bystrzejewski et al. 2009; Anjum et al. 2016).

7.1.2 POLLUTION ISSUE

Environmental pollution is a global challenge arising from the fast-growing population and technological advancement. The environment which comprises the soil, water, and air is contaminated through natural and anthropogenic activities. The wrong handling of agents of pollution such as sewage, pharmaceutical waste, agrochemicals, and industrial waste into water bodies and land has contributed immensely to hazardous release into the surroundings. Several facts have been documented with respect to this: over 1 million live without any form of sanitation, discharge of untreated sewage into water sources in most developing countries, oil spillage into water bodies and agricultural soils, discharge of industrial effluents into the environment. According to a report in 2012, almost 10 million chlorinated aliphatic hydrocarbons (CAH) were documented to be released in the United States. The commonly released are trichloroethylene (TCE), dichloromethane (DM), carbon tetrachloride (CT), chloroform (CF), and tetrachloroethylene (PCE). In addition, approximately

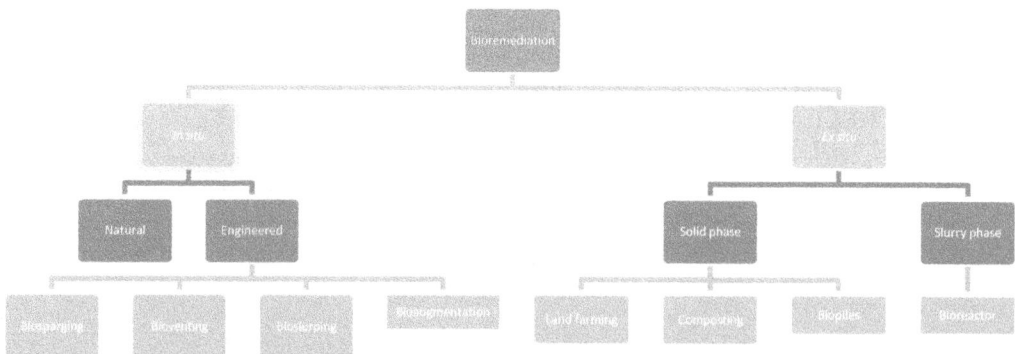

FIGURE 7.1 Different bioremediation approaches.

65% of these released pollutants are suspected to be cancerous (Xuhui Maoa et al. 2012). The pollution challenges involve climatic change, ozone layer depletion, reduced air and water quality, soil contamination, etc., and these are mostly caused by increasing human population and technological advancement. The advent of more industries

7.2 ENVIRONMENTAL POLLUTANT

Pollution occurs when waste materials mostly resulting from human activities (anthropogenic) and most seldom from natural occurrences are added to the environment. In a situation when the waste materials are not adequately absorbed, disintegrated, or otherwise eliminated through physical, natural, and biological means of the environment, the negative impact may result as the materials accrue and further metamorphosed into noxious substances. Therefore, any substances that can cause environmental pollution are called environmental pollutant. Similarly, it is referred to as deleterious substance in any form (solid, liquid, and gases) existing in a certain amount in the biosphere that can be so harmful to the entire ecosystem. Environmental pollutant contaminates the air, land, and water, putting man and the whole biota in danger. Pollutants are environmentally mobile and prevalent in addition; some are stored in animal fat cells and accumulate in the food chain. The three major ways of exposure to environmental pollutants are through ingestion, inhalation, and physical contact with skin (Möller et al. 1994; Wilson et al. 1996).

7.2.1 CLASSIFICATION OF ENVIRONMENTAL POLLUTANTS

Classification of environmental pollutants could possibly assist to be cognizant of its nature, hence, the right methods to manage it. It also assists in identifying the source, origin, causative material, chemical composition, state of matter, and area of impact. Based on foregoing, methods and procedures can be derived to remove or abolish such pollutants from the very onset and awareness can be created and rationalized. On the other hand, it must come to satisfactory conclusion that most of these classifications are in essence imitative and not objective so, the manner of adaptation is based on the pollutant purpose. In spite of these constraints, there are substantial values in having some standard methods provided that predictive environmental management tools for possible pollutant effects are taken into consideration. Table 7.1 shows some classes of environmental pollutants.

7.2.1.1 Pollutant Properties

As a particular valuable approach, pollutant qualities must be highly considered during evaluation of actual effects of pollution because the evaluation entails both the assessment of its overall qualities and the inhabitant environment. The properties include toxicity of the pollutant, as well as its bioaccumulation, mobility, persistence, and ease of control (BioWise, 2001).

7.2.1.2 Toxicity

The pollutant toxicity characterizes likely harmful effects (both short and long terms) of the pollutant to lives. It is connected with the amount or strength of pollutant and period or time of exposure to it; however, this connection is very difficult to establish. Naturally, extremely toxic substances can be put to death as fast as possible while slight toxic needs a very long time of exposure to do any harm. This is without uncertainty or ambiguity. Meanwhile, some pollutants possibly will kill rapidly in extreme concentrations and could also have an effect on organisms' behaviors or their exposure to environmental stress over its duration of life when exposed to low concentrations (BioWise, 2001).

7.2.1.3 Bioaccumulation

As widely known, many pollutants though exist in a small concentration in the environment, maybe absorbed by living organisms in the surroundings and become concentrated in their tissues over time.

TABLE 7.1

Classification of Pollutant

Classification	Type	Description	Example	References
Depending upon their source	Natural pollutants	Those which are released from nature	Particles from sea salt, dust from a volcanic eruption, photochemically formed ozone, etc.	Moller et al. (1994), Wilson et al. (1996), Ghio et al. (1999), Pooley et al. (1999)
	Man-made/ anthropogenic pollutants	Those which are released due to human activities (commercial, industrial, etc.)	SO_2 produced from the burning of fossil fuels, NO_x produced from internal combustion engine, CO and CO_2 emitted from fuel combustion, etc.	
Based on origin	Primary pollutants	Pollutants that are emitted from a direct source into the environment	SO_x, NO_x, CO, VOCs, Hg, particulate matter (PM), etc.	Moller et al. (1994), Wilson et al. (1996), Ghio et al. (1999), Pooley et al. (1999)
	Secondary pollutants	Pollutants that are formed due to interactions between the primary pollutants themselves or/and other surrounding components	Peroxyacyl nitrates (PANs), SO_3, H_3SO_4, O_3, HNO_3, etc.	
Based on persistent of the emitted substance	Persistent pollutants	Pollutants that remain or persist in the environment for a very long time without any alteration to its original form	Metals such as beryllium, lead, mercury, pesticides, etc.	Moller et al. (1994), Wilson et al. (1996), Ghio et al. (1999), Pooley et al. (1999)
	Nonpersistent pollutants	Pollutants that do not remain in the environment for a long time before they break down into a simple form	Food wastes, sewage, animal wastes, crop residues, etc.	
Based on chemical composition	Inorganic pollutants	Pollutants that are considered to be of mineral but not of carbon or biological origin	Arsenic, cadmium, lead, mercury, chromium, aluminum, nitrates, nitrites, fluorides, etc.	Moller et al. (1994), Wilson et al. (1996), Ghio et al. (1999), Pooley et al. (1999)
	Organic pollutants	Pollutants having molecules that contained carbon	Benzene, 1,3-Butadiene, PAH, dioxins, CO and CO_2 etc.	
Based on the state of matter	Particulate pollutants	Pollutants made of finely divided solids and liquids that will settle out under proper conditions.	Dust, fumes, smoke, fly ash, mist, and spray.	Moller et al. (1994), Wilson et al. (1996), Ghio et al. (1999), Pooley et al. (1999)
	Gaseous pollutants	Pollutants include vapors of substances that are liquid or solid at normal temperatures and pressures that are formless fluids completely occupying space, into which they are released, behave more like air, and do not settle out	CO, CO_2, SO_x, NO_x, HC, oxidants, etc.	
Based on area of impact	Ozone depletion substances	Pollutants that affect the stratosphere	CFCs, halons, HCFCs, etc.	Moller et al. (1994), Wilson et al. (1996), Ghio et al. (1999), Pooley et al. (1999)
	Greenhouse gases	Pollutants that occur in the troposphere	CO_2, CH_4, N_2O, CFCs and other hydrocarbons, O_3, etc.	
	Toxins	Pollutants that affect humans and other living creatures	Dioxins, furans, etc.	

They can accumulate in the organism to a level which constitutes environmental and health risk. The pollutant that acts in this particular way (be absorbed by living organisms) needs a serious attention due to the fact that a comparatively small related level of contamination may accumulate up the food chain.

7.2.1.4 Mobility

The peculiarity of a pollutant to scatter or be diluted is an essential characteristic in its entire effect because it influences concentration. A few pollutants are reluctantly mobile and have a tendency to remain in "hotspots" very close to their source point. Conversely, these pollutants can be transported by the environmental media to far sites where they have never been used or produced; most often their distribution may not be uniform.

7.2.1.5 Persistence

Persistence is the extent or length of time an effect lasts. It is an essential characteristic particularly in pollution and is regularly connected to bioaccumulation and mobility. Extremely toxic pollutants that are ecologically not stable and breakdown speedily is less dangerous than persistent pollutants, although it is possible that these are inherently less toxic.

7.2.1.6 Ease of Control

The efficient control of any pollution depends on several parameters which include the pollutant's mobility, its nature, the extent or length of time of pollution effect and specific consideration of the location. Obviously, source control is the best method forasmuch as it gets rid of the problem from the source. However, this is not always achievable, and in situations like this, containment could be the way out, though highly concentrated hotspots could also be formed.

7.2.2 Effects of Pollutant

The adverse health effects of environmental pollutants cannot be underestimated. It can affect life even at an early stage. The most important adverse health effects include infant mortality, malignancies, mental, perinatal, cardiovascular and respiratory disorders, allergy, high stress oxidative, endothelial dysfunction, and several other negative impacts (Kelishadi et al., 2009; Kelishadi and Poursafa, 2010). Although, the emphasis is always on short-term impact of these contaminants; however, a large selection of threats of pollution from childhood and their impact on chronic none communicable infections of adulthood should also be emphasized. According to literature exposure to environmental particulates has been connected to a rise in morbidity and mortality risks from various diseases, organ disturbances, cancers, and other chronic diseases (Kargarfard et al., 2011; Coogan et al., 2012). So, something must be done in order to control the pollutants else, they will degrade the environment. Action must be taken to embark on different kinds of interventions. However, these pollutants can be removed by both biological and physicochemical methods but the physicochemical ways are more costly than the former and demand high energy with excessive consumption of chemical reagents hence, the main reason for degrading of the pollutants using microorganisms (biological method) (Hamdi et al., 2007).

7.3 CONCEPT OF BIOAUGMENTATION

One of the biological (*in situ*) ways is bioaugmentation that enhances the power of polluted sites when single strains of microorganisms are introduced with needed biodegradable potentials. Bioaugmentation contains additional (augmented) specific microbial cultures that are naturally grown independently under well-defined conditions, to accomplish a particular function of remediation in a specific environment (Alvarez and Illman, 2006). Bioaugmentation is a process of applying a natural type of autochthonous or allochthonous or genetically improved microbes to a contaminated harmful waste site with the intention to fast-track the removal of undesirable compounds (Mrozik and Piotrowska-Seget, 2010). It is mostly applied in surroundings where bioactivities are insignificant or areas where reduction of pollutants is difficult but needs to be controlled. The principle behind this approach is to improve the extent or level of biodegradation of complex contaminants by adding pollutant degrading microbes (Leahy and Colwell, 1990; Adams et al., 2015). Improving the micro-biota of polluted area will not only improve the removal of the contaminants from the specific area but concurrently enhance the genomic potential of the desired area. Thus, bioaugmentation is equivalent to an improvement in the genetic consortium and hence, variation in the gene of the area or site to be decontaminated. In principle, the variation in a gene could be raised by augmenting the microbial diversity (Dejongbe et al., 2001; Shukla et al., 2010). More importantly, the effectiveness of a bioaugmentation process is dependent on certain abiotic and biotic factors.

Bioaugmentation process has been used in agriculture ever since 1800 as perceived in augmentation of nitrogen-fixing *Rhizobium* on the roots of leguminous plants (Gentry et al., 2004) but it is progressively in use now to improve biodegradation of incalcitrant organic contaminants. Presently, bioaugmentation is getting more attention as a method to increase the catabolic possibility at polluted sites and improve the biological degradation of resistant contaminants that are of priority.

7.3.1 Methods of Bioaugmentation

Two distinctive bioaugmentation methods have been technically established. The first method is founded on the injection of microbes with the required catabolic possibility to supplement or substitute indigenous microbial population. In this instance, the chosen bacteria have the potential to survive and out-compete the indigenous microbes then occupy a certain metabolic role in the polluted medium (Vogel and Walter, 2002). The second bioaugmentation method approach involves adding a substantial concentration of cells that function for a moment as bio-catalysts which biodegrade a considerable amount of targeted pollutants before becoming inactive and stopping the functioning (Duba et al., 1996; Krumme et al., 1994). In this instance, the microbes introduced are incapable of surviving due to genomic biological and abiotic stress built up in the new medium or environment. In such situations, regular biomass re-added is needed with time because the injected cells cannot flourish in situ. In the same way, other approaches that are similar to aforementioned methods above have been successfully employed in bioaugmentation. These include the addition of pre-adapted unadulterated bacterial strain, addition of pre-adapted consortium, the introduction of inherently engineered bacteria, and addition of biodegraded germane genes packed in a vector which is transferred to the native microorganisms through conjugation (El Fantroussi and Agathos, 2005). High potentially metabolized microorganisms should be chosen to maintain the important characteristics in severe environmental conditions(Thompson et al., 2005; van der Gast et al., 2004). In place of metal that is co-contaminated with high concentration of organic contaminants, the co-contaminants that impede the organic compounds should be degraded by the microbial consortia (Roane et al., 2001).

All the approaches stated above have been used for multi-component systems and is a better illustration of an existent environment compared to a model based on a single-component system (Ledin 2000). Using microbial population instead of a pure culture for bioaugmentation is usually more beneficial as it offers a diversity of metabolism required for field application (Alisi et al. 2009). Also, Alisi et al. (2009) effectively achieved total degradation of phenanthrene and diesel oil, 60% reduction in isoprenoids; and 70% overall reduction in hydrocarbons in a period of 42 days with the use of a microbial formula prepared with certain native strains. Likewise, Li et al. (2009) confirmed that native microorganisms possibly will degrade PAH in contaminated soil that is advanced in years. The result showed that addition of microbic consortia of five fungi (*Aspergillus niger*, *Cuuninghamella* sp., *Alternaria alternate (fr.) Keissler*, *Phanerochaete chrysosporium*, and *Penicillium chrysogenum*) and three bacteria (*Flavo bacterium*, *Bacillus* sp., and *Zoogloea* sp.) improved the biodegradation rate (41.3%) substantially (Li et al., 2009). Bioaugmentation in wastewater activated sludge system is a bit simple to achieve due to addition of microorganisms which is easily intermixed in the reactors moreover, conditions of the reaction can be engineered to improve its survival and operation. Conversely, bioaugmentation in an aquifer is difficult to some extent and should envisage difficulties attached to the existence of additional strains, their circulation through the polluted site, and low nutrient concentrations and target pollutants that assist as substrata to the augmented microorganisms (Da-Silva and Alvarez, 2010). The accomplishment of bioaugmentation rests mostly on the modification of the microbial consortia on the site where the decontamination is needed. Similarly, the achievement of the process depends on the capacity of the newly injected microbial consortia to contend with the native microbes, predators, and several other components (Godleads et al., 2015). Generally, bioaugmentation can be influenced by both abiotic (pH, temperature, humidity, aerobic conditions, and nutrients' availability) and biotic (toxic contaminant concentrations and competition with native microorganisms) factors. However, bio-stimulation is commonly applied alongside bioaugmentation to make the added cells to be viable and/or to boost their metabolic potentials, which leads to a vigorous long-term biodegradation efficiency (Gentry et al., 2004).

7.3.2 Inoculums Used in Bioaugmentation

Based on the abovementioned, inoculums used in bioaugmentation are typical of bacteria and occasionally, of fungi substances, these consist of the following:

a. Mixed cultures: These include a set of native bacteria which have been extremely improved on the contaminants of interest (Steffan et al., 1999; Ulrich and Edwards, 2003). The seed culture to be improved is typically collected from the polluted site of interest.

b. Pure cultures: These consist of improved single-strain bacteria suitable for degrading the contaminants. The specimens consist of *Desulfomonile tiedje* or *Dehalococcoides ethenogenes* (that respectively dechlorinate chlorobenzoate and tetrachloroethene) and *Pseudomonas stutzeri* that degrades carbon tetrachloride (Zheng et al., 2001).

c. Genetic elements: In this process, gene vectors are introduced to replace the bacteria (Mohan et al., 2009; Pepper et al., 2002). The plasmid coding for an enzyme with the required catabolic capability is introduced into the native bacteria (horizontal gene transfer). This process reduces problems related to the inoculation and circulation of (bigger) bacterial cells all through the polluted area. The native bacteria that obtain the plasmid could also to a lesser extent be more sensitive to abiotic or biotic stresses than injected (exogenous) microbes.

d. Genetically modified microorganisms (GMOs): They were first modified about two decades due to their simple genetics with a plan to inject some specific traits into microbial communities through genetic manipulation (Brokamp and Schmidt, 1991; Fulthorpe and Wyndham, 1991). This is with the aim to improve degradation including biological reduction of 2,4-D (Dejonghe et al., 2000), phenol (Watanabe et al., 2002), toluene, chlorobenzene, indole, and conversion of mercury ion to elemental mercury that is less harmful (Lange et al., 1998). Presently, the availability of GMOs is commercially widespread and can be obtained from many firms as freeze-dried organisms which can then be rehydrated and freshened in solution prior to inoculation. However, GMOs have been widely exploited in a research laboratory and agriculture, no sufficient study has been carried out to evaluate their strength and related long-term lifecycle effects, which may perhaps include circulation of the exogenous gene around species with potential impact on biodiversity and microbial population. Consequently, the effect of high speculation and polarization as a result of in vitro genetic manipulation presents an important political impediment to application of GMOs in remediation. In reality, several nations have positioned lawful impediments on distribution of GMOs for site cleanup purposes. Consequently, specific country legislation should be accessed before considering GMOs as an agent of bioaugmentation.

7.3.3 Techniques for Introduction of the Inoculums

There are different techniques for the introduction of microorganisms into the polluted area. The major injection approaches are as follows:

1. Liquid injection: In this technique, high concentrations of microbes in the range of 109–1,012 cells per milliliter of inoculum blended with the nutrient solution are injected with the aid of a liquid delivery pump. In order to improve microbial transportation followed by bioaugmentation, scientists have tried several methods (Gentry et al., 2004) which are potentially important but must deal with certain challenges like surfactant toxicity or the loss of biodegradation fitness as a result of starvation or mutation.
2. Immobilization and then injection: The method involves fusing the cells into vectors like biosolids, char-amended soil, clay, lignite, gel beads, and peat before mixing with a polluted area (Gardin and Pauss, 2001; Gentry et al., 2004). The vector body makes available an interim shelter and nutrients for the injected microbes, or else it may not be able to out-compete native microbes. Also, disinfection of the vector body generally enhances the inoculum's shelf-life and improves its survival in the environment.
3. Encapsulation and then injection: This involves encapsulation of microbes using micro-beads or porous ceramics to improve their survival (Gentry et al., 2004; Mertens et al., 2006; Moslemy et al., 2003). The product can then be introduced as a "hydraulic fracturing" fluid to remediate to a greater extent impermeable formation. The medium that is not toxic to the microbes enables the dispersion of gas and liquid. The capsule guards the cells against high concentrations of the toxic effect of the chemical generally found at the source. Furthermore, carbonaceous materials can augment the capsulate composition to stimulate growth of the inoculum all through its acclimatization to the new environment. On the other hand, if cells grow favorably at the expense of adding carbon to the capsule rather than the target contaminant, adverse effects may occur.

7.3.4 Pros and Cons of Bioaugmentation

7.3.4.1 Pros

i. The process is cost-effective.
ii. They are eco-friendly and hence have a less negative impact on the environment.
iii. Less labor demand results in less operational cost. The microbes do the work when introduced into the contaminated site.
iv. Biodegradation of oil into simple carbon compounds by microbes thereby hindering oil build up in them.
v. The process can also be carried out in the affected area.

7.3.4.2 Cons

i. Provision of appropriate growth conditions (such as temperature, nutrient, and humidity) for the microbes to thrive.
ii. Inability to metabolize certain kinds of waste.
iii. Waste generation during the process of ingestion and metabolization of waste.
iv. Microbial completion between the indigenous ones and those introduced can lead to bioaugmentation failure.
v. The bioaugmentation process depends highly on the right selection of suitable microbial strain, their continual survival, and activity.

7.4 APPLICATIONS OF BIOAUGMENTATION

i. Dimethylene glycol monobutyl ether (DGBE): This belongs to the group of glycol ethers as shown in Figure 2 which have a large application as a chemical solvent. They are toxic substances that are utilized in paint, ink, detergents. They are soluble in both polar and non-polar solvents and therefore can easily penetrate biological membranes. There are few research work in bioaugmented degradation of DGBE wastewater (Young-Mo et al., 2007; Hsi-Ling et al., 2013) because of frequent use of DGBE worldwide, its slow degradation rate in natural environments, low removal efficiency, and the high-concentration inhibition, method for enhancing the biodegradation of DGBE are imperative (Blažo et al., 2012; Mohammad et al., 2012). Recently, *Serratia* sp. was used in bioaugmentation of DGBE contained in wastewater discharged from silicon plate manufacturing company. The result shows high removal efficiency at both laboratory and full-scale levels (Chen et al. 2016).

ii. Polycyclic aromatic hydrocarbons and heterocyclic compounds: Industrial wastewater from coal coking process contains a high concentration of organic and inorganic contaminants among which nitrogen heterocyclic compounds and benzene compounds are harmful and has the potential of causing cancer (Bai et al. 2010; Fuller et al. 2014; Puyol et al. 2015). Bioaugmentation treatments with strains that can effectively reduce one specific organic contaminant have been shown to be less expensive and have been employed for treating coking wastewaters that contain phenol, naphthalene, or polycyclic nitrogen-containing aromatics (Bai et al., 2011, Park et al., 2009).

iii. Chlorinated and fluorinated compound removal: These compounds have wide applications in industries as lubricants, degreasing agents, pesticide and fungicides, wood preservatives, plastic components, adhesives, and solvents. They have been used for over three decades. Most of the chlorinated solvents have the following properties: less soluble in water, variable vapor pressure, and denser than water. Hence, a good number of chlorinated solvents are classified as dense non-aqueous phase liquids (DNAPLs). DNAPLs tend to sink and accumulate on the non-permeable layer at the bottom of a confined aquifer. The commonest among them are 1,2-dichloroethane (1,2-DCA), 1,1,1-trichloroethane (1,1,1-TCA), carbon tetrachloride, methylene chloride, chloroform, tetrachloroethene (PCE), and trichloroethene (TCE) (Sutherson 1997). The structures of some of these compounds are shown in Figure 7.3. Several research output has shown successful bioaugmentation of water and soil polluted with chlorinated compounds (Khondee et al., 2012, Okino et al., 2002, Hou et al., 2016).

iv. Quinoline and pyridine: These are toxic heterocyclic aromatic organic compounds as shown in Figure 7.4 that are usually used in dye production, solvent for resins and terpenes, nicotinic acid, pharmaceutical products (like antipsychotic drugs and blood-thinning agents), agrochemicals, rubber products, vitamins, textile water repellants, etc. They are possible carcinogenic agents. Studies have shown biodegradation of quinoline and pyridine present in contaminated wastewater and soil (Bai et al., 2010, Bai et al., 2011, Wang et al., 2002a, Wang et al., 2002b). And it was observed that microbes enhanced the biodegradation process.

v. Heavy metals: These are metallic chemical elements that have relatively high density and are hazardous even in a small amount. A list of them includes arsenic, lead, mercury, cadmium, chromium, iron, aluminum. They are significant contaminant and their toxicity

FIGURE 7.2 Structure of diethylene glycol monobutyl ether.

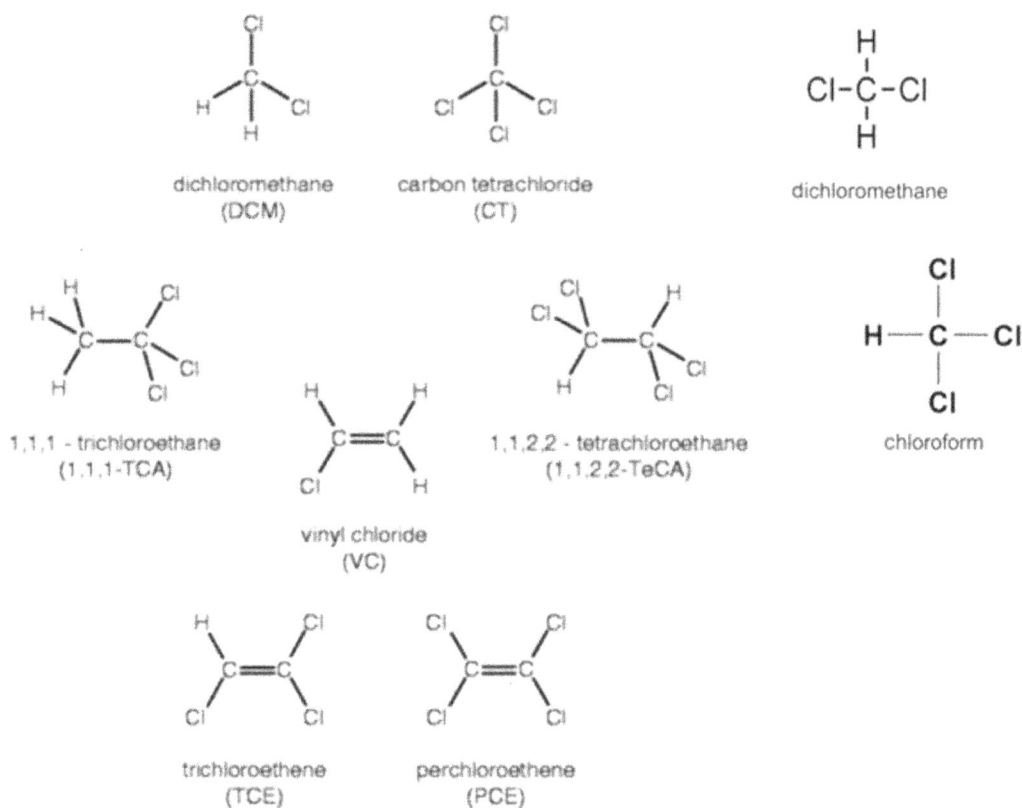

FIGURE 7.3 Structure of various forms of chlorinated and fluorinated compounds.

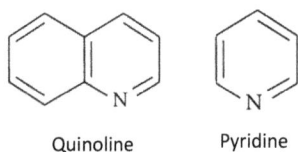

Quinoline   Pyridine

FIGURE 7.4 Structure of quinoline and pyridine.

adversely impacts the ecosystem. The major heavy metals common in wastewater are arsenic, cadmium, chromium, copper, lead, nickel, and zinc. Natural and anthropogenic sources of heavy metals are erosion, weathering, mining, industrial effluents, sewage discharge, agrochemicals. There is several research output that has shown successful bioaugmentation of heavy metals. Auwalu et al., (2020) reported efficient removal of heavy metals and metalloids from contaminated soil using consortia of filamentous fungi. Also, the bioaugmentation of soil contaminated with zinc minimized its adverse effects (Strachel et al 2020). Furthermore, soil contaminated with multiple heavy metals was successfully bioaugmented using *Bacillus subtilis* specie assisted by Novogro via adsorption reduce the negative impact of cadmium, chromium, mercury, and lead (wang et al., 2014).

vi. Cyanide: This is a chemical compound that contains the cyano group. They are toxic substances with environmental and health risks. They are used for several industrial purposes in photography, synthetic plastic, electroplating, mining, and clinical chemistry. Remediation of cyanides in polluted sites can be achieved using biological, chemical, and physical methods. Biological method (i.e., biodegradation) is cheap, eco-friendly, and efficient. There are several research outputs in his regards with respect to efficient removal of cyanides from polluted soil and water effluents cassava processing factories (Park et al., 2008, Akcil, 2003, Akcil et al., 2003, Dwivedi et al., 2011, Fatma et al., 2000, Kaewkannetra et al., 2009).

vii. Synthetic dyes: A dye is a colored organic compound that is used to impart the color of various substrates. They are made of harmful chemical compounds that pose an environmental risk. These chemical compounds comprise mercury, lead, chromium, toluene, sodium chloride, and benzene. Therefore effluents from the dye production process are potential water and soil pollutant if not properly treated. The report has shown microbes can be used for decoloration of dyes and their biodegradation process in polluted sites (Wang et al., 2012, Fan et al., 2009, Qu, et al., 2006, Qu et al., 2005, Qu et al., 2009).

7.5 UTILIZATION OF NANOMATERIAL TO BIOAUGMENTATION

Nanomaterial plays a vital role in the biological waste treatment process. This possesses unique properties such as high adsorption capacity and enhanced redox and photocatalytic properties which is necessary for waste treatment both in the catalysis process, toxic material degradation, and in improving the efficiency of the microbes in the degradation process. Application of nanomaterials to the bioremediation process offers the following benefits:

i. Better reactivity as the surface area per unit mass increases at a nanoscale level ensuring more contact with the surrounding material.
ii. Less activation energy is needed as for chemical reaction to take place nanomaterial possesses quantum effect property.
iii. Easy detection of toxic material as nanomaterials exhibits surface Plasmon resonance.
iv. Nanomaterial of any size or shape can be used for environmental cleanup.

There are several classes of nanomaterials each characterized by different morphological properties. Nanomaterials that have been utilized for environmental cleanup include nano-silver, nano-zero-valent iron, nano chitosan, nano-zinc oxide, nano-magnesium oxide, nano-zinc oxide, nano-cerium oxide, nano titanium dioxide, carbon nanotubes, fullerenes, and their derivates, etc. Nano-zero-valent nanoparticle is the commonly used in the environmental cleanup from inorganic salt, chlorinated organic pollutants, and heavy metals. However, studies show that it could be unsafe for the microbial community and the ecosystem at large. Many nanomaterials alongside nano-zero valent nanoparticles have been demonstrated to possess antimicrobial properties (Markova et al., 2013; Hsiao et al., 2006; Zhang et al., 2008; Lyon et al., 2008; Zhu et al., 2014; Henry et al., 2007; Stoimenov et al., 2002; Liu et al., 2019).

A report shows that carbon nanotube was applied in the treatment of water from the textile industry where it serves as redox mediators for azo dye decolorization and the outcome shows high reduction rates. Also, the presence of carbon nanotube improved the catalytic rates (Pereira et al., 2014). Other studies show that carbon nanotubes can improve or retard the biodegradation process for atrazine depending on the concentration used in the process (Zhang et al., 2015). Additionally, it has been used for heavy metal removal (Kaushal and Singh et al., 2017; Bhanjana et al., 2017a) like copper, cerium, gold, silver, chromium, etc. Readers can further consult the following research output for more studies (Bystrzejewski et al., 2009; Chen et al., 2017; Markova et al., 2013; Towell et al., 2011).

7.6 FUTURE PROSPECTS: CHALLENGES AND SOLUTION

Despite the promising benefits of nanotechnology application in environmental cleanup (bioremediation), the process is still limited by two major challenges which include the possibility of environmental toxicity and inhibition of microbial population. Also, the unique size of nanomaterials makes it easy to diffuse into the atmosphere when introduced into the environment and their form and properties can easily be influenced or changed. Continual exposure of the environment to nanomaterial can lead to contamination which may result in health issues. However, the research community is putting in more effort to proffer solutions to this challenge. Recent studies show that microbial inhibition by nanomaterials is dependent on its concentration level. This implies certain concentration level enhances microbial activities instead of inhibition. In a report by Zhang et al. (2015), it can be seen at a concentration below 25 mg/L microbial inhibition was not observed while at a concentration above 100 mg/L microbial population was affected negatively using carbon nanotube. Also, a similar report was documented during biodegradation of wastewater from dye-producing industry using carbon-based material (Pereira et al., 2014).

Therefore, in conclusion, the potential benefits of using nanomaterial in the bioaugmentation process can be harnessed fully but a more in-depth study is recommended on the interaction of nanomaterial with the environment during the bioremediation process. This is very vital to ascertain for every nanomaterial the suitable concentration level that favors high pollutant removal with less toxicity impact on the environment. Also, using nanomaterial produced from biological material could help in reducing their toxicity impact on the environment so the green synthesis of nanomaterials should be understudied for this purpose. Furthermore, growing microbes that are resistant to a high concentration of nanomaterial can be added advantage. Finally, if matching solutions can be proffered for these challenges then the future exploitation of nanomaterial for the environmental cleanup process can be established beyond laboratory experimentation.

REFERENCES

Adams, GO, Tawari-Fufeyin, P, Igelenyah, E (2014). Bioremediation of spent oil contaminated soils using poultry litter. *Res J Engineer Appl Sci.* 3(2): 124–30.

Akcil A, Karahan AG, Ciftci H, Sagdic O. 2003. Biological treatment of cyanide by natural isolated bacteria (Pseudomonas sp.). *Miner Eng* 16: 643–9.

Akcil A. 2003. Destruction of cyanide in gold mill effluents: biological versus chemical treatments. *Biotechnol Adv* 21: 501–11.

Alisi C, Musella R, Tasso F, Ubaldi C, Manzo S, Cremisini C, Sprocati AR (2009). Bioremediation of diesel oil in a co-contaminated soil by bioaugmentation with a microbial formula tailored with native strains selected for heavy metals resistance. *Sci Total Environ* 407:3024–32.

Alvarez, P.J. and Illman, W.A. (2006). *Bioremediation and Natural Attenuation: Process Fundamentals and Mathematical Models.* New York: Wiley.

Anjum, M., Miandad, R., Waqas, M., Gehany, F., & Barakat, M. A. (2016). Remediation of wastewater using various nano-materials. *Arab J Chem* 12(8): 4897–919.

Auwalu H, Agamuthu P., Innocent C.O., Fauziah S.H. (2020). Bioaugmentation of assisted mycoremediation of heavy metals and/metalloids landfill contaminated soil using consortia of filamentous fungi. *Biochem Eng J.* https://doi.org/10.1016/j.bej.2020.107550.

Bai, Y.H., Sun, Q.H., Sun, R.H., Wen, D.H., Tang, X.Y., 2011. Bioaugmentation and adsorption treatment of coking wastewater containing pyridine and quinoline using zeolite-biological aerated filters. *Environ Sci Technol.* 45(5), 1940–1948.

Bai, Y.H., Sun, Q.H., Zhao, C., Wen, D.H., Tang, X,Y. (2010) Bioaugmentation treatment for coking waste-water containing pyridine and quinoline in a sequencing batch reactor. *Appl Microbiol Biotechnol* 87:1943–1951

BioWise, UK Department of Trade and Industry (2001). Biotechnology Improves Product Quality, Crown Copyright.

Brokamp A, Schmidt FRJ (1991). Survival of Alcaligenes xylosoxidans degrading 2,2-dichloropropionate and horizontal transfer of its halidohydrolase gene in a soil microcosm. *Curr Microbiol* 22: 299–306.

Bystrzejewski, M., Pyrzyńska, K., Huczko, A., & Lange, H. (2009). Carbon-encapsulated magnetic nanoparticles as separable and mobile sorbents of heavy metal ions from aqueous solutions. *Carbon* 47(4): 1201–1204.

Chen, K., He, J., Li, Y., Cai, X., Zhang, K., Liu, T., et al. (2017). Removal of cadmium and lead ions from water by sulfonated magnetic nanoparticle adsorbents. *J Colloid Interf Sci* 494: 307–316.

Chen, M., Fan, R., Zou, W., Zhou, H., Tan, Z., Li, X. (2016). Bioaugmentation for treatment of full-scale diethylene glycol monobutyl ether (DGBE) wastewater by Serratia sp. BDG-2. *J Hazard Mater.* 309: 20–26.

Coogan, P.F., White, L.F. and Jerrett, M. (2012). Air pollution and incidence of hypertension and diabetes mellitus in black women living in Los Angeles, *Circulation* 125: 767–772.

Da Silva, M.L.B., Alvarez, P.J.J., (2010). Indole-based assay to assess the effect of ethanol on Pseudomonas putida F1 dioxygenase activity. *Biodegradation* 21: 425–430.

Dejonghe, W., Goris, J., Fantroussi, S.E., Höfte, M., De Vos, P., Verstraete, W., Top, E.M. (2000). Effect of dissemination of 2,4-dichlorophenoxyacetic acid (2,4-D) degradation plasmids on 2,4-d degradation and on bacterial community structure in two different soil horizons. *Appl Environ Microbiol* 66: 3297–3304.

Dejonghe, W., Boon, N., Seghers, D., Top, E.M., Verstraete, W. (2001). Bioaugmentation of soils by increasing microbial richness: missing links. *Environ Microb* 3:649–657.

Duba, A.G., Jackson, K.J., Jovanovich, M.C., Knapp, R.B., Taylor, R.T. (1996). TCE remediation using in situ, resting state bioaugmentation. *Environ Sci Technol* 30: 1982–1989.

Dwivedi, N., Majunder, C.B., Mondal, P., Dwivedi, S. (2011). Biological treatment of cyanide containing waste water. *Res J Chem Sci* 7: 15–21.

El-Fantroussi, S., Agathos, S. N. (2005) Is bioaugmentation a feasible strategy for pollutant removal and site remediation? *Curr Opin Microbiol* 8:268–27.

Fan, L., Ni, J., Wu, Y., Zhang, Y. (2009). Treatment of bromoamine acid wastewater using combined process of micro-electrolysis and biological aerobic filter. *J Hazard Mater* 162: 1204–1210.

Fatma, G, Hasan, C, Ata, A. (2000). Biodegradation of cyanide containing effluents by Scenedesmus obliquus. *J Hazard Mater* 162: 74–9.

Fuller, M.E. et al (2014) Laboratory evaluation of bioaugmentation for aerobic treatment of RDX in groundwater. *Biodegradation* 26:77–89

Fulthorpe, R.R., Wyndham, R.C. (1991). Transfer and expression of the catabolic plasmid pBRC60 in wild bacterial recipients in a freshwater ecosystem. *Appl Environ Microbiol* 57(5): 1546–1553.

Gardin, H, Pauss, A. (2001). K-carrageenan/gelatin gel beads for the co-immobilization of aerobic and anaerobic microbial communities degrading 2,4,6-trichlorophenol under air-limited conditions. *Appl Microbiol Biotechnol* 56: 517–523.

Gentry, T.J., Rensing, C., Pepper, I.L. (2004). New approaches for bioaugmentation as a remediation technology. *Crit Rev Environ Sci Technol* 34(5): 447–494.

Ghio, A. J., Stonehuerner, J., Dailey, L. A. and Carter, J. D. (1999). Metals associated with both the water-soluble and insoluble fractions of an ambient air pollution particle catalyze an oxidative stress. *Inhal Toxicol* 11: 37–49.

Godleads, O. A., Prekeyi, T. F., Samson, E. O., and Igelenyah, E. (2015). Bioremediation, biostimulation and bioaugmention: a review. *Int J Environ Bioremed Biodegrad* 1(3): 28–39.

Hamdi, H., Benzarti, S., Manusadzˇianas, L., Aoyama, I. and Jedidi, N. (2007). Bioaugmentation and biostimulation effects on PAH dissipation and soil eco toxicity under controlled conditions. *Soil Biol Biochem* 39(8): 1926–1935.

Henry, J., Balikdjian, D., Storme, G., Lustman-Maréachal, J., Degt, J.B. (2007). Photocatalytic and antibacterial activity of TiO2 and Au/TiO2 nanosystems. *Nanotechnology* 18: 14026–14029.

Hou, J., Liu, F., Wu, N., Ju, J., Yu, B. (2016). Efficient biodegradation of chlorophenols in aqueous phase by magnetically immobilized aniline-degrading Rhodococcus rhodochrous strain. *J Nanobiotechnol* 14: 5.

Hsiao, M.T., Chen, S.F., Shieh, D.B., Yeh, C.S. (2006). One-pot synthesis of hollow Au3Cu1 spherical-like and biomineral botallackite Cu2(OH)3Cl flowerlike architectures exhibiting antimicrobial activity. *J Phys Chem B* 110: 205–210.

Hsi-Ling, C., Yi-Tang, C., Yi-Fen, L., Ching-Hsing, L., 2013. Biodegradation of decabromodiphenyl ether (BDE-209) by bacterial mixed cultures in a soil/water system. *Int Biodeteriorat Biodegrad* 85: 671–682.

Kaewkannetra P, Imai T, Garcia GFJ, Chiu TY. 2009. Cyanide removal from cassava mill wastewater using Azotobacter vinelandii TISTR 1094 with mixed microorganisms in activated sludge treatment system. *J Hazard Mater* 172: 224–8.

Kargarfard, M., Poursafa, P., Rezanejad, S. and Mousavinasab, F. (2011). Effects of exercise in polluted air on the aerobic power, serum lactate level and cell blood count of active individuals. *Int J Prevent Med* 2(3): 145–150.

Kelishadi, R. and Poursafa, P. (2010). Air pollution and non-respiratory health hazards for children. *Arch Med Sci* 6(4):483–495.

Kelishadi, R., Mirghaffari, N., Poursafa, P. and Gidding S.S. (2009). Lifestyle and environmental factors associated with inflammation, oxidative stress and insulin resistance in children. *Atherosclerosis* 203(1):311–319.

Khondee, N., Tathong, S., Pinyakong, O., Powtongsook, S., Chatchupong, T., Ruangchainikom, C., Luepromchai, E. (2012). Airlift bioreactor containing chitosan-immobilized Sphingobium sp. P2 for treatment of lubricants in wastewater. *J Hazard Mater.* 213–214: 466–473.

Krumme, M.L., Smith, R.L., Egestorff, J., Thiem, S.M., Tiedje, J.M., Timmis, K.N., Dwyer, D.F. (1994). Behavior of pollutant-degrading microorganisms in aquifers: predictions for genetically engineered organisms. *Environ Sci Technol* 28: 1134–1138.

Lange, C.C., Wackett, L.P., Minton, K.W., Daly, M.J. (1998). Engineering a recombinant Deinococcus radiodurans for organopollutant degradation in radioactive mixed waste environments. *Nat Biotechnol* 16: 929–933.

Leahy, J.G. and Colwell, R.R. (1990). Microbial Degradation of Hydrocarbons in the Environment. *Microbiol Rev.* 54: 305–315.

Ledin, M. (2000). Accumulation of metals by microorganisms—processes and importance for soil systems. *Earth Sci Rev* 51:1–31.

Li, X.J., Lin, X., Li, P.J., Liu, W., Wang, L., Ma, F., Chukwuka, K.S. (2009). Biodegradation of the low concentration of polycyclic aromatic hydrocarbons in soil by microbial consortium during incubation. *J Hazard Mater* 172:601–605.

Liu, S., Wei, L., Hao, L., Fang, N., Chang, M.W., Xu, R., Yang, Y., Chen, Y. (2009). Sharper and faster nano darts kill more bacteria: a study of antibacterial activity of individually dispersed pristine single-walled carbon nanotube, *ACS Nano* 3:3891–3902.

Lyon, D.Y. and Alvarez, P.J. (2008). Fullerene water suspension (nC60) exerts antibacterial effects via ROS-independent protein oxidation. *Environ Sci Technol* 42: 8127–8132.

Markova, Z., Siskova, K.M., Filip, J., Cuda, J., Kolar, M., Safarova, K., Medrik, I., Zboril, R. (2013). Air stable magnetic bimetallic Fe-Ag nanoparticles for advanced antimicrobial treatment and phosphorus removal. *Environ Sci Technol* 47:5285–5293.

Mertens, B., Boon, N., Verstraete, W. (2005). Stereospecific effect of hexachlorocyclohexane on activity and structure of soil methanotrophic communities. *Environ Microbiol.* 7: 660–669.

Mohan, S.V., Rao, N.C., Prasad, K.K., Sarma, P.N., 2005. Bioaugmentation of an anaerobic sequencing batch biofilm reactor (AnSBBR) with immobilized sulphate reducing bacteria (SRB) for the treatment of sulphate bearing chemical wastewater. *Process Biochem.* 40: 2849–2857.

Möller, L. Schuetzle, D. Autrup, H. (1994). Future research needs associated with the assessment of potential human health risk from exposure to toxic ambient air pollutants. *Environ Health Perspect* 102(4): 193–210.

Moslemy, P., Guiot, S.R., Neufeld, R.J. (2002). Production of size-controlled gellan gum microbeads encapsulating gasoline-degrading bacteria. *Enzyme Microb Technol* 30: 10–18.

Mrozik, A. and Piotrowska-Seget, Z. (2010). Bioaugmentation as a strategy for cleaning up of soils contaminated with aromatic compounds. *Microb Res* 165: 363–375.

Okino, S., Iwasaki, K., Yagi, O., Tanaka, H. (2002). Removal of mercuric chloride by a genetically engineered mercury-volatilizing bacterium Pseudomonas putida PpY101/pSR134. *Bull Environ Contam Toxicol.* 68: 712–719.

Park, D., Lee, D.S., Kim, Y.M., Park, J.M., 2008. Bioaugmentation of cyanide-degrading microorganisms in a full-scale cokes wastewater treatment facility. *Bioresour Technol.* 99: 2092–2096.

Pepper, IL, Gentry, TJ, Newby, DT, Roane, TM, Josephson, K.L. (2002). The role of cell bioaugmentation and gene bioaugmentation in the remediation of co-contaminated soils. *Environ Health Perspect* 110(6): 943–946.

Pereira, R.A., Pereira, M.F.R., Alves, M.M., Pereira, L. (2014). Carbon based materials as novel redox mediators for dye wastewater biodegradation. *Appl Catalysis B* 144, 713–720.

Pooley, F. Mille, M. (1999). *Air Pollution and Health: Composition of Air Pollution Particles.* Academic Press, pp. 619–634.

Puyol, D., Monsalvo, V.M., Sanchis, S., Sanz, J.L., Mohedano, A.F., Rodriguez, J.J. (2015). Comparison of bioaugmented EGSB and GAC-FBB reactors and their combination with aerobic SBR for the abatement of chlorophenols. *Chem Eng J* 259:277–285

Qu, Y., Zhou, J., Wang, J., Fu, X., Xing, L. (2005). Microbial community dynamics in bioaugmented sequencing batch reactors for bromoamine acid removal. *FEMS Microbiol Lett.* 246: 143–149.

Qu, Y., Zhou, J., Wang, J., Song, Z., Xing, L., Fu, X. (2006). Bioaugmentation of bromoamine acid degradation with Sphingomonas xenophaga QYY and DNA fingerprint analysis of augmented systems. *Biodegradation* 17: 83–91.

Qu, Y.Y., Zhou, J.T., Wang, J., Xing, L.L., Jiang, N., Gou, M., Salah Uddin, M. (2009). Population dynamics in bioaugmented membrane bioreactor for treatment of bromoamine acid wastewater. *Bioresour Technol.* 100: 244–248.

Roane, T.M., Josephson, K.L., Pepper, I.L. (2001). Dual-bioaugmentation strategy to enhance remediation of cocontaminated soil. *Appl Environ Microbiol* 67:3208–3215.

Shukla, K.P., Singh, N.K., Sharma, S. (2010). Bioremediation: developments, current practices and perspectives. *Gen Eng Biotechnol J.* 3: 1–20.

Steffan, R.J., Sperry, K.L., Walsh, T., Vainberg, S, Condee, C.W. (1999). Field-scale evaluation of in situ bioaugmentation for remediation of chlorinated solvents in groundwater. *Environ Sci Technol* 33: 2771–2781.

Stoimenov, P.K., Klinger, R.L., Marchin, G.L., Klabunde, K.J. (2002). Metal oxide nanoparticles as bactericidal agents. *Langmuir* 18:6679–6686.

Strachel, R., Wyszkowska, J and Bacmaga, M (2020). Bioaugmentation of soil contaminated with zinc. *Water Air Soil Pollution* 231: 443. https://doi.org/10.1007/s11270-020-04814-5.

Sutherson, S. S. *Remediation Engineering: Design Concepts.* Lewis Publishers, 1997.

Thompson, I.P., van der Gast, C.J., Ciric, L., Singer, A.C. (2005). Bioaugmentation for bioremediation: the challenge of strain selection. *Environ Microbiol* 7:909–915.

Towell, M.G., Browne, L.A., Paton, G.I., Semple, K.T. (2011). Impact of carbon nanomaterials on the behaviour of 14C-phenanthrene and14C-benzo-[a]pyrene in soil. *Environ Pollut* 159 706–715.

Ulrich, AC, Edwards, EA (2003). Physiological and molecular characterization of anaerobic benzene-degrading mixed cultures. *Environ Microbiol* 5: 92–102.

Van der Gast, C.J., Whiteley, A.S., Thompson, I.P. (2004). Temporal dynamics and degradation activity of a bacterial inoculum for treating waste metal-working fluid. *Environ Microbiol* 6:254–263.

Vogel, T.M., Walter, M.V. (2002). Bioaugmentation. In *Manual of Environmental Microbiology.* 2nd edn. CJ Hurst, RL Crawford, GR Knudsen, MJ McIrney, LD Stezenback (eds.). ASM, Washington.

Wang, J. L., Quan, X. C., Han, L. P., Qian, Y., and Werner, H. (2002a). Kinetics of co-metabolism of quinoline and Glucose by *Burkholderia pickettii. Process Biochem* 37(8), 831–836.

Wang, J. L., Quan, X. C., Han, L. P., Qian, Y., and Werner, H. (2002b). Microbial degradation of quinoline by immobilized Cells of *Burkholderia pickettii. Water Res* 36: 288–296.

Wang, J., Liu, G.-F., Lu, H., Jin, R.-F., Zhou, J.-T., Lei, T.-M. (2012). Biodegradation of Acid Orange 7 and its auto-oxidative decolorization product in membrane-aerated biofilm reactor. *Int Biodeterior Biodegrad.* 67: 73–77.

Watanabe, K, Teramoto, M, Harayama, S. (2002) Stable augmentation of activated sludge with foreign catabolic genes harboured by an indigenous dominant bacterium. *Environ Microbiol* 4: 577–583.

Wilson, R., Spengler J. (1996) *Particles in Our Air.* Harvard University Press, USA.

Xuhui, M., James, W., A. Ciblaka, E. E. Cox, C. Riis, M. Terkelsend, D. B. Gent, A. N. Alshawabkeha. (2012). Electrokinetic-enhanced bioaugmentation for remediation of chlorinated solvents contaminated clay. *J Hazard Mater* 213–214.

Young-Mo, K., In-Hyun, N., Kumarasamy, M., Stefan, S., David, E. C., Yoon-Seok, C., 2007. Biodegradation of diphenyl ether and transformation of selected brominated congeners by Sphingomonas sp. PH-07. *Environ Biotechnol* 77: 187–194.

Zhang, Y., Peng, H., Huang, W., Zhou, Y., Yan, D. (2008). Facile preparation and characterization of highly antimicrobial colloid Ag or Au nanoparticles. *J Colloid Interf Sci* 325: 371–376.

Zhang, C., Li, M., Xu, X., Liu, N. 2015. Effects of carbon nanotubes on atrazine biodegradation by Arthrobacter sp. *J Hazard Mater.* 287: 1–6.

Zheng, D., Carr, C.S., Hughes, J.B. (2001) Influence of hydraulic retention time on extent of PCE dechlorination and preliminary characterization of the enrichment culture. *Biorem J* 5: 159–168.

Zhu, B., Xia, X., Xia, N., Zhang, S. Guo, X. (2014). Modification of Fatty acids in membranes of bacteria: implication for an adaptive mechanism to the toxicity of carbon nanotubes. *Environ Sci Technol* 48: 4086–4095.

8 Plasmid-Mediated Bioaugmentation

Elizabeth Mary John and M.S. Jisha
Mahatma Gandhi University

CONTENTS

8.1 INTRODUCTION

Bioaugmentation, the addition of indigenous or allochthonous, original wild-type or genetically modified organisms capable of degrading pollutant compounds, has always been proved as a useful approach to enhance detoxification of xenobiotics and other pollutants in the environment. These strains, highly efficient in laboratory conditions, failed to do so under natural conditions as they ought to compete with the native microbial community. Thus, this strategy is limited by the poor survival rate, the incapability of the introduced microorganisms to establish enough biomass, and their low activity to eliminate pollutants as well as the influence of various abiotic and biotic factors in the environment. The limitations can be brought under control by adding nutrient sources, changing operation parameters, or modifying the genetic construction of the microbes (Boon et al. 2000). The studies on horizontal gene transfer (HGT), the process of lateral movement of genetic material between organisms, had led to the idea of using mobile elements such as plasmids to develop microbes that can adapt, establish and degrade the target molecule faster in the medium than the wild organisms.

Plasmids are demonstrated to carry the gene or gene clusters of many xenobiotic-degrading capabilities. The association of naturally occurring plasmids and the HGT of catabolic genes in plasmids with the degradation have been extensively documented for synthetic compounds such as pesticides, hydrocarbons, wastewater, and sludge. In our recent curing and transformation study, a 4 kb plasmid responsible for chlorpyrifos degradation was found to be occurring in three bacteria from different genera. We speculate the natural HGT of plasmid between the isolates might be the reason for the acquisition of adaptive genes since they have been isolated from the same soil sample under selective pressure with the organophosphate pesticide; however this needs to be confirmed by sequencing of the plasmid. On further transformation of this isolated plasmid into non-degrading *Escherichia coli* JM109 cells, the transformed cells showed better degradation (>96%) of the

DOI: 10.1201/9781003187622-8

pesticide (John, Varghese, and Jisha 2020) compared to the primary isolates (John, Sreekumar, and Jisha 2016), indicating the practicability of employing engineered strains harboring catabolic plasmids as an alternative approach in bioaugmentation.

Such horizontal transfer of catabolic plasmids in laboratories had enabled the scientist to develop new strains capable of degrading xenobiotic compounds. Both the introduction of the original host microorganism harboring plasmids coding for enzymes for the degradation of target compounds or their transconjugants have been attempted as bioaugmentation agents or as a degradative plasmid donor. Additionally, HGT of self-transmissible plasmids could encourage the recombination of metabolic genes in the microbial population residing in the waste, sludge, or soil that has been bioaugmented, and thereby increase the catabolic potential of the natural microbial inhabitants for bioremediation. Hence, plasmid-mediated bioaugmentation into polluted sites of well-established and acclimatized indigenous bacteria might be an unconventional approach to beat the challenges of bioaugmentation and to persuade and augment the metabolic potential of the microbes to use as a bioremediation agent to degrade recalcitrant pollutants and/or to immobilize and treat heavy metals.

The main objective of this chapter is to underline the probabilities of plasmid gene transfer to establish and expand genetic information particularly emphasizing the bioaugmentation for recalcitrant by dissemination of catabolic genes through plasmids. The following sections summarize plasmid-mediated bioaugmentation studies that led to the persistent capacity for the biodegradation of xenobiotics and give an insight into the governing factors of the process as well as its advantages and disadvantages.

8.2 PLASMIDS AS BIOAUGMENTATION AGENT

Several methods have been investigated to implement bioaugmentation, varying from cell bioaugmentation to gene bioaugmentation. The former one, cell bioaugmentation, the addition of single or mixed cultures of microbes harboring the required catabolic genes (chromosomal or plasmid), non-capsulated or encapsulated to the environment, has great potential for the remediation of environments polluted with organic, inorganic, and/or synthetic compounds. Disappointingly, this strategy frequently turns ineffective, owing to the inability to compete and establish themselves with the indigenous microbial population and poor catabolic activity of inoculated microbial strains. Moreover, rapid decrease of viability, faster mortality, and limited dispersion of the inoculated strains necessitate the use of biostimulants, surfactants, foams, non-adhesive strains, and immobilization techniques (Franzetti et al. 2009; Wang and Mulligan 2004). Since the laboratory strains do not succeed in establishing in the added environment, scientists emerge with the idea of gene bioaugmentation (or plasmid-mediated bioaugmentation), the bioremediation strategy that enhances biodegradative potential *via* dissemination of catabolic genes from inoculated microorganisms to indigenous microorganisms. Genes encoding the essential machineries for the degradation of compounds are often located on mobile genetic elements such as integrons, plasmids, and transposons (Garbisu et al. 2017), hence can be employed to spread the contaminant degradation genes from a degrading/non-degrading donor to indigenous microbial population as illustrated in Figure 8.1. The plasmid-mediated bioaugmentation can be achieved by either of the following methods: (i) HGT of self-transmissible plasmids and (ii) cloning the gene into a plasmid having a broad host range.

The HGT allows the rearrangement and exchange of genetic information encoded in plasmids, both intergenus and interspecies, *via* transformation, transduction, or conjugation. HGT, in particular, plasmid conjugal transfer could be useful not only for bioremediation (hence degradative plasmids can be described as conjugative) but also to transfer the communities with resistance to metal toxicants and antibiotics (Ikuma and Gunsch 2012). The process has enabled the spread of catabolic genes even between distantly related taxonomic groups, allowing the recipient to adapt and commence degradation of the target compound (Figure 8.2). The natural HGT has played a vital role in the evolution of bacteria, benefiting them with important traits. The process of plasmid-mediated bioaugmentation involves the introduction of original host bacteria harboring

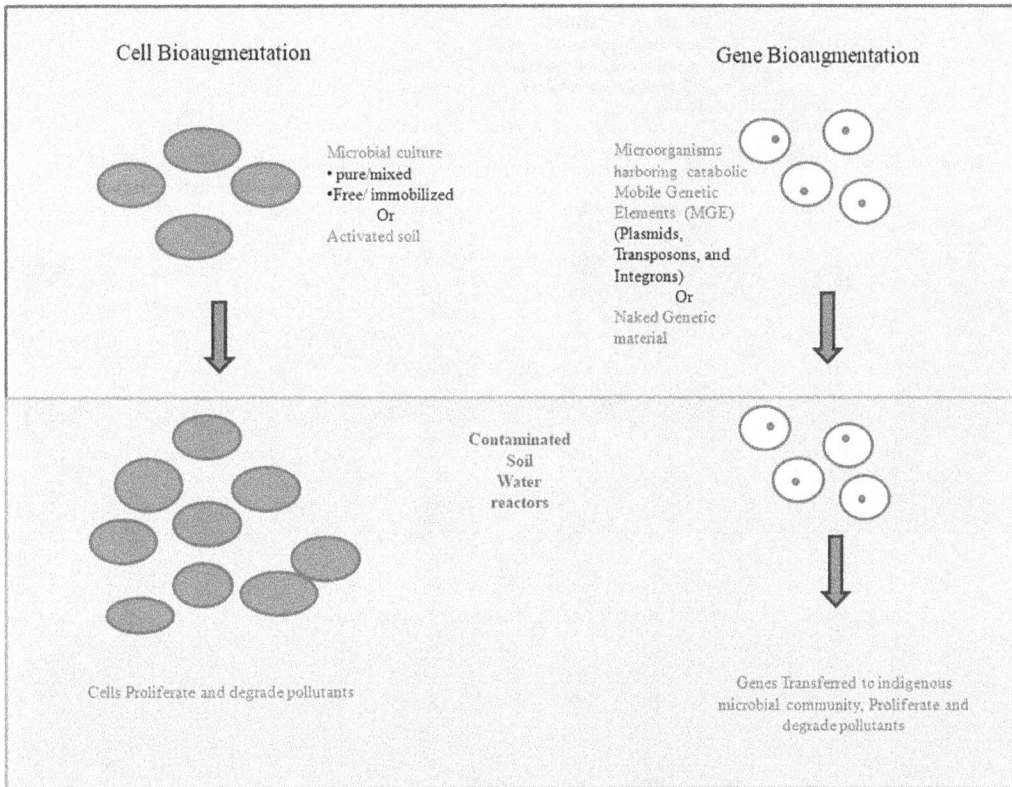

FIGURE 8.1   Comparison between cell bioaugmentation and gene bioaugmentation technique.

self-transmissible catabolic plasmids as a donor into the polluted sites, where the plasmids are trans-ferred to the existing bacterial populations by HGT. As a result, the recipient may start expressing a new beneficial catabolic trait or become an improved version of the wild type in terms of adapta-tion and catabolic activity, by the acquisition of plasmids from the donor. This is always advanta-geous that it causes *in situ* genetic modification by increasing and improving the genetic content of microorganisms that are already been adapted to the pollution by acquiring a plasmid carrying the desired gene, and hence the survival and proliferation of the augmented host microorganism is not a prerequisite to commencing biodegradation utilizing newly introduced biodegradative genes in the plasmid (DiGiovanni et al. 1996). This was evidenced when the bioaugmentation efficiency of two different bacterial donors, *Ralstonia eutropha* JMP134 and *Escherichia coli* D11, delivering the self-transmissible plasmid pJP4, carrying 2,4-dichlorophenoxyacetic acid (2,4-D) degradative genes to local soil bacteria was compared. Though both bacteria were good enough to deliver pJP4 plasmid to the soil, only the original host *Ralstonia eutropha* JMP134 was able to mineralize 2,4-D whereas the *Escherichia coli* D11 failed to do so as it lacked the necessary chromosomal genes to complete the detoxification of 2,4-D. Initial mineralization by the inoculants within 28 days was observed in the soil that received *Ralstonia eutropha* JMP134 bioaugmentation compared to the non-bioaugmented or soil inoculated with *Escherichia coli* D11 which took 49 days. However, on a second amendment, the more rapid degradation recorded in the soil that received the *Escherichia coli* D11 inoculums than the soil that received the original host *Ralstonia eutropha* JMP134 inocu-lums and non-bioaugmented soil was mediated by the numerous transconjugants, indicating a suc-cessful HGT from the *Escherichia coli* D11 to soil microcosm (Newby, Gentry, and Pepper 2000). Similar results were recorded by the studies of (Bathe 2004; Bathe, Schwarzenbeck, and Hausner 2005; Stephan Bathe, Schwarzenbeck, and Hausner 2009), using a donor strain *Pseudomonas*

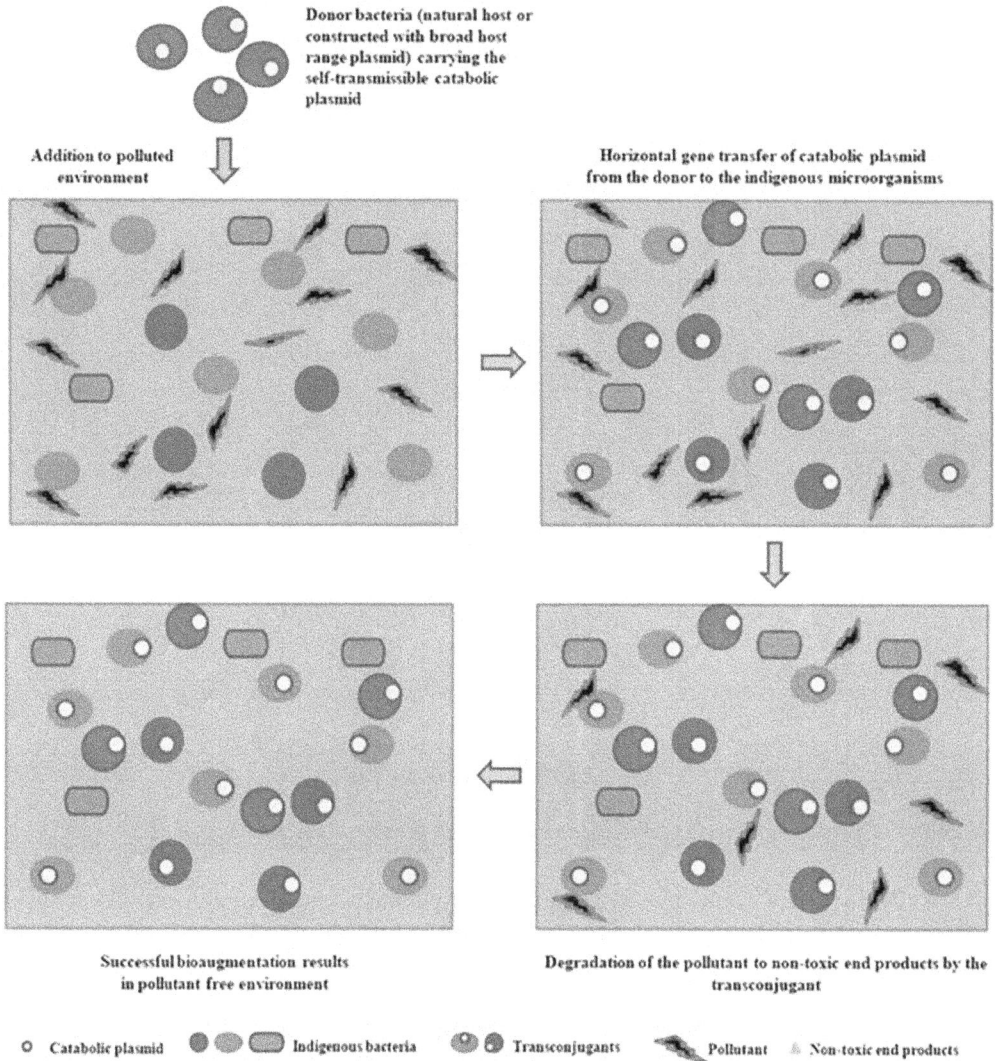

FIGURE 8.2 Plasmid-mediated bioaugmentation—a potential strategy for environmental bioremediation.

putida, that carries the plasmid pNB2 of 3-chloroaniline (3-CA), but lacked expression of the catabolic genes of plasmids or a few chromosomal genes required to complete the degradation pathway. When bioaugmented to activated sludge reactor, better adaptation and successful degradation of 3-CA was contributed by the transconjugant populations, specifically the strains of *Comamonas testosterone,* that acquired the degradation ability *via* horizontal transfer of the plasmid pNB2 within the bioreactor. However, the functioning and reproducibility of the HGT mechanism rely on several factors such as the host's fitness and environmental factors as discussed in the coming sessions: demanding in-depth research and understanding of several parameters.

Alternatively, precise/selective remediation genes can be introduced into the chromosome of the microorganism of interest or plasmids in the laboratory and can then be used as bioaugmentation agents. This transfer can be accomplished with the use of original transmissible plasmids by mating between donor and target recipient microorganisms or by cloning the gene into a suitable broad-host-range plasmid, in the absence of transmissible plasmid, followed by the conjugation or transformation to the donor microorganism. In both cases, the recipient aids the dissemination of catabolic plasmids to indigenous bacteria which had already been adapted within the environment to be bioremediated.

8.3 EXAMPLES OF PLASMID-MEDIATED BIOAUGMENTATION

8.3.1 PESTICIDES

Genes for the biodegradation of synthetic pesticides such as pEMT1 (Dejonghe et al. 2000), pMCP424 of monocrotophos (Bhadbhade et al. 2002), pV2 of dichlorvos (Tang et al. 2009), pVAG33 of γ-HCH (Zhang et al. 2010), pJP4 of 2,4-D (Aspray, Hansen, and Burns 2005; Inoue et al. 2012; Tsutsui et al. 2013) and pDOC of chlorpyrifos (Zhang et al. 2012) have been extensively investigated and successfully employed for the bioremediation of pesticides through transformation, transduction, and conjugation (Cycoń, Mrozik, and Piotrowska-Seget 2017). In general, the transformants exhibit a higher degradation rate than the parent isolates suggesting an effective approach for improving the pesticide-degrading capability of microorganisms. Along with the enhanced survival rate and faster acquisition of remediation activity of the bacteria, persistent capacity was observed for the biodegradation of many pesticides (Table 8.1) in soil. For example, the plasmid pDOC isolated from a chlorpyrifos degrading bacterium *Bacillus laterosporus* DSP, when transferred into the soil, the members of the genera *Staphylococcus* and *Pseudomonas* in soil acquired the chlorpyrifos degradation ability in 5 days post transfer of pDOC and presented an enhanced degradation (Zhang et al. 2012). Additionally, the engineered strain harboring plasmids for degradation can be applied and monitored in soil without exerting large impacts on the indigenous microbial community (Inoue et al. 2012) or in the form of a biocatalyst in a bioreactor for the effective degradation of pesticides (Tsutsui et al. 2013). However, gene bioaugmentation has the potential to alter the indigenous microbial gene pool in the soil as evidenced by (Newby, Gentry, and Pepper 2000). They reported the potential for indigenous microorganisms that did not have the genes, to degrade 2,4-D when furnished with the self-transmissible plasmid pJP4 *via* gene bioaugmentation from *Escherichia coli* D11. The rapid degradation of 2,4-D was due to the transconjugant microcosms that received plasmid from the *Escherichia coli* D11, which would not be otherwise possible as the donor lacked the necessary chromosomal and plasmid genes to allow for complete mineralization of 2,4-D.

8.3.2 EXPLOSIVE COMPOUNDS

Bioremediation is always challenging in the case of explosive compounds as the genes involved in the remediation are highly conserved, and most of the naturally occurring microorganisms lack this specific gene function. Genetic engineering and HGT are proved to be benefiting bioaugmentation efforts to remediate the explosives such as RDX. Previously, (Jung et al. 2011) have demonstrated successful conjugation of 182 kb mobilizable plasmid pGKT2 carrying the xplAB locus responsible for RDX degradation to related bacteria, serving as a mechanism of allocation and persistence of degradation genes in the environment. Recently, transconjugant *Rhodococcus jostii* RHA1 with the plasmid pGKT2 carrying the RDX xplAB genes gained by lateral transfer, have been shown a good fitness cost and greater degradation of the explosive than the donor *Gordonia* sp. KTR9 strain in flow-through soil columns (Jung et al. 2019).

8.3.3 WASTEWATER AND ACTIVATED SLUDGE

Bioaugmentation with specialized microorganisms could be a powerful tool to improve the quality of wastewater after the transfer of catabolic plasmids to bacteria in activated sludge. Several studies have demonstrated the natural exchange of catabolic plasmids between the bacterial species of the sludge *via* HGT, thereby modifying microbial population with these functional genes and improving biodegradation of compounds such as 3-ClOBz (McClure, Fry, and Weightman 1991), 3-chlorobenzoate, and 4-methyl benzoate (Nusslein et al. 1992), 2,4-D (Stephan Bathe and Hausner 2010). However, a relatively long initial adaptation period and extended cycle durations may be expected since the degradation of the target compound starts only after the easily degradable carbon sources are exhausted, thereby forcing the transconjugants populations to proceed in

TABLE 8.1
Examples of Plasmid-Mediated Bioaugmentation

Xenobiotics	Plasmid	Donor/Host/Degradative Cell	Recipient/Transconjugants	Medium	References
3-Chlorobenzoate (3 ClOBz)	pD10 (constructed by cloning the SstI-C fragment of the 2,4-dichlorophenoxyacetic-acid-degradative plasmid pJP4 into the broad-host-range IncQ vector pKT231)	*Pseudomonas putida* UWC1	Autochthonous sludge bacterium	Plate-filter mating	McClure et al. (1989), McClure et al. (1991)
3-Chlorobenzoate and 4-Methyl benzoate	pWWO-EB62	-	*Pseudomonas putida* UWC1	Reactors	Nusslein et al. (1992)
Biphenyl and chlorobiphenyl	RP4::Tn4371	*Enterobacter agglomerans*	*Pseudomonas chlorophis, Pseudomonas corrugate* and *Comamonas* sp.	Sandy soil	De Rore et al. (1994)
2,4-Dichloro-phenoxyacetic acid (2,4-D)	pJP4	*Alcaligenes eutrophus* JMP134	-	Soil	Digiovanni et al. (1996)
2,4-Dichloro-phenoxyacetic acid (2,4-D)	pJP4	*Alcaligenes eutrophus* JMP134	*Variovorax paradoxus*	Model system	Digiovanni et al. (1996)
Phenolics	pheBA operon in a conjugative plasmid	*Pseudomonas putida* PaW85	-	River water	Peters et al. (1997)
3-Phenoxybenzoic acid (3-POB)	pPOB	*Pseudomonas pseudoalcaligenes* POB310, *Pseudomonas* sp. B13ST1	-	Sterilized and non-sterilized soil	Halden et al. (1999)
	pD30.9	*Pseudomonas* sp. B13-D5			
2,4-Dichloro-phenoxyacetic acid (2,4-D)	pJP4	*Ralstonia eutropha* JMP134 *Escherichia coli* D11	-	Soil	Newby, Gentry, and Pepper (2000)
2,4-Dichloro-phenoxyacetic acid (2,4-D)	pEMT1 and pJP4	*Pseudomonas putida* UWC3	-	Soil	Dejonghe et al. (2000)

(Continued)

TABLE 8.1 (*Continued*)
Examples of Plasmid-Mediated Bioaugmentation

Xenobiotics	Plasmid	Donor/Host/Degradative Cell	Recipient/Transconjugants	Medium	References
4-Chlorobenzoate and 4-chlorobiphenyl (4-CB)	Broad-host-range plasmid containing the 4-CBA degradation operon (fcb) isolated from *Arthrobacter globiformis* strain KZT1	*Rhodococcus* sp. strain RHA1		Medium and soil	Rodrigues et al. (2001)
2,4-Dichloro-phenoxyacetic acid (2,4-D)	pJP4	*Ralstonia eutropha* JMP134 *Escherichia coli* D11		Brazito soil and Madera soil	Pepper et al. (2002)
3-Chlorobenzoate (3-CB)	pBRC60	*Comamonas testosteroni* BR60			Pepper et al. (2002)
3-Chloroaniline (3-CA)	pNB2	*Pseudomonas putida*	*Comamonas testosteroni* strain 12	Mineral medium	Bathe (2004)
2,4-Dichloro-phenoxyacetic acid (2,4-D)	pJP4::gfp	*Pseudomonas putida* SM1443	*Burkholderia*	Soil suspension	Aspray, Hansen, and Burns (2005)
3-Chloroaniline (3-CA)	pNB2	*Pseudomonas putida* SM1443::gfp2x *Comamonas testosteroni* pNB2-transconjugant	-	Sequencing batch moving bed reactor or semi-continuous activated sludge reactors	Bathe et al. (2005), Stephan Bathe et al. (2009)
Naphthalene	pNF142	*Pseudomonas putida* KT2442 (pNF142:: TnMod-OTc)	-	Model soil system	Filonov et al. (2005)
2,4-Dinitrotoluene	pJS1 megaplasmid, contains the *dnt* genes for 2,4-dinitrotoluene degradation from *Burkholderia* sp. strain DNT	*Pseudomonas fluorescens* MP	-	-	Monti et al. (2005)
γ-Hexachlorocyclo-hexane (γ-HCH)	pLB1	*Sphingobium japonicum* UT26	Other α-Proteobacterial strains		Miyazaki et al. (2006)
Oil	pWW0	*Pseudomonas putida* PaW85	*Pseudomonas oryzihabitans* 29	Polluted soil	Jussila et al. (2007)

(Continued)

TABLE 8.1 (*Continued*)
Examples of Plasmid-Mediated Bioaugmentation

Xenobiotics	Plasmid	Donor/Host/Degradative Cell	Recipient/Transconjugants	Medium	References
Benzyl alcohol	pWWO (TOL plasmid)	*Pseudomonas putida* strain KT2442	-	Synthetic wastewater containing benzyl alcohol in sequencing batch biofilm reactors	Venkata Mohan et al. (2009)
Dichlorvos	pV2	-	*Trichoderma atroviride*	-	Tang et al. (2009)
γ-Hexachlorocyclo-hexane (γ-HCH)	pVAG33	Constructed	*Pseudomonas nitroreducens* J5-1	-	Zhang et al. (2010)
RDX	pGKT2	*Gordonia* sp. KTR9	*Gordonia polyisoprenivorans*, *Rhodococcus jostii* RHA1 and *Nocardia* sp. TW2	-	Jung et al. (2011)
Chlorpyrifos	pDOC	*Bacillus laterosporus* DSP	*Pseudomonas* and *Staphylococcus*	Soil	Zhang et al. (2012)
Toluene	pWW0-derivative TOL	*Pseudomonas putida* BBC443	-	Soil column	Ikuma and Gunsch (2012)
2,4-Dichloro-phenoxyacetic acid (2,4-D)	pJP4	*Pseudomonas putida* or *Escherichia coli*	-	Soil	Inoue et al. (2012)
2,4-Dichloro-phenoxyacetic acid (2,4-D)	pJP4	*Cupriavidus necator* JMP134 and *Escherichia coli* HB101	*Achromobacter*, *Burkholderia*, *Cupriavidus*, and *Pandoraea*	Wastewater	Tsutsui et al. (2013)
Dichlorodiphenyltrichloroethane (DDT)	pDOD	*Sphingobacterium* sp. D-6, *Escherichia coli* TG I, *Klebsiella* TZ	-	Soil	Gao et al. (2015)
Toluene and xylene	-	*Alcaligenes feacalis* and *Alcaligenes* sp. DN25	*Escherichia coli* JM109	Phyllosphere	Undugoda et al. (2018)
RDX	pGKT2	*Gordonia* sp. KTR9	*Rhodococcus jostii* RHA1	Flow-through soil columns	Jung et al. (2019)
Chlorpyrifos	Not characterized	-	*Escherichia coli* JM109	Laboratory condition	John, Varghese, and Jisha (2020)

reactors (Bathe et al. 2004). Plasmid-mediated bioaugmentation was demonstrated by introducing the bacteria JMP134 and *Escherichia coli* HB101 harboring plasmid pJP4 for the enhanced removal of 2,4-D in sequencing batch reactors. Though the former bacteria could mineralize 100% of 2,4-D at 200 mg/L concentration immediately post its introduction they failed to sustain the activity on 7th day due to the decline in the viability of *Cupriavidus necator* JMP134. The latter cannot assimilate 2,4-D because of the absence of the chromosomal genes coding for the degradation of the intermediates, but acted as a donor. Complete removal was observed in both reactors after 16 days with the manifestation of 2,4-D-degrading transconjugants belonging to the genera *Achromobacter, Cupriavidus, Pandoraea,* and *Burkholderia.* On further increase of 2,4-D concentration to 500 mg/L, enhanced (>95%) stable removal was recorded in both reactors despite a progressive depression in the reactor that received *Cupriavidus necator.* The overall result recommended plasmid-mediated bioaugmentation to improve the degradation activities in the activated sludge unrelatedly to the survival and 2,4-D degradation capacity of the introduced strain (Tsutsui et al. 2013). Alternatively, bioaugmentation with plate mated transconjugants mixture can also be employed for activated sludge systems treating wastewaters in reactors. To cite an example, successful application of plasmid pNB2 through *Comamonas testosteroni* transconjugants for bioaugmentation of wastewater to remediate 3-CA in laboratory-scale sequencing batch moving bed reactors (SBMBRs) have been investigated (Bathe 2004; Bathe, Schwarzenbeck, and Hausner 2005).

8.3.4 Metal Toxicants

The gene bioaugmentation process by utilizing plasmids may also have applications to transport genetic information for the remediation of toxic metal where the inhabitant species that are already adapted to the metal-contaminated sites uptake such engineered broad-host-range plasmids and carry out the transformation of toxic metals. Plasmid-dependent intracellular mechanism of detoxification was elucidated in a *Pseudomonas* sp. strain H1 and *Bacillus* sp. strain H9 for the metal cadmium. This phenomenon has been employed to enhance the biodegradation of 2,4-D in the soil co-contaminated with cadmium, which cannot otherwise be happening with *Ralstonia eutropha* JMP134 alone. However, dual bioaugmentation of soil microcosm with the cadmium-resistant isolates *Pseudomonas* and *Bacillus* appeared to be a feasible approach for the remediation of co-contaminated soils since the metal was thought to exert an inhibitory effect on the ability of *Ralstonia eutropha* JMP134 to degrade 2,4-D (Roane, Josephson, and Pepper 2001). The most studied plasmid is the mega-sized, natural, self-transferable plasmid pSinA (109 kb), first described in *Sinorhizobium* sp. M14 (Alphaproteobacteria), that involved in the complete oxidation of the metalloid arsenic and resistance to heavy metals (cadmium, cobalt, zinc, and mercury). This broad-host-range plasmid could successfully transfer arsenic metabolism genes into the members of Alpha- and Gammaproteobacteria in arsenic-contaminated soils *via* conjugation, and stably express the enzyme arsenite oxidase in natural transconjugant cells even after closely 60 generations of growth under non-selective conditions (Drewniak et al. 2013). Two recombinant broad-host-range plasmid vectors named pAIO1 and pARS1 have been constructed by cloning arsenite oxidation and arsenic resistance modules of the plasmid pSinA into vectors pBBR1MCS-2 and pCM62 respectively and were introduced into phylogenetically and physiologically distant bacteria. These genetically modified strains displayed not only an improved resistance and oxidation abilities, but also were able to grow on mine waters contaminated with arsenic without any nutrient supplementation (Drewniak et al. 2015). Studies also revealed that these metal tolerating capabilities of the M14 strain are also conferred by eight modules in a narrow-host-range repABC-type replicon pSinB of approximately 300 kb size that encoding transporters, efflux pumps, and enzymes copper oxidases and permeases possibly involved in resistance to metals arsenic, cadmium, cobalt, copper, iron, mercury, nickel, silver, and zinc (Romaniuk et al. 2017). A recent study by (diCenzo et al. 2018), validates and reports a novel pilot-scale two-stage installation, coupling the microbiological module of the megaplasmids pSinA and pSinB of *Ensifer* (*Sinorhizobium*) sp. M14, for the

bioremediation of water contaminated with arsenic. Although these studies are fruitful in laboratory microcosms, the application in actual environments has yet to be attempted.

8.3.5 OTHER COMPOUNDS

Direct measurement of HGT of catabolic plasmids and marked biodegradation of respective contaminant has been carried out under both well-defined laboratory conditions and in *in situ* situations for phenolics (Peters et al. 1997), 3-phenoxybenzoic acid (3-POB) (Halden et al. 1999),
3-chlorobenzoate (3-CB) (Pepper et al., 2002), naphthalene, 2-metylnaphthalene and salicylate
(Filonov et al. 2005), oil-contaminated soil (Jussila et al. 2007), toluene (Ikuma and Gunsch 2012),
nitrobenzene and triazine (Segura, Molina, and Ramos 2014), and xylene and its related monoaromatic hydrocarbons (Undugoda et al. 2018). Table 8.1 provides a summary of the plasmid-
mediated bioaugmentation studies.

8.4 FACTORS INFLUENCING PLASMID-MEDIATED BIOAUGMENTATION

Successful implementation of genetic bioaugmentation is not only relying on relocating the plasmid
to the indigenous inhabitants but also on the phenotypic outcome as an overall amplification of the
catabolic activity of the *in situ* microbial community following these augmentation events. It has
been observed that an array of abiotic and biotic factors influence the transfer and catabolic effectiveness of the bioaugmentation process. The most significant abiotic selective barriers that regulate
the efficiency of bioaugmentation are temperature, moisture (Miller, Stratton, and Murray 2004;
Aminov 2011; Gao et al. 2015), pH, aeration, organic matter, nutrient type and concentration, soil
type, and the method of dispersion (Ikuma and Gunsch 2013; Król et al. 2011; Zhang et al. 2012).
Temperature is a crucial environmental factor affecting plasmid transfer that the process is low-
est under cold conditions whereas a progressive surge in temperature may result in raised transfer
frequencies (Bale, Fry, and Day 1987; Rochelle, Fry, and Day 1989; Johnsen and Kroer 2007).
Along with that, soil depth, the presence of plant roots, and plant types also tend to have important
effects on the transfer of the plasmid in soil and on the activities of transconjugants. The transfer
frequency is significantly affected by depth of the soil due to gradual decrease of microbial abundance and microbial metabolic activity with increasing soil depth; even though the upper layers of
soil is believed to have more microbial activity than do the lower layers, is not optimum for plasmid
transfer because of the impact of external environmental disturbances (Maila et al. 2005; Król et al.
2011). For example, the growth of transconjugant bacteria was recorded more rapidly in soil rich in
clay and organic matter than in sandy soil and the transfer frequency is high in loam and autoclaved
soil but absent in non-autoclaved soil (Halden et al. 1999). The transfer efficiency of the plasmid
pDOC, and the competence and the number of chlorpyrifos degraders in the soil were shown to be
greatly influenced by the soil type, temperature, and moisture level, where the best performance
for the transfer of pDOC was observed at temperature 30°C and water-holding capacity at 60%
and the degradation rate varied with the soil type (Zhang et al. 2012). To cite another example, the
transfer of a TOL-like plasmid harboring genes of catechol 2, 3-dioxygenase (C23O) between two
soil bacteria has been investigated under diverse conditions. Temperature played an important role
in that the transfer frequency increased with temperature (10–35°C), with the generation of 6×10^{-4}
transconjugants/donor cells at 35°C. The transfer frequency was also influenced by soil depth which
was significantly higher at 5–10 cm than other depths, where the activity coincides with the oxygen
gradient in soils. The highest transfer frequency up to about 1.3×10^{-3} transconjugants/donor cell
was recorded in the presence of tomato seedlings followed by corn and wheat seedlings compared
to soil without plants, which led to the conclusion that the presence and type of soil plants contribute to the plasmid-mediated bioaugmentation. However, temperature, soil and plant types had
a negligible influence on the activities of transconjugants, whereas the depth of the soil played a
significant role. Additionally, more efficient expression of C23O was observed by topsoil (0–5 cm)

transconjugants under normal incubation conditions, than when incubated with additional nutrient sources in soil (Wang et al. 2014).

The nutrient composition and concentration and bioavailability of the contaminant in the medium can affect the transfer frequency as well as the activity. Nitrogen is a vital parameter that in its absence HGT event was found increasing by three orders of magnitude in case of transfer of plasmid pGKT2 from the donor *Gordonia* sp. KTR9 to *Rhodococcus jostii* RHA1, *Nocardia* sp. TW2 and *Gordonia polyisoprenivorans*. Furthermore, such repressive effects of external nitrogen sources in transconjugants were reported to be strain-specific while studying RDX degradation (Jung et al. 2011). On the contrary, the nitrogen sources such as sodium nitrate and ammonium sulfate had been demonstrated to exert small and variable effects on the degradation rates of toluene by the trans-conjugant strains (Jung et al. 2019). The addition of easily degradable alternative carbon sources such as glucose and Luria-Bertani (LB) broth at a specific concentration (0.5–5 g/L and 10–100% respectively) into the experimental system is shown to improve the transfer frequency, the expression of the acquired genes in some transconjugants, and thereby specific degradation rates compared with no addition as described in the following paragraphs (Ikuma and Gunsch 2010); however, excess concentrations have an inhibitory effect on percentage and functionality of transconjugants (Ikuma and Gunsch 2013; Ikuma, Holzem, and Gunsch 2012). Moreover, the relation of plasmid transfer frequency to pollution is depicted by the likely loss of the plasmid when the donors were grown without exposure to the targeted pollutants for a long time. But the effect of the presence and concentration of the contaminant on the transfer and degradation is still conflicting that the selective pressure to force for the conjugation of catabolic plasmids differs considerably depending on the contaminant and the plasmid (De Rore et al., 1994; Neilson et al., 1994; Wuertz et al., 2004; Ikuma and Gunsch, 2012).

The primary biotic factor that affects HGT and plasmid-mediated bioremediation is the type of donor bacteria and its plasmid and the host bacteria. The choice of the donor strain is a governing factor in generating a transconjugant population and in turn in the success of bioaugmentation (Gao et al. 2015). This effectiveness has been elucidated in a comparative bioaugmentation study of plasmid transfer and subsequent soil degradation of 2,4-D using two different donors—the natural host *Ralstonia eutropha* JMP134 and constructed *Escherichia coli* D11—of the plasmid pJP4. Bioaugmentation of soil with the laboratory-generated D11 has resulted in a more diversified and better degradative transconjugant population than of the soil augmented with *Ralstonia eutropha* pointing towards the significant role of the type of the donor in the bioaugmentation effort (Newby, Gentry, and Pepper 2000). The type of the host cells is as equally decisive as the donor cells since plasmid bioaugmentation objects to transfer degradative genes into native microorganisms having a greater fitness for survival in the polluted environments. Some destructive effects can be exerted on the recipient bacterial chromosome since the plasmids harbor their own DNA restriction enzymes. In contrast, the DNA restriction and anti-restriction systems present in recipient bacteria determine the fate of the newly added plasmids inside recipient cells. In this struggle to thrive inside the cells, plasmids may integrate into the chromosome of the recipient bacteria, thus, in turn, affect HGT success. Additionally, the binding of a donor to recipient bacteria is affected by the pilus specificity and surface exclusion phenomenon (Thomas and Nielsen 2005).

Despite the new adaptive benefits conferred by plasmids, this acquisition can unconstructively affect cellular functions in recipient cells and entails a metabolic burden in the host, termed "fitness (metabolic) cost", when it first appears in a new bacterial host. According to (Garbisu et al. 2017), the energy expenditure due to consumption of molecular building blocks, chromosomal disturbances caused by the horizontally acquired genes, sequestration of cellular processes, and related molecular machineries, and the large size of the plasmid are the key reasons behind the metabolic cost. In the absence of selection for any plasmid-encoded traits, this manifests reducing the reproductive rate and competitiveness of the plasmid-carrying strain and thereby hindering the endurance of plasmids in bacterial communities. Nonetheless, the host bacteria overpower these detrimental effects through necessary compensatory mutations in the plasmid and/or the host chromosome over time; hence, playing a decisive role in bacterial evolution by engendering selection against plasmid-carrying strains (San Millan and MacLean 2017). The same plasmid has different fitness costs in different hosts as depicted by the noticeable decrease in the retention of the plasmid pGKT2 carrying the xplAB genes for RDX degradation in the transconjugant *Nocardia* sp. TW2 to 50 generations against the high stability of the same in transconjugants *Gordonia polyisoprenivorans* and *Rhodococcus jostii* RHA1 for up to 100 generations (Jung et al. 2011). The authors speculated that the reason for this two-fold stability difference occurred without any selective pressure in *Nocardia* sp. could be due to the large metabolic expense experienced by the incorporation of pGKT2 plasmid. Later, in another study, *Rhodococcus jostii* RHA1 transconjugants were demonstrated to have a fitness cost while the *Gordonia polyisoprenivorans* and TW2 showed minimal cost (Jung et al. 2019). Thus, it is said impractical to use the fitness cost of a plasmid measured in one bacterial host to predict the impact of the same plasmid in another bacterial host (San Millan and MacLean 2017); and a prior characterization of the possible recipient cells would potentially advantageous.

Studies suggest that both the guanine and cytosine (G+C) content of the recipient genome and its phylogenetic relationships to the donor strains are strongly interrelated that it is difficult to separate the controlling effects of them on the functional expression of the transferred plasmid. Their effect was explained when the function of TOL*gfp*mut3b plasmid in eight transconjugant bacterial strains and the donor strain belonging to *Enterobacteriaceae* and *Pseudomonadaceae*, respectively, were investigated (Ikuma and Gunsch 2013). Despite all the studied transconjugants were able to uptake the TOL*gfp*mut3b plasmid and had the genomic G+C content between 50% and 60%, the very limited TOL plasmid functions displayed by the six *Enterobacteriaceae* transconjugant strains (*Escherichia* sp. TOL, *Escherichia coli* RP-1, *Escherichia coli* TOL, *Pantoea agglomerans* TOL, *Enterobacter cloacae* TOL, and *Serratia marcescens* TOL), when grown using toluene as the sole source of carbon, was attributed to the difference in G+C content (49.9%–58.8% compared to an average TOL plasmid G+C content of 59%) and the relatively high phylogenetic distances from the donor strain *Pseudomonas putida* BBC443. On the other hand, the other two transconjugants *Pseudomonas fluorescens* and *Pseudomonas putida* being *Pseudomonadaceae* strains were phylogenetically related to the donor and showed better functioning of the plasmid. However, amendment with easily biodegradable carbon source tackled the limited rates of degradation and improved the success rates of gene bioaugmentation by intensifying the functional metabolic process even in transconjugant communities having a 10% difference in G+C content compared to the donor strain. Though, work through different mechanisms, the addition of both glucose and LB broth had exerted larger positive effects on the expression of TOL plasmid in strains with lower G+C contents, where glucose exhibited a stronger connection with genomic G+C content compared to LB broth. Both glucose and LB broth enhanced overall toluene biodegradation by transconjugant strains irrespective of the correlation with the phylogenetic distance between the transconjugants and the donor strains. These results opposed the previous findings that in nature HGT occurs only when the difference in G+C content between strains falls <5% (Popa et al. 2011). Although the authors believe that the stability of the TOL plasmid is unrelated to the genetic construction of the transconjugant, further research is recommended.

Nevertheless, a strong interdependence is elucidated, the influence of the initial recipient to donor cell density on plasmid transfer is paradoxical (Pei and Gunsch 2009; Pinedo and Smets 2005). Plasmid spread is restricted in bacterial communities that are initially not well-mixed and such a spatial separation between donor and recipient strains in the medium because of factors such as physical isolation, cell–cell signals, fertility inhibition, oxygen availability, cell density, cell-to-cell contact mechanism, and juxtaposition can be addressed with the help of biofilm (Stalder and Top 2016; Król et al. 2011) and by employing a solid surface for conjugation. The former approach can be exploited for enhancing catabolic processes in an environment by introducing mobile genetic elements carrying desired genes into an existing microbial population. The biofilm structure determines how far a donor can penetrate and facilitates the transfer of plasmids *via* bacterial conjugation through creating open channels and pores allowing the efficient transport of donor cells and further collision to recipient cells (Wuertz, Okabe, and Hausner 2004). Such a biofilm-mediated better adherence of donor cells, transfer of TOL plasmid through conjugation to the native biofilm population, survival and increased persistence of both the donor strain, the transconjugant as well as enhanced biodegradation was demonstrated for 3-CA while treating synthetic wastewater in a laboratory-scale sequencing batch moving bed reactor (Bathe et al., 2005) and for benzyl alcohol (Venkata Mohan et al. 2009). However, a too high mixing rate in the reactors may impair biofilm architecture and bacterial interactions within it, thereby reducing the gene transfer rates in reactors.

The rhizosphere or phyllosphere grant the microorganism a complex environment for the detoxification of harmful molecules. Additionally, phytosphere-associated plasmids confer a specific fitness advantage to host bacteria (Lilley and Bailey 1997). But, for this, the introduced strains have to fight with the enormous phytosphere microbial population as well as the toxic stress induced by the plants. In planta, HGT is an alternative to this and such plasmid-mediated bioremediation of contaminants have been explored in natural plant-associated microbial endophytes to improve the phytoremediation of environmental pollutants. For example, the efficiency of lupin plants to phytoremediate the volatile organic contaminant toluene increased when the seeds were inoculated with *Burkholderia cepacia* L.S.2.4, a natural endophyte of yellow lupin, genetically engineered to carry toluene degradation plasmid pTOM of *Burkholderia cepacia* G4 (Barac et al. 2004). Similarly, when the Poplar plants were inoculated with the endophyte *Burkholderia cepacia* VM1468, the HGT of the plasmid pTOM-Bu61 carrying genes for toluene degradation to the different endogenous endophytic communities resulted in exerting a positive impact on plant growth and reduced toluene evapotranspiration (Taghavi et al. 2005). Through *in situ* horizontal transfer from endophytic bacteria, the transconjugants rhizosphere bacteria acquired the plasmid with the C23O gene coding for the degradation of phenol (Wang et al. 2007). The findings of Wang et al. (2014) on the bioaugmentation under the effect of soil with different plants also support the rhizospheric effect on HGT.

8.5 ADVANTAGES AND THE SUCCESS CONSTRAINS

Plasmid-mediated bioaugmentation has several advantages over cell bioaugmentation especially for the cleanup of pollutants in environmental conditions of poor nutrients and microbial activity (Dejonghe et al. 2000). First of all, the problem of poor survival and low metabolic activity of the cells in cell bioaugmentation is solved when plasmids are used. Here, plasmids are transferred from a donor to the indigenous microbial population, which is already been surpassed the various biotic and abiotic stresses in that particular environment. The advantage of this adaptation has a direct impact on genetic complementation and in turn reflected in significantly increased catabolic capabilities (Figure 8.3). Secondly, the self-transmissible nature of the plasmids greatly reduces the efforts of gene bioaugmentation, where conjugation results in the widespread of many catabolic genes. The low copy number and the presence of genes required for the conjugation process and stabilization within the host, making it well maintained and consistent over succeeding generations are additional benefits contributing to HGT. Unlike the cell bioaugmentation, the introduced host cells

FIGURE 8.3 Increased catabolic activity results in successful plasmid-mediated bioaugmentation.

are not expected to have that long-term competitive endurance to execute their function as a donor, since the catabolic activity truly is relied on the number and diversity of the recipient population. The genetically constructed strains may express stable and competitive existence of the plasmid and a higher specific growth rate than that of its natural host (Filonov et al. 2005).

Plasmid-encoded metabolic pathways are generally genetically flexible and can be modified and/ or constructed in a host cell. Moreover, the possibilities of interspecies and even intergeneric HGT highlights big advantages (Mrozik and Piotrowska-Seget 2010) in extending the metabolic potential of microbial communities over large taxonomic distances. Lastly, this approach overcomes the barriers associated with the inoculation and spread of larger bacterial cells throughout the polluted area (Da Silva and Alvarez 2010).

Unfortunately, outnumbered limitations constrain the application of gene bioaugmentation to treat environmental niches. The incompetence of donor bacteria to persist in the environment and to perform plasmid transfer into recipient population, inadequate number of donor and recipient bacteria, and poor stability of plasmids once in the recipient bacteria are the proffered reasons for HGT failure (Thomas and Nielsen 2005). Starting from the donor bacteria, its ability to outcompete the indigenous population and process inhibition is a limiting factor in the plasmid-mediated bioaugmentation success. To propose an optimal plasmid-mediated bioaugmentation, a study to decide the appropriate plasmid and to analyze the transfer efficiency of the microorganisms and stability and persistence of the plasmid in the recipient bacteria is needed. Moreover, the genetic incompleteness of the plasmids is another constrain in the bioaugmentation process where the plasmid of interest encodes only a part of the genes required for complete degradation of a xenobiotic. This requirement for genetic complementation, concealed in the host bacteria where the host chromosome carries the missing part to accomplish full degradation of the compound, becomes visible in the recipient lacking the necessary genes on their chromosomes. Such futile bioaugmentation could be rectified by the identification of plasmids or by using a diverse recipient community instead of pure cultures.

Lack of knowledge about the phylogenetic profile of the indigenous microbial communities of the contaminated sites under study causes a huge debt since the expression of the acquired catabolic pathway is strongly bonded to the fitness efficiency, the G+C content, and phylogenic individuality of the recipient population limit (Ikuma and Gunsch 2012). The chosen genetically modified recipient must be capable of maintaining the plasmid, executing their assigned functions, and safe for bioaugmentation (Aquino, Barbieri, and Oller Nascimento 2011). Also, it is mandatory to have a thorough examination of the above-mentioned abiotic parameters that stimulate HGT events and efficient expression of transferred metabolic genes in predicting the feasibility of gene bioaugmentation for real environmental applications (Ikuma and Gunsch 2012). Additionally, co-contamination with other pollutants can delay or even inhibit indigenous microbial activity and the intensity of this inhibition is dramatic at a higher level of contamination. And for many toxicants, a reamendment with additional pollutants is necessary for transconjugant growth and the greatest rates of degradation (Pepper et al. 2002).

The phenotypic outcome of HGT events in the real contaminated environment is always unconvinced because many strains that showed catabolic activity in the laboratory failed to do so in the actual polluted environment. This might be because of the greatly limited consistency and compatibility of plasmid in the donor and recipient. An important concern with the use of genetically modified microorganisms (GMOs) is the instability of the plasmids and the subsequent loss of the desired phenotype. Furthermore, conjugal transfer of plasmid in the actual environment is limited to few recipient bacterial strains. Therefore, continuous monitoring of the survival of the introduced strain and transfer of plasmid along with the transconjugant occurrence, their identification, plasmid persistence, and performance in the reactor or environment is inevitable. Like many other GMOs, the use of genetically engineered strains is legally regulated for bioremediation purposes. For some reason, many consider gene transfer from introduced organisms to indigenous microorganisms a threat or risk especially for GMOs when the transfer is not the original goal. The European Union

(EU) regulations govern the testing and release of GMOs in the EU and they demand a compulsory long-term monitoring of the GM inoculants to address the issues with the impact of the inoculums and the fate of modified genes in the release sites (Morrissey et al. 2002). According to the Toxic Substances Control Act (TSCA) by United State Environmental Protection Agency (US EPA), any microorganism produced by combining genetic components from organisms in different genera is a "new" organism; likewise, the microorganisms that acquired mobile genetic elements such as plasmids is must be treated as "new" if the mobile genetic element was first identified in a microorganism belonged to a different genus. Some countries may legally restrict the application of such new stains unless there is an inescapable need for field bioaugmentation, even if this transfer of plasmid from a donor to a recipient occurs naturally (US EPA 2012).

8.6   FUTURE

To date, only a handful of studies have investigated the application in aquatic environments. Further investigations involving the occurrence of different transconjugants and the onset of xenobiotics in real aquatic conditions such as a coastline or marine environment and sediments have to be conducted. Improvement of *in situ* monitoring techniques to detect the transfer frequencies and maintenance of plasmids within the application sites is needed. Our knowledge of plasmid-mediated bioaugmentation is limited so far to those from cultivable bacteria. A new approach "transposon-aided capture (TRACA) of plasmids" with plasmid-safe DNase, which has been successfully used for the identification of plasmids residents in the human gut metagenome, has to be attempted to fill this gap regarding plasmids from non-cultivable bacteria. The recent reports on non-homologous end-joining mechanisms open a new window to spread genes across distantly related prokaryotes (G+C difference <5%) through bypassing the donor-recipient sequence similarity barrier (Popa et al. 2011) Furthermore, the impact of catabolic plasmids on bacterial cell physiology is another area of ongoing research. Combined genomic and transcriptomic approaches to develop a prior understanding of the molecular basis of the origin of cost, fitness effects of HGT, the transcriptional effects of plasmids, and the ability of compensatory evolution in the recipient population to reduce the cost can be expanded to get a full picture for improving the biotechnological utilization of bacteria for bioaugmentation. Such a detailed multidisciplinary research approach on natural host bacteria, its inappropriateness for bioaugmentation, incomplete catabolic pathway, and further possible modifications at the molecular level would be recommended to solve the following issues with plasmid-mediated bioaugmentation: (i) the limitations in using the natural host for plasmid-mediated bioaugmentation, (ii) host cells catabolic plasmids are stable in their natural host even in the absence of selective pressure, but easily eliminated from a recipient cell if appropriate cautions are not taken, (iii) incomplete catabolic pathway where the host or recipient bacteria lack necessary chromosomal genes, and (iv) legal restrictions in real-life applications where a recombinant or genetically engineered bacteria with the catabolic plasmid is treated as a "new strain".

8.7   CONCLUSION

Although the efficiency of the process is greatly affected by many physiologic and environmental conditions, the introduction of the degradative donor plasmid into the polluted sites is an effective bioaugmentation strategy for the bioremediation of many contaminants. Moreover, the degradation capacity is further expanded once the plasmid is successfully transferred into the local microbial community, thereby eliminating all the prerequisites for the survival and persistence of the introduced strain in the environment for a long-term application. In conclusion, the introduction of self-transmissible plasmids encoding enzymes and other machineries that can degrade xenobiotics or donor strains harboring catabolic plasmids can lead to disseminating xenobiotic degradation ability to the native microbial community thereby is an alternative approach in bioaugmentation. However, for a feasible bioaugmentation strategy, more understanding of several parameters is needed on a wider frame; the advancing multidisciplinary approaches and techniques enable more accurate prediction of the necessary operational parameters in a given contaminated environment.

REFERENCES

Aminov, R.I. 2011. "Horizontal Gene Exchange in Environmental Microbiota." *Frontiers in Microbiology*. Frontiers Research Foundation. https://doi.org/10.3389/fmicb.2011.00158.

Aquino, E., C. Barbieri, and C.A. Oller Nascimento. 2011. "Engineering Bacteria for Bioremediation." In *Progress in Molecular and Environmental Bioengineering - From Analysis and Modeling to Technology Applications*. InTech. https://doi.org/10.5772/19546.

Aspray, T.J., S.K. Hansen, and R.G. Burns. 2005. "A Soil-Based Microbial Biofilm Exposed to 2,4-D: Bacterial Community Development and Establishment of Conjugative Plasmid PJP4." *FEMS Microbiology Ecology* 54 (2): 317–27. https://doi.org/10.1016/j.femsec.2005.04.007.

Bale, M. J., J. C. Fry, and M. J. Day. 1987. "Plasmid Transfer between Strains of Pseudomonas Aeruginosa on Membrane Filters Attached to River Stones." *Journal of General Microbiology* 133 (11): 3099–3107. https://doi.org/10.1099/00221287-133-11-3099.

Barac, T., S. Taghavi, B. Borremans, A. Provoost, L. Oeyen, J.V. Colpaert, J. Vangronsveld, and D. Van Der Lelie. 2004. "Engineered Endophytic Bacteria Improve Phytoremediation of Water-Soluble, Volatile, Organic Pollutants." *Nature Biotechnology* 22 (5): 583–88. https://doi.org/10.1038/nbt960.

Bathe, S. 2004. "Conjugal Transfer of Plasmid PNB2 to Activated Sludge Bacteria Leads to 3-Chloroaniline Degradation in Enrichment Cultures." *Letters in Applied Microbiology* 38 (6): 527–31. https://doi.org/10.1111/j.1472-765X.2004.01532.x.

Bathe, S., and M. Hausner. 2010. "Plasmid-Mediated Bioaugmentation of Wastewater Microbial Communities in a Laboratory-Scale Bioreactor." *Methods in Molecular Biology* 599: 185–200. https://doi.org/10.1007/978-1-60761-439-5_12.

Bathe, S., N. Schwarzenbeck, and M. Hausner. 2005. "Plasmid-Mediated Bioaugmentation of Activated Sludge Bacteria in a Sequencing Batch Moving Bed Reactor Using PNB2." *Letters in Applied Microbiology* 41 (3): 242–47. https://doi.org/10.1111/j.1472-765X.2005.01754.x.

Bathe, S., N. Schwarzenbeck, and M. Hausner. 2009. "Bioaugmentation of Activated Sludge towards 3-Chloroaniline Removal with a Mixed Bacterial Population Carrying a Degradative Plasmid." *Bioresource Technology* 100 (12): 2902–9. https://doi.org/10.1016/j.biortech.2009.01.060.

Bhadbhade, B.J., P.K. Dhakephalkar, S.S. Sarnaik, and P.P. Kanekar. 2002. "Plasmid-Associated Biodegradation of an Organophosphorus Pesticide, Monocrotophos, by Pseudomonas Mendocina." *Biotechnology Letters* 24 (8): 647–650. https://doi.org/10.1023/A:1015099409563.

Boon, N., J. Goris, P. De Vos, W. Verstraete, and E.M. Top. 2000. "Bioaugmentation of Activated Sludge by an Indigenous 3-Chloroaniline- Degrading Comamonas Testosteroni Strain, I2gfp." *Applied and Environmental Microbiology* 66 (7): 2906–13. https://doi.org/10.1128/AEM.66.7.2906-2913.2000.

Cycoń, M., A. Mrozik, and Z. Piotrowska-Seget. 2017. "Bioaugmentation as a Strategy for the Remediation of Pesticide-Polluted Soil: A Review." *Chemosphere*. https://doi.org/10.1016/j.chemosphere.2016.12.129.

Dejonghe, W., J. Goris, S.E. Fantroussi, M. Hofte, P. D. Vos, W. Verstraete, and E. M. Top. 2000. "Effect of Dissemination of 2,4-Dichlorophenoxyacetic Acid (2,4-D) Degradation Plasmids on 2,4-d Degradation and on Bacterial Community Structure in Two Different Soil Horizons." *Applied and Environmental Microbiology* 66 (8): 3297–3304. https://doi.org/10.1128/AEM.66.8.3297-3304.2000.

diCenzo, G.C., K. Debiec, J. Krzysztoforski, W. Uhrynowski, A. Mengoni, C. Fagorzi, A. Gorecki, et al. 2018. "Genomic and Biotechnological Characterization of the Heavy-Metal Resistant, Arsenic-Oxidizing Bacterium Ensifer Sp. M14." *Genes* 9 (8). https://doi.org/10.3390/genes9080379.

DiGiovanni, G. D., J. W. Neilson, I. L. Pepper, and N. A. Sinclair. 1996. "Gene Transfer of Alcaligenes Eutrophus JMP134 Plasmid PJP4 to Indigenous Soil Recipients." *Applied and Environmental Microbiology* 62 (7): 2521–26. https://doi.org/10.1128/aem.62.7.2521–2526.1996.

Drewniak, L., M. Ciezkowska, M. Radlinska, and A. Sklodowska. 2015. "Construction of the Recombinant Broad-Host-Range Plasmids Providing Their Bacterial Hosts Arsenic Resistance and Arsenite Oxidation Ability." *Journal of Biotechnology* 196–197: 42–51. https://doi.org/10.1016/j.jbiotec.2015.01.013.

Drewniak, L., L. Dziewit, M. Ciezkowska, J. Gawor, R. Gromadka, and A. Sklodowska. 2013. "Structural and Functional Genomics of Plasmid PSinA of Sinorhizobium sp. M14 Encoding Genes for the Arsenite Oxidation and Arsenic Resistance." *Journal of Biotechnology* 164 (4): 479–88. https://doi.org/10.1016/j.jbiotec.2013.01.017.

Filonov, A. E., L. I. Akhmetov, I. F. Puntus, T. Z. Esikova, A. B. Gafarov, T. Yu Izmalkova, S. L. Sokolov, I. A. Kosheleva, and A. M. Boronin. 2005. "The Construction and Monitoring of Genetically Tagged, Plasmid-Containing, Naphthalene-Degrading Strains in Soil." *Microbiology* 74 (4): 453–58. https://doi.org/10.1007/s11021-005-0088-6.

Franzetti, A., P. Caredda, C. Ruggeri, P. La Colla, E. Tamburini, M. Papacchini, and G. Bestetti. 2009. "Potential Applications of Surface Active Compounds by Gordonia sp. Strain BS29 in Soil Remediation Technologies." *Chemosphere* 75 (6): 801–7. https://doi.org/10.1016/j.chemosphere.2008.12.052.

Gao, C., X. Jin, J. Ren, H. Fang, and Y. Yu. 2015. "Bioaugmentation of DDT-Contaminated Soil by Dissemination of the Catabolic Plasmid PDOD." *Journal of Environmental Sciences (China)* 27 (C): 42–50. https://doi.org/10.1016/j.jes.2014.05.045.

Garbisu, C., O. Garaiyurrebaso, L. Epelde, E. Grohmann, and I. Alkorta. 2017. "Plasmid-Mediated Bioaugmentation for the Bioremediation of Contaminated Soils." *Frontiers in Microbiology* 8 (OCT): 1966. https://doi.org/10.3389/fmicb.2017.01966.

Halden, R.U., S.M. Tepp, B.G. Halden, and D.F. Dwyer. 1999. "Degradation of 3-Phenoxybenzoic Acid in Soil by Pseudomonas Pseudoalcaligenes POB310(PPOB) and Two Modified Pseudomonas Strains." *Applied and Environmental Microbiology* 65 (8): 3354–59. https://doi.org/10.1128/aem.65.8.3354–3359.1999.

Ikuma, K., and C. Gunsch. 2010. "Effect of Carbon Source Addition on Toluene Biodegradation by an Escherichia Coli DH5α Transconjugant Harboring the TOL Plasmid." *Biotechnology and Bioengineering* 107 (2): 269–77. https://doi.org/10.1002/bit.22808.

Ikuma, K., and C.K. Gunsch. 2012. "Genetic Bioaugmentation as an Effective Method for in Situ Bioremediation." *Bioengineered* 3 (4): 236–41. https://doi.org/10.4161/bioe.20551.

Ikuma, K., and C.K. Gunsch 2013. "Functionality of the TOL Plasmid under Varying Environmental Conditions Following Conjugal Transfer." *Applied Microbiology and Biotechnology* 97 (1): 395–408. https://doi.org/10.1007/s00253-012-3949-8.

Ikuma, K., R.M. Holzem, and C.K. Gunsch. 2012. "Impacts of Organic Carbon Availability and Recipient Bacteria Characteristics on the Potential for TOL Plasmid Genetic Bioaugmentation in Soil Slurries." *Chemosphere* 89 (2): 158–63. https://doi.org/10.1016/j.chemosphere.2012.05.086.

Inoue, D., Y. Yamazaki, H. Tsutsui, K. Sei, S. Soda, M. Fujita, and M. Ike. 2012. "Impacts of Gene Bioaugmentation with PJP4-Harboring Bacteria of 2,4-D-Contaminated Soil Slurry on the Indigenous Microbial Community." *Biodegradation* 23 (2): 263–76. https://doi.org/10.1007/s10532-011-9505-x.

John, E.M., E.M. Varghese, and M.S. Jisha. 2020. "Plasmid-Mediated Biodegradation of Chlorpyrifos and Analysis of Its Metabolic By-Products." *Current Microbiology Icrobiology* 77: 3095–3103. https://doi.org/https://doi.org/10.1007/s00284-020-02115-y.

John, E.M., J. Sreekumar, and M.S. Jisha. 2016. "Optimization Of Chlorpyrifos Degradation By Assembled Bacterial Consortium Using Response Surface Methodology." *Soil and Sediment Contamination: An International Journal*, June, 00–00. https://doi.org/10.1080/15320383.2016.1190684.

Johnsen, A.R., and N. Kroer. 2007. "Effects of Stress and Other Environmental Factors on Horizontal Plasmid Transfer Assessed by Direct Quantification of Discrete Transfer Events." *FEMS Microbiology Ecology* 59 (3): 718–28. https://doi.org/10.1111/j.1574-6941.2006.00230.x.

Jung, C. M., Matthew Carr, G. Alon Blakeney, and K. J. Indest. 2019. "Enhanced Plasmid-Mediated Bioaugmentation of RDX-Contaminated Matrices in Column Studies Using Donor Strain Gordonia Sp. KTR9." *Journal of Industrial Microbiology and Biotechnology* 46 (9–10): 1273–81. https://doi.org/10.1007/s10295-019-02185-3.

Jung, C. M., F. H. Crocker, J. O. Eberly, and K. J. Indest. 2011. "Horizontal Gene Transfer (HGT) as a Mechanism of Disseminating RDX-Degrading Activity among Actinomycete Bacteria." *Journal of Applied Microbiology* 110 (6): 1449–59. https://doi.org/10.1111/j.1365-2672.2011.04995.x.

Jussila, M.M., J. Zhao, L. Suominen, and K. Lindström. 2007. "TOL Plasmid Transfer during Bacterial Conjugation in Vitro and Rhizoremediation of Oil Compounds in Vivo." *Environmental Pollution* 146 (2): 510–24. https://doi.org/10.1016/j.envpol.2006.07.012.

Król, J.E., H. Duc Nguyen, L.M. Rogers, H. Beyenal, S.M. Krone, and E.M. Top. 2011. "Increased Transfer of a Multidrug Resistance Plasmid in Escherichia Coli Biofilms at the Air-Liquid Interface." *Applied and Environmental Microbiology* 77 (15): 5079–88. https://doi.org/10.1128/AEM.00090-11.

Lilley, A.K., and M.J. Bailey. 1997. "Impact of Plasmid PQBR103 Acquisition and Carriage on the Phytosphere Fitness of Pseudomonas fluorescens SBW25: Burden and Benefit." *Applied and Environmental Microbiology* 63 (4): 1584–87. https://doi.org/10.1128/aem.63.4.1584-1587.1997.

Maila, M.P., P. Randima, K. Surridge, K. Drønen, and T.E. Cloete. 2005. "Evaluation of Microbial Diversity of Different Soil Layers at a Contaminated Diesel Site." *International Biodeterioration and Biodegradation* 55 (1): 39–44. https://doi.org/10.1016/j.ibiod.2004.06.012.

McClure, N. C., J. C. Fry, and A. J. Weightman. 1991. "Survival and Catabolic Activity of Natural and Genetically Engineered Bacteria in a Laboratory-Scale Activated-Sludge Unit." *Applied and Environmental Microbiology* 57 (2): 366–73. https://doi.org/10.1128/aem.57.2.366-373.1991.

Miller, M. N., G. W. Stratton, and G. Murray. 2004. "Effects of Soil Moisture and Aeration on the Biodegradation of Pentachlorophenol Contaminated Soil." *Bulletin of Environmental Contamination and Toxicology* 72 (1): 101–8. https://doi.org/10.1007/s00128-003-0246-3.

Miyazaki, R., Y. Sato, M. Ito, Y. Ohtsubo, Y. Nagata, and M. Tsuda. 2006. "Complete Nucleotide Sequence of an Exogenously Isolated Plasmid, PLB1, Involved in γ-Hexachlorocyclohexane Degradation." *Applied and Environmental Microbiology* 72 (11): 6923–33. https://doi.org/10.1128/AEM.01531-06.

Monti, M.R., A.M. Smania, G. Fabro, M.E. Alvarez, and C.E Argaraña. 2005. "Engineering Pseudomonas Fluorescens for Biodegradation of 2,4-Dinitrotoluene." *Applied and Environmental Microbiology* 71 (12): 8864–72. https://doi.org/10.1128/AEM.71.12.8864-8872.2005.

Morrissey, J.P., U.F. Walsh, A. O'Donnell, Y. Moënne-Loccoz, and F. O'Gara. 2002. "Exploitation of Genetically Modified Inoculants for Industrial Ecology Applications." *Antonie van Leeuwenhoek, International Journal of General and Molecular Microbiology* 81 (1–4): 599–606. https://doi.org/10.1023/A:1020522025374.

Mrozik, A., and Z. Piotrowska-Seget. 2010. "Bioaugmentation as a Strategy for Cleaning up of Soils Contaminated with Aromatic Compounds." *Microbiological Research* 165 (5): 363–75. https://doi.org/10.1016/j.micres.2009.08.001.

Neilson, J. W., K. L. Josephson, I. L. Pepper, R. B. Arnold, G. D. Di Giovanni, and N. A. Sinclair. 1994. "Frequency of Horizontal Gene Transfer of a Large Catabolic Plasmid (PJP4) in Soil." *Applied and Environmental Microbiology* 60 (11): 4053–58. https://doi.org/10.1128/aem.60.11.4053-4058.1994.

Newby, D. T., T. J. Gentry, and I. L. Pepper. 2000. "Comparison of 2,4-Dichlorophenoxyacetic Acid Degradation and Plasmid Transfer in Soil Resulting from Bioaugmentation with Two Different PJP4 Donors." *Applied and Environmental Microbiology* 66 (8): 3399–3407. https://doi.org/10.1128/AEM.66.8.3399-3407.2000.

Nusslein, K., D. Maris, K. Timmis, and D. F. Dwyer. 1992. "Expression and Transfer of Engineered Catabolic Pathways Harbored by Pseudomonas Spp. Introduced into Activated Sludge Microcosms." *Applied and Environmental Microbiology* 58 (10): 3380–86. https://doi.org/10.1128/aem.58.10.3380-3386.1992.

Pei, R., and C. K. Gunsch. 2009. "Plasmid Conjugation in an Activated Sludge Microbial Community." *Environmental Engineering Science* 26 (4): 825–31. https://doi.org/10.1089/ees.2008.0236.

Pepper, I.L, T.J. Gentry, D.T. Newby, T.M. Roane, and K.L. Josephson. 2002. "The Role of Cell Bioaugmentation and Gene Bioaugmentation in the Remediation of Co-Contaminated Soils." *Environmental Health Perspective* 10 (6): 943–46.

Peters, M., E. Heinaru, E. Talpsep, H. Wand, U. Stottmeister, A. Heinaru, and A. Nurk. 1997. "Acquisition of a Deliberately Introduced Phenol Degradation Operon, PheBA, by Different Indigenous Pseudomonas Species." *Applied and Environmental Microbiology* 63 (12).

Pinedo, C.A., and B.F. Smets. 2005. "Conjugal TOL Transfer from Pseudomonas Putida to Pseudomonas Aeruginosa: Effects of Restriction Proficiency, Toxicant Exposure, Cell Density Ratios, and Conjugation Detection Method on Observed Transfer Efficiencies." *Applied and Environmental Microbiology* 71 (1): 51–57. https://doi.org/10.1128/AEM.71.1.51-57.2005.

Popa, O., E. Hazkani-Covo, G. Landan, W. Martin, and T. Dagan. 2011. "Directed Networks Reveal Genomic Barriers and DNA Repair Bypasses to Lateral Gene Transfer among Prokaryotes." *Genome Research* 21 (4): 599–609. https://doi.org/10.1101/gr.115592.110.

Roane, T. M., K. L. Josephson, and I. L. Pepper. 2001. "Dual-Bioaugmentation Strategy to Enhance Remediation of Cocontaminated Soil Downloaded From." *Applied and Environmental Microbiology* 67 (7): 3208–15. https://doi.org/10.1128/AEM.67.7.3208-3215.2001.

Rochelle, P. A., J. C. Fry, and M. J. Day. 1989. "Factors Affecting Conjugal Transfer of Plasmids Encoding Mercury Resistance from Pure Cultures and Mixed Natural Suspensions of Epilithic Bacteria." *Journal of General Microbiology* 135 (Pt 2): 409–24. https://doi.org/10.1099/00221287-135-2-409.

Rodrigues, J. L.M., O. V. Maltseva, T. V. Tsoi, R. R. Helton, J. F. Quensen, M. Fukuda, and J. M. Tiedje. 2001. "Development of a Rhodococcus Recombinant Strain for Degradation of Products from Anaerobic Dechlorination of PCBs." *Environmental Science and Technology* 35 (4): 663–68. https://doi.org/10.1021/es001308t.

Romaniuk, K., L. Dziewit, P. Decewicz, S. Mielnicki, M. Radlinska, and L. Drewniak. 2017. "Molecular Characterization of the PSinB Plasmid of the Arsenite Oxidizing, Metallotolerant Sinorhizobium Sp. M14- Insight into the Heavy Metal Resistome of Sinorhizobial Extrachromosomal Replicons." *FEMS Microbiology Ecology* 93 (1). https://doi.org/10.1093/femsec/fiw215.

Rore, H.D., K. Demolder, K.D. Wilde, E. Top, F. Houwen, and W. Verstraete. 1994. "Transfer of the Catabolic Plasmid RP4::Tn4371 to Indigenous Soil Bacteria and Its Effect on Respiration and Biphenyl Breakdown." *FEMS Microbiology Ecology* 15 (1–2): 71–77. https://doi.org/10.1111/j.1574-6941.1994.tb00231.x.

San Millan, A., and R. Craig MacLean. 2017. "Fitness Costs of Plasmids: A Limit to Plasmid Transmission." In *Microbial Transmission*, 5:65–79. https://doi.org/10.1128/microbiolspec.mtbp–0016–2017.

Segura, A., L. Molina, and J. Luis Ramos. 2014. "Plasmid-Mediated Tolerance Toward Environmental Pollutants." *Microbiology Spectrum* 2 (6). https://doi.org/10.1128/microbiolspec.plas-0013-2013.

Silva, M. L. B. D, and P. J. J. Alvarez. 2010. "Bioaugmentation." In *Handbook of Hydrocarbon and Lipid Microbiology*, 4531–4544. Springer, Berlin Heidelberg. https://doi.org/10.1007/978-3-540-77587-4_356.

Stalder, T., and E. Top. 2016. "Plasmid Transfer in Biofilms: A Perspective on Limitations and Opportunities." *Npj Biofilms and Microbiomes*. Nature Publishing Group. https://doi.org/10.1038/npjbiofilms.2016.22.

Taghavi, S., T. Barac, B. Greenberg, B. Borremans, J. Vangronsveld, and D. Van Der Lelie. 2005. "Horizontal Gene Transfer to Endogenous Endophytic Bacteria from Poplar Improves Phytoremediation of Toluene." *Applied and Environmental Microbiology* 71 (12): 8500–8505. https://doi.org/10.1128/AEM.71.12.8500–8505.2005.

Tang, J., L. Liu, S. Hu, Y. Chen, and J. Chen. 2009. "Improved Degradation of Organophosphate Dichlorvos by Trichoderma Atroviride Transformants Generated by Restriction Enzyme-Mediated Integration (REMI)." *Bioresource Technology* 100 (1): 480–483. https://doi.org/10.1016/j.biortech.2008.05.022.

Thomas, C.M., and K.M. Nielsen. 2005. "Mechanisms of, and Barriers to, Horizontal Gene Transfer between Bacteria." *Nature Reviews Microbiology*. https://doi.org/10.1038/nrmicro1234.

Tsutsui, H., Y. Anami, M. Matsuda, K. Hashimoto, D. Inoue, K. Sei, S. Soda, and M. Ike. 2013. "Plasmid-Mediated Bioaugmentation of Sequencing Batch Reactors for Enhancement of 2,4-Dichlorophenoxyacetic Acid Removal in Wastewater Using Plasmid PJP4." *Biodegradation* 24 (3): 343–352. https://doi.org/10.1007/s10532-012-9591-4.

Undugoda, LJS, RV Kandisa, S Kannangara, and DM Sirisena. 2018. "Plasmid Encoded Toluene and Xylene Degradation by Phyllosphere Bacteria." https://doi.org/10.4172/2161–0525.1000559.

US EPA. 2012. "Microbial Products of Biotechnology Summary of Regulations under the Toxic Substances Control Act." https://www.epa.gov/sites/production/files/2015-08/documents/biotech_fact_sheet.pdf.

Venkata Mohan, S., C. Falkentoft, Y. V. Nancharaiah, B. S.Mc Swain Sturm, P. Wattiau, P.A. Wilderer, S. Wuertz, and M. Hausner. 2009. "Bioaugmentation of Microbial Communities in Laboratory and Pilot Scale Sequencing Batch Biofilm Reactors Using the TOL Plasmid." *Bioresource Technology* 100 (5): 1746–53. https://doi.org/10.1016/j.biortech.2008.09.048.

Wang, S., and C.N. Mulligan. 2004. "An Evaluation of Surfactant Foam Technology in Remediation of Contaminated Soil." *Chemosphere* 57 (9): 1079–1089. https://doi.org/10.1016/j.chemosphere.2004.08.019.

Wang, Y., S. Kou, Q. Jiang, B. Xu, X. Liu, J. Xiao, Y. Tian, C. Zhou, C. Zhang, and M. Xiao. 2014. "Factors Affecting Transfer of Degradative Plasmids between Bacteria in Soils." *Applied Soil Ecology* 84 (December): 254–261. https://doi.org/10.1016/j.apsoil.2014.07.009.

Wang, Y., M. Xiao, X. Geng, J. Liu, and J. Chen. 2007. "Horizontal Transfer of Genetic Determinants for Degradation of Phenol between the Bacteria Living in Plant and Its Rhizosphere." *Applied Microbiology and Biotechnology* 77 (3): 733–739. https://doi.org/10.1007/s00253-007-1187-2.

Wuertz, S., S. Okabe, and M. Hausner. 2004. "Microbial Communities and Their Interactions in Biofilm Systems: An Overview." http://iwaponline.com/wst/article-pdf/49/11-12/327/420335/327.pdf.

Zhang, H., H. Wan, L. Song, H. Jiang, H. Wang, and C. Qiao. 2010. "Development of an Autofluorescent Pseudomonas Nitroreducens with Dehydrochlorinase Activity for Efficient Mineralization of Gamma-Hexachlorocyclohexane (Gamma-HCH)." *Journal of Biotechnology* 146 (3): 114–119. https://doi.org/10.1016/j.jbiotec.2010.01.020.

Zhang, Q., B. Wang, Z. Cao, and Y. Yu. 2012. "Plasmid-Mediated Bioaugmentation for the Degradation of Chlorpyrifos in Soil." *Journal of Hazardous Materials* 221–222 (June): 178–184. https://doi.org/10.1016/j.jhazmat.2012.04.024.

9 Bioaugmentation in Rhizoengineering for Xenobiotic Biodegradation

N.D. Dhanraj, Edna Mary Varghese, and M.S. Jisha
Mahatma Gandhi University

CONTENTS

9.1 INTRODUCTION

Plant rhizospheres are highly complex. It consists of various living and nonliving constituents that interact with each other in a particular way. From olden days itself humans tried to manipulate rhizosphere in order to get the desired result like improvement in crop production. But those manipulations, done in early ages, were based on inferences rather than a thorough understanding about the rhizospheric interactions. With advancements in science and technology, mechanistic aspects of complex rhizospheric interactions could be studied. Environmental contamination with xenobiotic compounds is a worldwide problem. Biodegradation of xenobiotic contaminants within the plant rhizosphere gained increased attention with the emergence of phytoremediation. With the increased sophistication in scientific tools and techniques, interests in rhizoengineering are in its ascendancy. This has led to the disclosure of the complexity of rhizosphere and hence enabled us to manipulate it for good causes (Dzantor, 2007).

Bioaugmentation is the process of addition or augmentation of desired microorganisms which are grown on well-defined conditions for performing a remediation process in a given environment (Alvarez and Illman, 2006). One approach to bioaugmentation is the addition of potential strains to complement or replace the native microbial strains. Another approach is the addition of large concentrations of microbial cells that will momentarily act as biocatalysts that will metabolize a large amount of the target contaminant (Duba et al., 1996; Krumme et al., 1994).

Researchers have exploited the symbiotic relationship between the plant and microbes in the rhizosphere for the bioremediation of the contaminants within it. This rhizoremediation has been proved

DOI: 10.1201/9781003187622-9

to degrade various organic carbon contaminants such as parathion (Andersonet al., 1993), atrazine (Andersonand Coats, 1995), trichloroethylene (Shim et al., 2000, Yee et al., 1998), and polychlorinated biphenyls (PCBs) (Brazil et al., 1995, Villacieros et al., 2005). Rhizoremediation can be achieved through rhizoengineering, in which the rhizospheric microbiomes can be manipulated according to the purpose (Kumari et al., 2019). Engineering the rhizosphere of the plant *Cucurbita pepo* by the addition of microbial consortia had increased the uptake of DDT (dichlorodiphenyltrichloroethane) by that plant. This showed that engineering the rhizosphere by bioaugmenting the potent microbial strains to a contaminated site will increase the biodegradation of that particular xenobiotic compound (Tartanus et al., 2017). In this chapter, the prospects of bioaugmentation in rhizoengineering and rhizosphere engineering processes in bioaugmentation for the degradation of xenobiotic compounds are discussed in detail. Also, the different bioaugmentation processes and types of remediation like phytoremediation and rhizoremediation are explained in view with rhizoengineering.

9.2   PHYTOREMEDIATION

Some organic compounds can be transported through the plant body. Mostly, the low molecular weight compounds are transported to the leaves from where it is removed from the plant through evapotranspiration (phytovolatilization). Nonvolatile compounds are degraded and transformed into non-toxic compounds within the plant (phytodegradation or phytoextraction). The compounds which are stable are sequestered in the biomass which can be removed physically. All these could be included under the category of phytoremediation, where plants are used for the remediation of xenobiotic contaminated sites (Gerhardt et al., 2009).

Boyajian and Carreira (1997) suggested that plants have greater potential for phytoremediation in the natural environment as plants have approx. 100 million miles of roots per acre. But high concentration of contaminants in the soil may inhibit the growth of roots and in turn the plants. In such situations, synergistic interactions of plants and their associated microbes come into play. It is these plant–microbe interactions that carry forward the xenobiotic remediation processes when a plant alone cannot perform it due to the contaminant stress. However, a deeper understanding of these interactions is required to be accepted as a predictable degradation alternative (Thijs et al., 2016).

Microbes from the contaminated soil are used for this microbe-assisted biodegradation. The mechanism adopted would probably be a give-and-take system in which rhizosphere microbes degrade the xenobiotic contaminant into simpler molecules which are easily taken up by the plant and the plant in turn feed its microbes through its root exudates (Gerhardt et al., 2009). Thijs et al. (2016) studied phytoremediation at the metaorganism (host+microbiome) level and discussed a competition-driven model to explain rhizosphere colonization in contaminated soil. They proved that a catabolic plant growth-promoting microbiome could be established in the rhizosphere, with the right level and mix of root exudates, as long as they are competent. Thus, microbe-assisted phytoremediation is primarily rhizoremediation.

9.3   RHIZOREMEDIATION

Several xenobiotic compounds need to be removed from the soil as mentioned earlier. Many microbes have the ability to interact with these xenobiotics and biotransform them. In agricultural soils, high microbial load is observed in the root vicinity areas called the rhizosphere. Rhizospheres foster these interactions, removing xenobiotics within the rhizosphere itself (Anderson and Coats, 1995). Hence, increased biodegradation of xenobiotics occurs in the rhizosphere known as rhizoremediation (Anderson et al., 1995). Rhizoremediation is a blessing in pesticide and herbicide-contaminated agricultural soil as a low-cost alternative for soil cleanup.

As mentioned earlier, rhizosphere microbes partnered with their host plants can bring about efficient rhizoremediation of xenobiotics. Tartanus et al. (2017) carried out a field trial for the rhizoremediation of DDT. They used fungal and bacterial consortium associated with *Cucurbita pepo* L. for the biodegradation of the DDT within the rhizosphere. This favored the uptake of DDT by the plant. Also, genetically engineered varieties are used in rhizoremediation like that of *Pseudomonas fluorescens,* used for the biodegradation of PCBs (Villacieros et al., 2005), 2,4-dinitrotoluene (Monti et al., 2005), and trichloroethylene (Yee et al., 1998). The success of rhizoremediation depends on the capacity of the contaminant degraders and plant growth-promoting rhizobacteria (PGPR) that can colonize the roots. The relation between PGPR and rhizoremediation is quite indirect. Certain organic compounds in the root exudates of plants are required by PGPR for root colonization. These organic compounds are structural analogs to certain xenobiotic compounds and hence could be metabolized by PGPR (discussed in detail in later sections). Thus, the biodegradation of xenobiotic compounds could be naturally occurring in the rhizosphere (Gerhardt et al., 2009; Chaudhry et al., 2005; Holden and Firestone, 1997).

9.4   BIOAUGMENTATION – A KEY STRATEGY IN BIOREMEDIATION

Bioaugmentation is the addition of desired microorganisms (which can be autochthonous or allochthonous wild type or genetically modified microorganisms) which are grown in well-defined conditions to an environment for beneficial purposes such as plant growth promotion and bioremediation (Da Silva and Alvarez, 2010). Figure 9.1 schematically represents the two bioaugmentation processes. Bioaugmentation enhances the degree of degradation of complex organic pollutants. This not only increases the biodegradation rate but also enhances the genetic diversity of a site as a new genetic variation is introduced to the gene pool of a particular site through bioaugmentation (Goswami et al., 2018).

In bioaugmentation, the success rate highly depends on the survival and degradation capability of the introduced microorganisms. These two factors are in turn dependent on the various biotic factors such as competition and predation and abiotic factors such as prevailing environmental

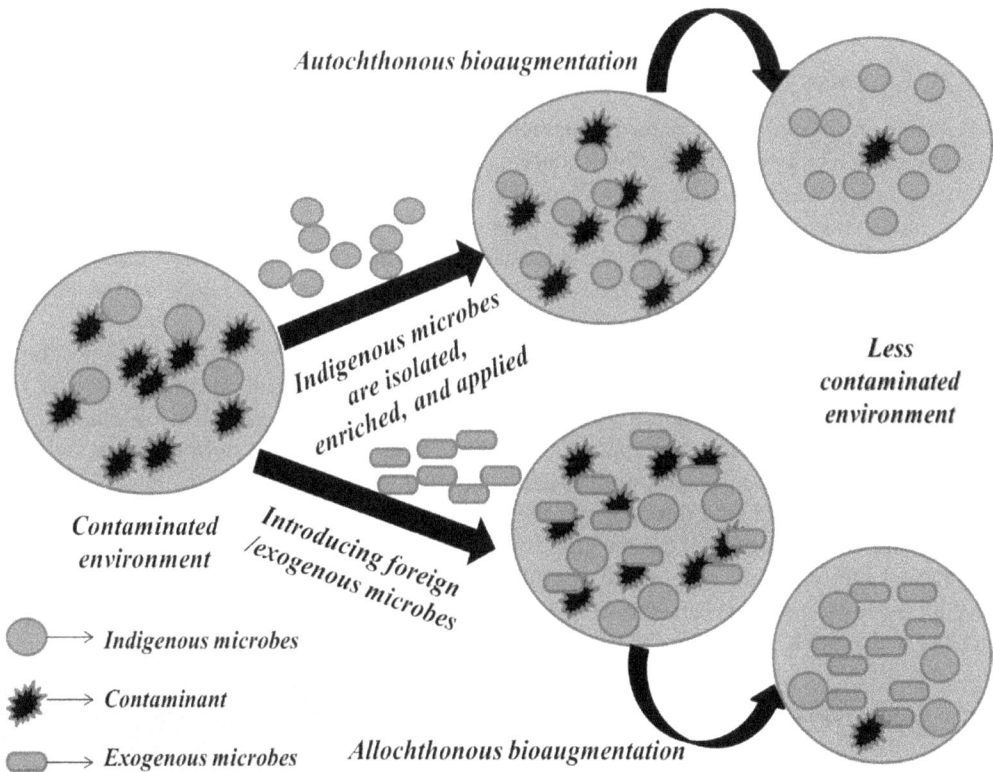

FIGURE 9.1   An overview of the bioaugmentation process.

conditions. Hence, allochthonous bioaugmentation will prove less effective as it involves addition of exogenous microbes which are foreign to the contaminant environment. However, autochthonous bioaugmentation can overcome these problems as well-characterized indigenous microorganisms, which are already acclimatized to the xenobiotic environments, are employed. These microbes are enriched under the same conditions prior to re-inoculation (Hosokawa et al., 2009).

9.4.1 FACTORS INFLUENCING BIOAUGMENTATION

Success of a bioaugmentation method chiefly depends on how much the organisms, which are newly introduced, will adapt to a particular environment. Also, it depends on the ability of the introduced microorganisms to compete with the indigenous microorganisms, predators, and various abiotic factors (Godleads et al., 2015). Apart from these, the other factors affecting bioaugmentation are pH, temperature, moisture, organic matter content, aeration, nutrient content, and soil type (Mrozik and Piotrowska-Seget, 2010; Hong et al., 2007). Among these, the most important soil parameter that influences the biodegradation activity of the inoculated microorganisms is the organic matter content of the soil. Organic matter content of the soil plays a key role in the bioavailability of the soil pollutants and thus influences the survival of microbes and the effectiveness of the bioaugmentation process (Mrozik and Piotrowska-Seget, 2010; Greer and Shelton, 1992).

Rhizosphere competence is the first and the most important factor to be looked upon for efficient rhizoremediation of xenobiotics using bioaugmentation. Many scientists have reported several microbial selection procedures to combat rhizosphere competence. These include root colonization ability, *in vitro* plant growth promoting (PGP) abilities, use of suitable plant–microbe combinations, employing bacterial consortium, and inoculation of competent plant endophytes (Kuiper et al., 2002; Kuiper et al., 2004; Thijs et al., 2014; Thijs and Vangronsveld, 2015).

9.4.2 CHOOSING THE RIGHT AGENT FOR BIOAUGMENTATION

When choosing microorganisms for bioaugmentation, we should consider factors *viz.* fast growth, ease of culture, ability to withstand high concentrations of contaminants, and the ability to survive in a wide range of environmental conditions (Goswami et al., 2018). Several studies have proven the bioaugmenting capability of various strains of microbes to degrade a wide range of xenobiotics including Quinoline (Wang et al., 2004), chlorinated benzenes (Tchelet et al., 1999), and polyethylene succinate (Tribedi et al., 2012). Table 9.1 lists the microorganisms used in bioaugmentation for the biodegradation of various xenobiotics.

Certain high-level PCBs and chloroethylenes might not be able to be degraded by the common naturally occurring degrading strains due to the lack of complex microbial catabolic pathways (Thijs and Vangronsveld, 2015). In such situations, genetically engineered or genetically modified microorganisms (GEM/GMO) find their applications (Brazil et al., 1995). Also, during co-contamination circumstances, resistant microbial strains need to be used for the survival of the introduced strain, which can be developed only through genetic engineering. However, this technique confronts several drawbacks. Introduction of foreign genes into the environment is the major disadvantage of this technique. Limited success rates with new synthetic vectors, high costs, and the huge labor involved have all made the microbial genetic modification processes enormously challenging (Hernandez-Sanchez and Wittich, 2012; Thijs and Vangronsveld, 2015).

9.5 RHIZOSPHERE ENGINEERING

The success of various agricultural and environmental practices depends hugely upon the establishment of large numbers of beneficial, metabolically active populations of soil microbes (Metting, 1992). Their high adaptability in varied environments, faster growth rate, and their

TABLE 9.1
Microbes Used in Bioaugmentation for the Biodegradation of Various Xenobiotics

S. No.	Pollutant Degraded	Organisms Used	References
1.	4 Chlorobenzoic acid and Chlorobenzoates	*Pseudomonas putid* PaW340/pDH5	Massa et al., 2009
2.	Aroclor 1221 and 1232	*Cupriavidus necator* RW112	Wittich and Wolff, 2007
3.	Aroclor 1242	*Burkholderia xenovorans* LB400 (ohb)	Rodrigues et al., 2006
		Pseudomonas aeruginosa TMU56	Hatamian-Zarmi et al., 2009
4.	2,4-Dinitrotoluene	*Pseudomonas fluorescens*	Monti et al., 2005
		Burkholderia sp.	So et al., 2004
5.	Naphthalene	*Pseudomonas putida*	Filonov et al., 2005
		Bacillus naphthovorans sp. nov.	Zhuang et al., 2002
6.	Atrazine	*Eschirichia coli*	Strong et al., 2000
		Agrobacterium tumefaciens, *Caulobacter crescentus,* *Pseudomonas putida,* *Sphingomonas yaniokuyae, Nocardia* sp., *Rhizobium* sp., *Flavobacterium oryzihabitans* and *Variovorax paradoxus*	Smith et al., 2005
7.	Biphenyl ethylbenzene	*Pseudomonas fluorescens* CS2	Parameswarappa et al., 2008
8.	Pentachlorophenol	*Sphingobium chlorophenolicum*	Dams et al., 2007
		Pseudomonas spp.	Leung et al., 1997
		Citrobacter freundii	WerheniAmmeri et al., 2016
9.	1,2-Dichloroethane	*Desulfitobacterium dichloroeliminans*	Maes et al., 2006
10.	Chloronitroaromatics	*Comamonas* sp. strain CNB-1	Liu et al., 2007
		Cupriavidus sp. strain a3	Tiwari et al., 2017
		Burkholderia sp. strain SJ98	Pandey et al., 2012
11.	Anthracene	*Mycobacterium fortuitum, Bacillus cereus,* phenanthrene and pyrene *Microbacterium* sp., *Gordonia polyisoprenivorans, Microbacteriaceae bacterium*	Jacques et al., 2008
12.	PAH (polycyclic aromatic hydrocarbon)	*Pseudomonas aeruginosa* strain 64	Straube et al., 2003
		Pseudomonas sp., *Stenotrophomonas* sp., and *Enterobacter* sp.	González et al., 2011
13.	Chlorinated ethenes	*Burkholderia cepacia*	Steffan et al., 1999
14.	Phenol, trichloroethane	*Pseudomonas putida* BCRc14349	Chen et al., 2007
15.	TCE (*trichloroethylene*), *cis*-DCE (*cis-1, 2-dichloroethene*), and vinyl chloride	*Dehalococcoides* sp. and *Desulfuromona* sp.	Lendvay et al., 2001
16.	TCE	*Burkholderiacepacia* *Pseudomonas* sp. strain ASA86 and	Bourquin et al., 1997
		Burkholderia sp. strain TAM17	Chee, 2011
		Burkholderiakururiensis sp. nov.,	Zhang, 2000

(Continued)

TABLE 9.1 (*Continued*)
Microbes Used in Bioaugmentation for the Biodegradation of Various Xenobiotics

S. No.	Pollutant Degraded	Organisms Used	References
17.	Petroleum	*Betaproteobacterium, Enterobacter* sp.	Graj et al., 2013
		Acinetobacter calcoaceticus, Comamonas sp.,	
		Pseudomonas alcaligenes	
		Ochrobactrum intermedium, Klebsiella oxytoca,	
		Sphingobacterium multivorum,	
		Pseudomonas putida	
		Chryseobacterium sp., *Stenotrophomonas*	
		maltophilia,	
		Pseudomonas stutzeri, Alcaligenes	
		xylosoxidans,	
		Sphingobacterium sp., *Citrobacter freundii,*	
		Sphingobacterium sp.	
18.	Chlorpyrifos	*Staphylococcus warneri Pseudomonas Putida,*	
		Stenotrophomonas maltophilia	John et al., 2016
		Flavobacterium sp.	Mallick et al., 1999
		Enterobacter strain B-14	Singh et al., 2004
		Stenotrophomonas sp. YC-1	Yang et al., 2006
		Sphingomonas sp. Dsp-2	Li et al., 2007
		Bacillus pumilus	Anwar et al., 2009
		Bacillus subtilis	El-Helow et al., 2013
		Mesorhizobium sp. HN3	Jabeen et al., 2014
		Sphingomonas sp. and *Brevundimonas* sp.	Santillan et al., 2020
19.	Linear alkylbenzenesulfonates (LAS)	*Pseudomonas nitroreducens* and *Pseudomonas aeruginosa*	Asok and Jisha, 2012

ability to metabolize diverse natural and xenobiotic compounds have assisted them in their establishment in the soil ecosystem (Narasimhan et al., 2003). Majority of such population in the soil is found associated with plant roots, in huge numbers (Whipps, 1990; Metting, 1992). Plant root exudation has caused this abundance of beneficial soil microbes in vegetated soils (Brimecombe, 2001), particularly in the immediate vicinity of the roots known as the rhizosphere. Hence, rhizosphere can be considered as the ideal site to alter the soil microbial populations to suit various applications (O'Connell, 1996). Rhizosphere is the zone around the root system which is influenced by the root activity. This area is the zone of beneficial interactions between the plant and soil-borne microbes. These interactions make plants vulnerable to a variety of biotic and abiotic stress. Hence, plants have inherent mechanisms for modifying the rhizosphere to withstand these kinds of stresses. Understanding such processes involved will pave the way for rhizosphere modification for various purposes. Rhizosphere can be modified by the addition of beneficial microbes, biodegradable biostimulants, or transgenic plants. This concept is called rhizosphere engineering/rhizoengineering (Ryan et al., 2009). Rhizoengineering has been employed for plant growth promotion (Ryan et al., 2009), biodegradation of various xenobiotic compounds within the rhizosphere (rhizoremediation) (Dzantor, 2007), and bioremediation of heavy metals (Braud et al., 2009).

9.5.1 RHIZOENGINEERING FOR XENOBIOTIC BIODEGRADATION

Phytoremediation of organic contaminants takes place in three ways: (i) through uptake, translocation into plant shoots, and metabolism (*phytodegradation*), (ii) volatilization (*phytovolatilization*), and (iii) indirectly through plant–microbe–contaminant interactions within plant root zones (*rhizospheres*). In the biodegradation of organic compounds within the rhizosphere, generally, the rhizospheric microbes will get stimulated by the plant exudates such as carbohydrates, carboxylic and amino acids leading to higher rates of degradation in the rhizospheric region. But for some persistent organic compounds, this process is not enough. They require the presence of some substrate which can bind competent microbes or their consortia. Plant–rhizospheric microbe interactions have been studied extensively aiming for manipulating them for accelerating the rhizodegradation of various recalcitrant xenobiotic components. Attempts were done by bioaugmenting competent microbes to the contaminated sites or by substrate amendment approach. But these attempts did not meet the expectations as some of them failed to give good results (Dzantor, 2007). These failures may be due to the competition by the indigenous species, low concentrations of pollutants in the site, low bioavailability of the xenobiotic pollutants, alternative substrate, predators, etc. (Goldstein et al., 1985).

For successful adaptation and survival of microbes introduced in the rhizosphere, addition of enzyme-inducing substrate is not sufficient as it does not always favor the growth of introduced microbes. Manipulations that favor the introduced microbes should provide a selective advantage to them. Those advantages include nutritional favoritism or suppression of indigenous populations (Dzantor, 2007). For example, salicylate can be used as a carbon source that will help selectively increase the growth and activity of *Pseudomonas putida* PpG7 in soil (Colbert et al., 1993) and, for enriching detergent-adapted populations of *P. fluorescence* ANP15 and *P. aeruginsa* 7NSK2 in maize rhizosphere, the detergents Igepal CO-720 and dicotylsulfosuccinate were used (Devliegher et al., 1995).

Some natural products such as flavonoids and opines produced by the plant itself will support the growth of a particular bacteria in the rhizosphere. Flavonoids, naringin, and apigenin have been found to enhance the growth and activity of bacteria that can biodegrade PCBs (Fletcher and Hedge, 1995). Thus, production of plant products serves two benefits: it will increase the growth of particular microbes and induce enzymes for biodegradation. The understanding of this fact offers very attractive opportunities for rhizoengineering for enhancing rhizodegradation of xenobiotics (Dzantor, 2007). Apart from bacteria, fungi also help in bioremediation of xenobiotics in rhizosphere. Arbuscular mycorrhiza fungi (AMF) are being used for rhizoremediation, and this is termed as mycorrhizoremediation. The oxidative enzymes produced by fungi aid in the biodegradation of xenobiotic compounds (Khan, 2006).

9.5.2 BIOAUGMENTATION IN RHIZOENGINEERING

Here in rhizoengineering by bioaugmentation, the gene pool of the rhizosphere will be enriched. Exogenous microorganisms with better degradation capabilities can be introduced or microorganisms previously isolated from the same rhizosphere can be added. Among these, the second option gives more guarantee for the success of bioaugmentation, as the adaptability and competitiveness of the introduced microorganisms in this case would be higher (Tyagi et al., 2011; El Fanroussi and Agathos, 2005). Microbe-delivery methods into the rhizosphere include seed coating, root dipping, soil drenching, etc. Such inoculated root systems act as the 'bioinjector' of strains with degradation capability into the rhizosphere (Thijs and Vangronsveld, 2015) leading to rhizoengineering.

Several studies have been performed on the application of bioaugmentation on rhizospheric level for the biodegradation of various xenobiotic compounds. Rhizospheric bacteria previously isolated from the rhizosphere of plant *Brassica nigra* was inoculated to its rhizosphere again showed 87% PCB removal after 12 weeks of bioaugmentation (Singer et al., 2003). *Rhizobium* strain bioaugmented in the rhizosphere of alfalfa plants showed increased biodegradation of PCBs (Xu et al., 2010). Bioaugmentation with *Flavobacterium* sp. in rhizospheres has also shown increased PCB degradation. Many studies have not passed the field trials successfully, as many biotic and abiotic factors influence the growth and establishment of the introduced strain. So, the introduction of indigenous microbes is the best strategy. Also, carrier immobilization of the inoculum has been found fruitful in some studies (Pajuelo et al., 2014; Mrozik and Piotrowska-Seget, 2010).

Genetically modified microorganisms can be bioaugmented in the rhizosphere for the biodegradation of xenobiotics. Genetic modifications were done on *Pseudomonas* and *Rhizobium* that enhanced the capacity to degrade trichloroethylene in the plant rhizosphere (Yee et al., 1998). The *bph* operon from *Burkholderia xenovorans* LB400, was inserted into a mutant *P. fluorescens* F113 and showed significant PCB degradation (Brazil et al., 1995). Fast and efficient degradation is possible with genetically modified organisms. Arsenic-resistant operon *arsRDABC* was inserted into *Pseudomonas* F113rif PCB strain and hence were able to colonize the root and degrade biphenyl in the presence of sodium arsenate (Ryan et al., 2007).

9.5.3 Rhizoengineering Approaches in Bioaugmentation

Successful bioaugmentation in the rhizosphere depends upon various factors. Primarily, the introduced microbes are concerned mainly about their existence or survival, for which, the nutrition factor comes to play. Hence, the major rhizospheric nutrient source – the root exudates or rhizodeposits – need to be manipulated for successful microbial establishment in the rhizosphere for biodegradation. The major factors contributing to this manipulation would be contaminant concentration and rhizospheric competition. Rhizoengineering involving all these concepts could increase the survivability and efficiency of the introduced microbial strains eventually leading to enhanced xenobiotic degradation.

It is well known that the rhizosphere microbiome is largely influenced by root exudates (Berg and Smalla, 2009). Microbes selectively get repelled or attracted towards specific root exudates (Neal et al., 2012). The aromatic compounds in the root exudates, *viz.* flavonoids, terpenoids, phenols, and other lignin derivatives, play a key role in the plant selection of microbes. They act as co-substrates for complex xenobiotics like high chlorinated biphenyls and polycyclic aromatic hydrocarbons (PAHs) probably because of the similarity in chemical structures. They also act as xenobiotic degradation pathway inducers (Thijs and Vangronsveld, 2015). Employing transgenic plants, in the contaminated environment, can release such secondary metabolites as root exudates, specific to the introduced degradative microbial communities. This can promote their rhizospheric establishment thereby leading to contaminant degradation. Narasimhan et al. (2003) identified a wide range of phenylpropanoid compounds, including flavonoids, in the root exudates of *Arabidopsis* mutant plants with the help of the rhizosphere metabolomics approach. The mutants capable of overproducing flavonoids were then used for the colonization of *Pseudomonas putida* PML2 strain, which is a PGPR that can metabolize flavonoids and PCBs. A significantly higher degradation of PCBs was observed with strains associated with flavonoid-expressing *Arabidopsis* mutants than with the flavonoid-null ones.

Rhizoengineering exploits favoritism of the microbes towards certain exotic nutrients provided by plants through their root exudates (Jha et al., 2015). This bias towards a particular nutrient component is advantageous in bioaugmentation as it provides a competitive advantage to certain microbes due to the formation of a nutrient-limiting rhizosphere. Studies have revealed that the rhizosphere of many plant species is nutrient-limiting leading to the shortage of rhizobacterial populations (Normander, 1999). One of the earlier reports based on this was the opine concept. Opines are one of the exotic nutrients favored by many microbes. Based on opine metabolism, root colonization by a particular bacteria could be controlled (Savka and Farrand, 1997). Yergeau et al. (2014) stated that the selective pressure in a contaminated rhizosphere is double compared to a non-contaminated one. They studied the rhizosphere microbiome of willow in polluted and non-polluted soils. They found that there was a considerable upregulation in genes involved in nutrient acquisition, quorum sensing, bacterial competition, and biofilm formation in the contaminated rhizosphere than in the non-polluted one through metatranscriptomics studies. This is due to the presence of specialized carbon sources of the root exudates and the contaminants. Only those microorganisms that can metabolize these could get established in the rhizosphere (Yergeau et al., 2014) (Figure 9.2). Abundance of catabolic genes in the contaminated rhizosphere was also studied earlier by Sipila et al. (2008) based on qPCR results.

Siciliano et al. (2001) reported that contaminant concentration in soil is a major factor determining the structure and function of the rhizosphere and root endophytic microbiome. Thijs et al. (2016) hypothesized that plants favor contaminant degradation by changing their root exudates in response to it and termed this as the 'call for support' (Figure 9.2). Moreover, there are other reports displaying the plants' ability to alter their root exudates in response to contaminants (Zhou et al., 2011). Kawasaki et al. (2011) demonstrated that certain plant species indirectly respond to PAH contamination using a split-root model. They suggested that the root exudates of these plants can alter the microbial composition of their rhizosphere at the genus level and favor rhizoremediation. Hence, bioaugmentation using such plant systems will prove successful in remediating rhizospheres contaminated with xenobiotics like PAHs.

To sum up, the establishment of a 'degradative' microbiome will only be possible when there is a moderate to low concentration of contaminant and competitive advantage in the rhizosphere. Pollutant-tolerant microbes would be selected at high contaminant levels with low disturbances such as chemical treatment, fertilization, excavation, and tillage. High disturbances in the bulk soil would favor ruderal microbes in the rhizosphere to migrate to free niches where easily available carbon sources are abundant. This would hamper the biodegradation processes. Hence, for efficient rhizoremediation, microbes capable to compromise the selection pressures resulting from competition, stress, and disturbances (Thijs et al., 2016) should be introduced through bioaugmentation methods.

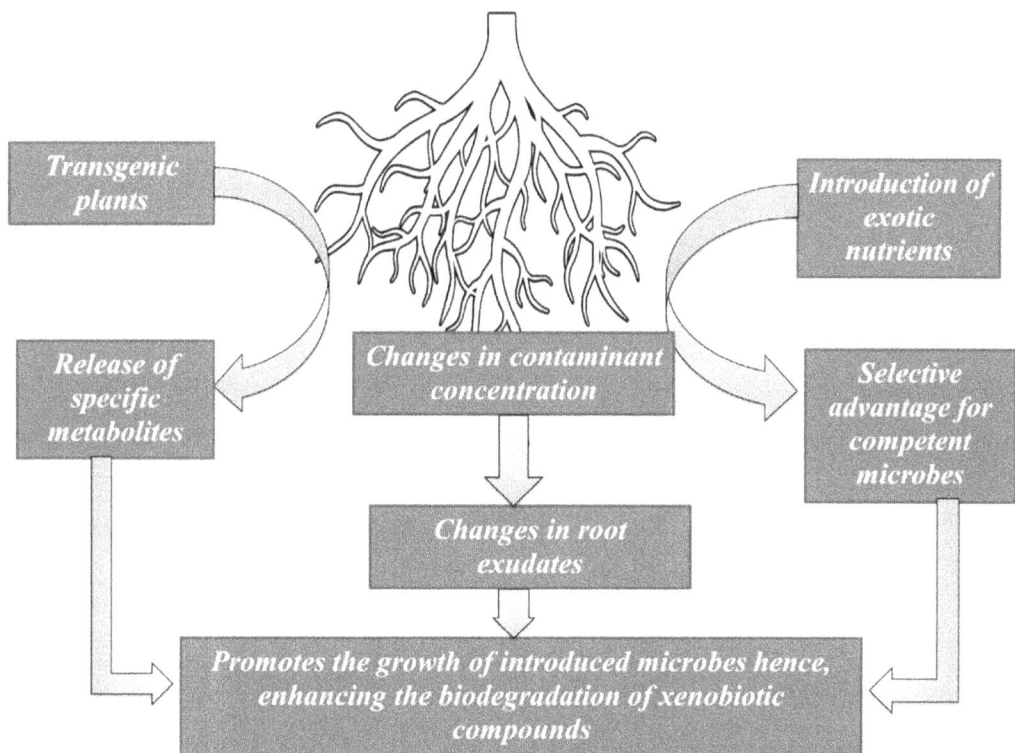

FIGURE 9.2 Root exudate-mediated rhizoengineering approaches in bioaugmentation. (Based on Thijs et al., 2016.)

9.6 CONCLUSION AND PERSPECTIVES

There isn't any doubt regarding the prime importance of xenobiotic degradation in vegetative soils. The question of 'how to carry about' has now attained clarity. Bioremediation supported by bioaugmentation is the excellent cost-effective green approach towards this. As far as arable lands are considered, rhizoremediation is crucial to avoid the deleterious effects of xenobiotics on plants. Both autochthonous and allochthonous bioaugmentation in the rhizosphere could be implemented depending upon the type of xenobiotic to be degraded. Autochthonous degrading microbes have the upper hand but if highly efficient exogenous candidates can thrive well in the xenobiotic environment, allochthonous rhizosphere bioaugmentation can be adopted. Rhizoengineering, an upcoming technology, could be implemented along with bioaugmentation to speed up the remediation processes. But the question still persists, is it bioaugmentation in rhizoengineering or rhizoengineering approach in bioaugmentation? In both cases, finally, the rhizosphere is established with degrading microbial strains. With such an engineered catabolic rhizosphere, xenobiotic degradation could never be a worry.

ACKNOWLEDGMENT

The authors would like to acknowledge DST-PURSE (No. SR/PURSE/Phase 2/26 (C)) and DST-FIST (No. SR/FST/LSI-660/2016 (C)) for the support rendered.

REFERENCES

Alvarez, P. J. J. and W. A. Illman. 2006. *Bioremediation and Natural Attenuation-Process Fundamentals and Mathematical Models*. New York: Wiley.

Anderson, T. A., E. A. Guthrie, and B. T. Walton. 1993. Bioremediation in the rhizosphere. *Environmental Science & Technology* 27:2630–2636.

Anderson, T. A., and J. R. Coats. 1995. An overview of microbial degradation in the rhizosphere and its implication for bioremediation. In H. D. Skipper and R. F. Turco (ed.), *Bioremediation: Science and Applications*. vol. 43. Madison, WI: Soil Science Society of America. pp. 135–143.

Anderson, T.A., D.C. White, and B.T. Walton. 1995. Degradation of hazardous organic compounds by rhizosphere microbial communities. In V. Singh, (ed.), *Biotransformations: Microbial Degradation of Health-Risk Compounds*. Elsevier, Amsterdam. pp. 205–225.

Anwar, S., F. Liaquat, Q. M. Khan, Z. M. Khalid, and S. Iqbal. 2009. Biodegradation of chlorpyrifos and its hydrolysis product 3, 5, 6-trichloro-2-pyridinol by *Bacillus pumilus* strain C2A1. *Journal of Hazardous Materials* 168:400–405.

Asok, A. K., and M. S Jisha. 2012. Biodegradation of the anionic surfactant linear alkylbenzenesulfonate (LAS) by autochthonous *Pseudomonas* sp. *Water, Air, & Soil Pollution* 223(8):5039–5048.

Berg, G. and K. Smalla. 2009. Plant species and soil type cooperatively shape the structure and function of microbial communities in the rhizosphere. *FEMS Microbiology Ecology* 68:1–13.

Boyajian G.E, and L.H. Carreira. 1997. Phytoremediation: a clean transition from laboratory to marketplace. *Nature Biotechnology* 15:127–128.

Braud, A., K. Jézéquel, S. Bazot, and T. Lebeau. 2009. Enhanced phytoextraction of an agricultural Cr-and Pb-contaminated soil by bioaugmentation with siderophore-producing bacteria. *Chemosphere* 74(2): 280–286.

Brazil, G. M., L. Kenefick, M. Callanan, A. Haro, V. de Lorenzo, D. N. Dowling, and F. O'Gara. 1995. Construction of a rhizosphere pseudomonad with potential to degrade polychlorinated biphenyls and detection of *bph* gene expression in the rhizosphere. *Applied and Environmental Microbiology* 61:1946–1952.

Brimecombe, M. J., F. A. De Leij, and J. M. Lynch. 2001. The effect of root exudates on rhizosphere microbial populations. In R. Pinton, Z. Varanini, P. Nannipieri, eds, *The Rhizosphere*. New York: Marcel Dekker, pp. 95–1.

Chaudhry, Q., M. Blom-Zandstra, S. Gupta, and E.J. Joner. 2005. Utilizing the synergy between plants and rhizosphere microorganisms to enhance breakdown of organic pollutants in the environment. *Environmental Science and Pollution Research* 12:34–48.

Chee, G. J. 2011. Biodegradation analyses of trichloroethylene (TCE) by bacteria and its use for biosensing of TCE. *Talanta* 85(4):1778–1782.

Colbert, S. F., T. Isakeit, M. Ferri, A. R. Weinhold, M. Hendson, and M. N. Schroth. 1993. Use of an exotic carbon source to selectively increase metabolic activity and growth of *Pseudomonas putida* in soil. *Applied and Environmental Microbiology* 59:2056–2063.

Da Silva, M. L. B., and P. J. J. Alvarez. 2010. Bioaugmentation. In: Timmis K.N. (ed) *Handbook of Hydrocarbon and Lipid Microbiology.* Springer, Berlin, Heidelberg. pp. 4531–4544.

Devliegher, W., S. M. A. Arif, and W. Verstraete. 1995. Survival and plant growth promotion of detergent-adapted *Pseudomonasfluorescence* ANP15 and *Pseudomonas aeruginosa*7NSK2. *Applied and Environmental Microbiology* 61:3865–3871.

Duba, A. G., K. J. Jackson, M. C. Jovanovich, R. B. Knapp, and R. T. Taylor. 1996. TCE remediation using in situ, resting state bioaugmentation. *Environmental Science & Technology* 30:1982–1989.

Dzantor, E. K. 2007. Phytoremediation: the state of rhizosphere 'engineering' for accelerated rhizodegradation of xenobiotic contaminants. *Journal of Chemical Technology & Biotechnology: International Research in Process, Environmental & Clean Technology* 82(3):228–232.

El Fanroussi, S., and S. N. Agathos. 2005. Is bioaugmentation a feasible strategy for pollutant removal and site remediation? *Current Opinion in Microbiology* 8:268–75.

El-Helow, E. R., M. E. I. Badawy, M. E. M. Mabrouk, E. A. H. Mohamed, and Y. M. El-Beshlawy. 2013. Biodegradation of Chlorpyrifos by a Newly Isolated *Bacillus subtilis* Strain, Y242. *Bioremediation* 17:113–123.

Fletcher, J. S. and R. S. Hedge. 1995. Release of phenols by perennial plant roots and their potential importance in bioremediation. *Chemosphere* 31:3009–3016.

Gerhardt, K. E., X. D. Huang, B. R. Glick, and B. M. Greenberg. 2009. Phytoremediation and rhizoremediation of organic soil contaminants: potential and challenges. *Plant Science* 176(1): 20–30.

Godleads, O. A., T. F. Prekeyi, E. O. Samson, and I. Ehinomen. 2015. Bioremediation, Biostimulation and Bioaugmention: A Review. *International Journal of Environmental Bioremediation & Biodegradation* 3(1):28–39.

Goldstein, R. M., L. M. Mallory, and M. Alexander.1985. Reasons for possible failure of inoculation to enhance biodegradation. *Applied and Environmental Microbiology* 50:977–983.

González, N., R. Simarro, M. C. Molina, L. F. Bautista, L. Delgado, and J. A. Villa. 2011. Effect of surfactants on PAH biodegradation by a bacterial consortium and on the dynamics of the bacterial community during the process. *Bioresource Technology* 102(20):9438–9446.

Goswami, M., P. Chakraborty, K. Mukherjee, G. Mitra, P. Bhattacharyya, S. Dey, and P. Tribedi. 2018. Bioaugmentation and biostimulation: a potential strategy for environmental remediation. *Journal of Microbiology & Experimentation* 6(5):223–231.

Greer, L. E., and D. R. Shelton. 1992. Effect of inoculants strain and organic matter content on kinetics of 2,4-dichloro-phenoxyacetic acid degradation in soil. *Applied and Environmental Microbiology* 58(5):1459–1465.

Hatamian-Zarmi, A., S. A. Shojaosadati, E. Vasheghani-Farahani, S. Hosseinkhani, and A. Emamzadeh. 2009. Extensive biodegradation of highly chlorinated biphenyl and Aroclor 1242 by *Pseudomonas aeruginosa* TMU56 isolated from contaminated soils. *International Biodeterioration & Biodegradation* 63(6):788–794.

Hernández-Sánchez, V., and R. M. Wittich. 2012. Possible reasons for past failures of genetic engineering techniques for creating novel, xenobiotics-degrading bacteria. *Bioengineered* 3(5):260–261.

Holden, P.A, and M. K. Firestone. 1997. Soil microorganisms in soil cleanup: how can we improve our understanding? *Journal of Environmental Quality* 26:32–40.

Hong, Q., Z. Zhang, Y. Hong, and S. Li 2007. A microcosm study on bioremediation of fenitrothion-contaminated soil using Burkholderia sp.FDS-1. *International Biodeterioration & Biodegradation* 59(1): 55–61.

Hosokawa, R., M. Nagai, M. Morikawa, and H. Okuyama 2009. Autochthonous bioaugmentation and its possible application to oil spills. *World Journal of Microbiology Biotechnology* 25:1519–1528.

Jabeen, H., S. Iqbal, and S. Anwar. (2014). Biodegradation of chlorpyrifos and 3, 5, 6-trichloro-2-pyridinol by a novel rhizobial strain *Mesorhizobium*sp. HN3. *Water and Environment Journal* 29:151–160.

Jha, P., J. Panwar, and P. N. Jha. 2015. Secondary plant metabolites and root exudates: guiding tools for polychlorinated biphenyl biodegradation. *International Journal of Environmental Science and Technology* 12:789–802.

John, E. M., J. Sreekumar, and M. S. Jisha. 2016. Optimization of chlorpyrifos degradation by assembled bacterial consortium using response surface methodology. *Soil and Sediment Contamination: An International Journal* 25(6):668–682.

Kawasaki, A., E. R. Watson, and M. A. Kertesz. 2011. Indirect effects of polycyclic aromatic hydrocarbon contamination on microbial communities in legume and grass rhizospheres. *Plant Soil* 358:169–182.

Khan, A. G. 2006. Mycorrhizore mediation, an enhanced form of phytoremediation. *Journal of Zhejiang University, Science B* 7(7):503–14.

Krumme, M. L., R. L. Smith, J. Egestorff, S. M. Thiem, J. M. Tiedje, K. N. Timmis, and D. F. Dwyer. 1994. Behavior of pollutant-degrading microorganisms in aquifers:predictions for genetically engineered organisms. *Environmental Science & Technology.* 28: 1134–1138.

Kuiper, I., E. L. Lagendijk, G. V. Bloemberg, and B. J. Lugtenberg. 2004. Rhizoremediation: a beneficial plant-microbe interaction. *Molecular Plant Microbe Interactions* 17(1):6–15.

Kuiper, I., L. V. Kravchenko, G. V. Bloemberg, and B. J. Lugtenberg. 2002. *Pseudomonas putida* strain PCL1444, selected for efficient root colonization and naphthalene degradation, effectively utilizes root exudate components. *Molecular Plant Microbe Interactions* 15(7):734–41.

Kumari, B., M. A. Mallick, M. K. Solanki, A. C. Solanki, A. Hora, and W. Guo. 2019. Plant growth promoting rhizobacteria (PGPR): modern prospects for sustainable agriculture. In *Plant Health under Biotic Stress*. Singapore: Springer. pp. 109–127.

Leung, K. T., M. B. Cassidy, K. W. Shaw, H. Lee, J. T. Trevors, E. M. Lohmeier-Vogel, and H. J. Vogel. 1997. Pentachlorophenol biodegradation by *Pseudomonas* spp. UG25 and UG30. *World Journal of Microbiology and Biotechnology.* 13(3):305–313.

Li, X., J. He, and S. Li. 2007. Isolation of a chlorpyrifos-degrading bacterium, *Sphingomonas* sp. strain Dsp-2, and cloning of the mpd gene. *Research in Microbiology* 158:143–149.

Mallick, K., K. Bharati, A. Banerji, N. A. Shakil, and N. Sethunathan. 1999. Bacterial degradation of chlorpyrifos in pure cultures and in soil. *Bulletin of Environmental Contamination and Toxicology* 62:48–54.

Metting, F. B. Jr. 1992. Structure and physiological ecology of soil microbial communities. In FB Metting Jr, ed, *Soil Microbial Ecology*. New York: Marcel Dekker, pp. 3–25.

Monti, M. R., A. M. Smania, G. Fabro, M. E. Alvarez, and C. E. Argaraña. 2005. Engineering *Pseudomonas fluorescens* for biodegradation of 2, 4-dinitrotoluene. *Applied and Environmental Microbiology* 71(12):8864–8872.

Mrozik A., and Z. Piotrowska-Seget. 2010. Bioaugmentation as a strategy for cleaning up of soils contaminated with aromatic compounds. *Microbiological Research* 165(5):363–375.

Narasimhan, K., C. Basheer, V. B. Bajic, and S. Swarup. 2003. Enhancement of plant-microbe interactions using a rhizosphere metabolomics-driven approach and its application in the removal of polychlorinated biphenyls. *Plant Physiology* 132:146–153.

Neal, A. L., S. Ahmad, R. Gordon-Weeks, and J. Ton. 2012. Benzoxazinoids in root exudates of maize attract *Pseudomonas putida* to the rhizosphere. *PLoS One* 7:e35498.

Normander, B., N. B. Hendriksen, and O. Nybroe. 1999. Green fluorescent protein marked *Pseudomonas fluorescens*: localization, viability, and activity in the natural barley rhizosphere. *Applied and Environmental Microbiology* 65:4646–4651.

O'Connell, K. P., R. M. Goodman, and J. Handelsman. 1996. Engineering the rhizosphere: expressing a bias. *Trends in Biotechnology* 14:83–88.

Pajuelo, E., I. D. Rodríguez-Llorente, A. Lafuente, P. Perez-Palacios, B. Doukkali, and M. A. Caviedes. 2014. Engineering the rhizosphere for the purpose of bioremediation: an overview. *CAB Reviews* 9(001): 1–17.

Pandey, J., N. K. Sharma, F. Khan, A. Ghosh, J. G. Oakeshott, R. K. Jain, and G. Pandey. 2012. Chemotaxis of Burkholderia sp. strain SJ98 towards chloronitroaromatic compounds that it can metabolise. *BMC Microbiology* 12(1): 1–9.

Ryan, P. R., Y. Dessaux, L. S. Thomashow, and D. M. Weller. 2009. Rhizosphere engineering and management for sustainable agriculture. *Plant and Soil* 321(1):363–383.

Ryan, R. P., D. Ryan, and D. N. Dowling 2007. Plant protection by the recombinant, root-colonizing *Pseudomonas fluorescens* F113rifPCB strain expressing arsenic resistance: improving rhizoremediation. *Letters of Applied Microbiology* 45:668–74.

Santillan, J. Y., N. L. Rojas, P. D. Ghiringhelli, M. L. Nóbile, E. S. Lewkowicz, and A. M. Iribarren. 2020. Organophosphorus compounds biodegradation by novel bacterial isolates and their potential application in bioremediation of contaminated water. *Bioresource Technology* 317:124003.

Savka, M. A., and S. K. Farrand. 1997. Modification of rhizobacterial populations by engineering bacterium utilization of novel plant-produced resource. *Nature Biotechnology* 15:363–368.

Shim, H., S. Chauhan, D. Ryoo, K. Bowers, S. M. Thomas, K. A. Canada, J. G. Burken, and T. K. Wood. 2000. Rhizosphere competitiveness of trichloroethylene-degrading poplar-colonizing recombinant bacteria. *Applied and Environmental Microbiology* 66:4673–4678.

Siciliano, S. D., N. Fortin, A. Mihoc, G. Wisse, S. Labelle, D. Beaumier, D. Ouellette, R. Roy, L. G. Whyte, M. K. Banks, P. Schwab, K. Lee, and C. W. Greer 2001. Selection of specific endophytic bacterial genotypes by plants in response to soil contamination. *Applied and Environmental Microbiology* 67:2469–2475.

Singer, C. A., D. Smith, W. A. Jury, K. Hathuc, and D. E. Crowley. 2003. Impact of the plant rhizosphere and augmentation on remediation of polychlorinated biphenyl contaminated soil. *Environmental Toxicological Chemistry* 22:1998–2004.

Singh, B. K., A. Walker, J. A. W. Morgan, and D. J Wright. 2004. Biodegradation of chlorpyrifos by *Enterobacter* strain B-14 and its use in bioremediation of contaminated soils. *Applied and Environmental Microbiology* 70:4855–4863.

Sipila, T. P., A. K. Keskinen, M. L. Akerman, C. Fortelius, K. Haahtela, and K. Yrjala. 2008. High aromatic ring-cleavage diversity in birch rhizosphere: PAH treatment-specific changes of I.E.3 group extradioldioxy genases and 16S rRNA bacterial communities in soil. *ISME Journal* 2:968–981.

Smith, D., S. Alvey, and D. E. Crowley. 2005. Cooperative catabolic pathways within an atrazine degrading enrichment culture isolated from soil. *FEMS Microbiology Ecology* 53:265–273.

So, J., D. A. Webster, B. C. Stark, and K. R. Pagilla. 2004. Enhancement of 2, 4-dinitrotoluene biodegradation by Burkholderia sp. in sand bioreactors using bacterial hemoglobin technology. *Biodegradation* 15(3):161–171.

Tartanus, M., E. Malusá, B. H. Łabanowska, A. Miszczak, and E. Szustakowska. 2017. DDT content in polish soils-current state and attempts of rhizo-bioremediation. *Journal of Research and Applications in Agricultural Engineering* 62(4).

Tchelet, R., R. Meckenstock, P. Steinle, and J. R. van der Meer. 1999. Population dynamics of an introduced bacterium degrading chlorinated benzenes in a soil column and in sewage sludge. *Biodegrdation* 10(2):113–125.

Thijs, S., and J. Vangronsveld. 2015. Rhizoremediation. In *Principles of Plant-Microbe Interactions.* Springer International Publishing.

Thijs, S., N. Weyens, W. Sillen, P. Gkorezis, R. Carleer, and J. Vangronsveld. 2014. Potential for plant growth promotion by a consortium of stress-tolerant 2,4-dinitrotoluene-degrading bacteria: isolation and characterization of a military soil. *Microbial Biotechnology* 7(4):294–306.

Thijs, S., W. Sillen, F. Rineau, N. Weyens, and J. Vangronsveld. 2016. Towards an Enhanced understanding of plant–microbiome interactions to improve phytoremediation: engineering the metaorganism. *Frontiers in Microbiology.* 7:341.

Tiwari, J., P. Naoghare, S. Sivanesan, and A.Bafana. 2017. Biodegradation and detoxification of chloronitroaromatic pollutant by Cupriavidus. *Bioresource Technology* 223:184–191.

Tribedi, P., S. Sarka, K. Mukherjee, and A. K. Sil. 2012. Isolation of a novel *Pseudomonas* sp. from soil that can efficiently degrade polyethylene succinate. *Environmental Science and Pollution Research* 19(6):2115–2124.

Tyagi, M., M. Manuela, R. Da Fonseca, and C. C. C. R. De Carvalho. 2011. Bioaugmentation and biostimulation strategies to improve the effectiveness of bioremediation processes. *Biodegradation* 22:231–41.

Villacieros, M., C. Whelan, M. Mackova, J. Molgaard, M. Sanchez-Contreras, J. Lloret, D. Aguirre de Carcer, R. I. Oruezabal, L. Bolanos, T. Macek, U. Karlson, D. N. Dowling, M. Martin, and R. Rivilla.2005. Polychlorinated biphenyl rhizoremediation by *Pseudomonas fluorescens* F113 derivatives, using a *Sinorhizobium meliloti nod* system to drive *bph* gene expression. *Applied and Environmental Microbiology* 71:2687–2694.

Wang, J. L., Z. Y. Mao, L. P. Han, and Y. Qian. 2004. Bioremediation of quinoline-contaminated soil using bioaugmentation in slurry-phase reactor. *Biomedical and Environmental Sciences* 17(2):187–195.

WerheniAmmeri, R., S. MokniTlili, I. Mehri, S. Badi, and A. Hassen. 2016. Pentachlorophenol biodegradation by *Citrobacter freundii* isolated from forest contaminated soil. *Water, Air, & Soil Pollution* 227(10):1–12.

Whipps, J. M. 1990. Carbon economy. In JM Lynch, ed, *The Rhizosphere.* New York: Wiley, pp. 59–97.

Xu, L., Y. Teng, Z. G. Li, J. M. Norton, and Y. M. Luo. 2010. Enhanced removal of polychlorinated biphenyls from alfalfa rhizosphere soil in afield study: the impact of a rhizobial inoculum. *The Science of the Total Environment* 408(5):1007–13.

Yang, C., N. Liu, X. Guo, and C. Qiao. 2006. Cloning of mpd gene from a chlorpyrifos degrading bacterium and use of this strain in bioremediation of contaminated soil. *FEMS Microbiology Letters* 265:118–125.

Yee, D. C., J. A. Maynard, and T. K. Wood. 1998. Rhizoremediation of trichloroethylene by a recombinant, root-colonizing *Pseudomonas fluorescens* strain expressing toluene *ortho*-monooxygenase constitutively. *Applied and Environmental Microbiology* 64:112–118.

Yergeau, E., S. Sanschagrin, C. Maynard, M. St-Arnaud, and C. W. Greer. 2014. Microbial expression profiles in the rhizosphere of willows depend on soil contamination. *The ISME Journal* 8:344–358.

Zhang, H., S. Hanada, T. Shigematsu, K. Shibuya, Y.Kamagata, T. Kanagawa, and R. Kurane. 2000. *Burkholderiakururiensis* sp. nov., a trichloroethylene(TCE)-degrading bacterium isolated from an aquifer polluted with TCE. *International Journal of Systematic and Evolutionary Microbiology* 50(2):743–749.

Zhou, N., P. Liu, Z. Y. Wang, and G. D. Xu. 2011. The effects of rapeseed root exudates on the forms of aluminum in aluminum stressed rhizosphere soil. *Crop Protection* 30:631–636.

Zhuang, W. Q., J. H. Tay, A. Maszenan, and S. Tay. 2002. *Bacillus naphthovorans* sp. nov. from oil-contaminated tropical marine sediments and its role in naphthalene biodegradation. *Applied Microbiology and Biotechnology* 58(4):547–554.

10 Bioaugmentation of Municipal Waste

Recycling of Electronic Wastes through Biohydrometallurgical Technology

Deviany Deviany
Institut Teknologi Sumatera

Siti Khodijah Chaerun
Institut Teknologi Bandung

CONTENTS

10.1 INTRODUCTION

Waste generation has been set off extensively by the human population, which grows rapidly coupled with accelerated industrialization and urbanization. Waste properties are determined by the GDP status of nations and their geographical positions (Das et al. 2019). Globally, 1.3 billion tons of municipal solid waste are produced per year, which is expected to escalate to 2.2 billion tons by 2025 (Hoornweg and Bhada-Tata 2012). In the case of least developed countries, waste generation rates are expected to increase twice for the next few decades. Organization for Economic Co-operation and Development (OECD) lists waste from households, institutions, commerce and trade, office buildings, and small business collected and managed by municipalities as municipal waste (European Commission 2016). Additionally, food waste, plastics, paper, textiles, glass and pottery, garden and park waste, rubber, electronic waste, and other materials such as dirt, soil, dust, and so on are listed as municipal waste by the Intergovernmental Panel on Climate Change (IPCC) (Hoornweg and Bhada-Tata 2012). This chapter will focus on the biological treatment (specifically by bioaugmentation) of electronic waste as an example of municipal solid waste that needs immediate attention due to its substantial impacts on society and the ecosystem. The biological treatment of electronic waste is known as biomining (urban biomining) or biohydrometallurgy technology.

Electronic waste, otherwise called a waste of electrical and electronic equipment (WEEE), was aroused due to the Industrial Revolution from the progressively advanced electronics industry. United Nations Environment Programme (UNEP) reported escalating illegal import of electronic

DOI: 10.1201/9781003187622-10

waste in South Africa with as much as 200%–400% and India up to 500% (Schluepet et al. 2009). The inorganic and organic constituents of WEEE could generate a negative impact if they were not correctly handled. Incineration and landfill came as a proposed method, although those were deemed unsuitable for their emerging problem affecting human health and the environment (Robinson 2009; Tsydenova and Bengtsson 2011). Conversely, precious and base metals such as palladium (Pd), gold (Au), silver (Ag), and copper (Cu) contained in the electronic waste make them a valuable resource for minerals and metals (Oguchi 2013). For the last decade, extensive studies have been conducted to improve technologies for the recovery of metals from WEEE, particularly with pyrometallurgy and hydrometallurgy, which use thermal treatment and extraction solutions, respectively, for valuable metals recovery from mineral and ores. However, these techniques have proven to threaten the environment (Korteet et al. 2000; Mecucci and Scott 2002). The application of biohydrometallurgy or biomining utilizing the ability of microbes to generate leaching agents has emerged as an alternative to the previous conventional methods. Biohydrometallurgy is considered an attractive solution owning to its advantages being appreciably low operational cost and reducing environmental contamination risk (Ehrlich 2001; Brierley 2010).

Electronic waste/WEEE compositions vary depending on the type of waste, divided into five categories: non-ferrous metals, ferrous metals, plastics, glass, and other materials. Ongondo et al. (2011) reported the type of components in WEEE such as 60% metals, 15% plastics, 12% CRT and LCD screens, 5% metal–plastic mixture, and other materials. The metals component in WEEE also varied depending on the type of electronic waste. For example, printed circuit boards (PCBs) are comprised of copper (Cu: 20%), iron (Fe: 8%), tin (Sn: 4%), lead (Pb: 2%), nickel (Ni: 2%), and zinc (Zn: 1%) with the precious metal content of silver (Ag: 0.2%), gold (Au: 0.1%), and palladium (Pd: 0.005%) (He et al. 2006).

Electronic waste recycling is deemed vital for a variety of reasons, including:

1. to lessen the burden of electronic waste as a form of toxic solid waste on the environment
2. to recover precious metals. Considering in 1 ton of used mobile phones, base, and precious metals can be extracted, including approximately 130 kg of Cu, 3 kg Ag, 300–500 g of Au, and 140 g of Pd. Estimation of Au content in WEEE is roughly 10–1,000 g/ton, significantly higher than in natural ores in the range of 0.5–13 g/ton
3. as an alternative resource for natural ores, which are limited in supply (non-renewable resources).

Recently, responsible management of WEEE related to the final disposal of electronic devices has been a global concern. Safe disposal of various electronic waste is a pressing concern requiring the global community's awareness. Regulation issued by the European Union, Directive 2008/98/EC, states that the management of WEEE follows principles of waste hierarchy which consist of (i) prevention, (ii) preparation for reuse, (iii) recycling, (iv) other use of recovery such as energy recovery, and (v) disposal.

Landfill disposal and incineration are often the predominant treatments for waste management, with few differences between countries. These practices have long been considered unsafe, raising problems for humans and nature. WEEE discarded to a landfill can release metals through leaching, contaminating the aquatic and terrestrial ecosystem. Incineration aids in reducing waste volume and acts as toxic chemical compounds such as dioxin and furan. These hazardous pollutants could be released from the incinerator when appropriate cleaning of the exhaust gas system was not supervised. Incineration also contributes to annual emission levels of cadmium (Cd) and mercury (Hg) (Crowe et al. 2003).

Extraction of precious and base metals such as Au, Ag, Cu, and Zn from electronic wastes is an alternative solution for waste management to reduce the application of incineration and landfill disposal. This waste recycling for metals recovery can be considered as an alternative to mineral extraction from low-grade resources. Extraction of metals from minerals and ores extensively consumes land and energy, generates waste and mining byproducts, and releases a large amount of sulfur dioxide (SO_2) and carbon dioxide (CO_2) compared to the extraction of metals from electronic wastes.

The first step in the recycling management scheme for electronic waste is the removal of toxic compounds/chemicals such as chlorofluorocarbons (CFC), polychlorinated biphenyl, and mercury (Hg) using pretreatments which include manual removal of pollutant/contaminant, crushing/shredding/grinding, air classifiers/hoods, and electrostatic and magnetic separations. According to Bigum et al. (2012), the foremost pretreatments step for electronic waste processing comprises 29% manually separated components, 33% ferromagnetic materials, and 26% plastics residue fraction with other material fractions amounting to 2%–3% of the stream. Pre-treatment plant of high-grade WEEE dust per 1,000 kg commonly produces ferromagnetic iron and steel (381 kg), plastic parts (265 kg), copper and precious metals (165 kg), materials with particular regulated treatments needed (114 kg), residual waste (53 kg) and aluminum (22 kg).

Commonly, the next step of treatment for WEEE is based on pyrometallurgy, hydrometallurgy, and electrometallurgy process. WEEE is heated at a high temperature in pyrometallurgy, sometimes above 1,000°C, separating into different phases to extract the precious metals from impurities. High energy consumption and toxic gas production are linked to this process. Hydrometallurgy utilized strong acids and bases as leaching agents combined with complexing agents in an aqueous solution to recover metals. These agents include sulfuric acid, acetic acid, carbonic acid, oxalic acid, thiosulfate, thiourea, and halide. Compared with pyrometallurgy, hydrometallurgy requires lower energy and operational cost for the plant with relatively low capacity. Electrometallurgy uses electrical power to recover metal from its compounds/ores, for example, electrowinning/electroextraction and electrorefining of copper, zinc, and other elements. This process can also be considered energy-consuming. Rocchetti et al. (2013) suggested a combination of treatment with a chemical approach that is usually more efficient and a biological approach that is relatively safe for the environment taking advantage of both leaching applications. To summarize, until today, the management of WEEE treatment is done to recover and recycle resources by pyrometallurgy and hydrometallurgy, regeneration of spent catalyst, and landfill disposal.

A safe environment innovation of WEEE management is needed more than ever, considering the global climate emergency. Biohydrometallurgy, or biomining, exploiting the ability of microbes to extract metals should be considered to be developed for treatment management of WEEE. The term urban biomining is also used to describe the process of metals recovery from electronic wastes. Some advantages of urban biomining are environmentally friendly with no secondary pollutant/toxic byproducts, considerably low energy consumption and operational cost, detoxification of hazardous and toxic wastes, and the possibility of recovering base and precious metals. However, optimization is required considering the processing time and the toxicity level of metals in WEEE to grow microbes used in urban biomining.

10.2   RECOVERY OF GOLD (AU) FROM ELECTRONIC WASTES BY ADDITION OF EXOGENOUS MICROBES (THE SO-CALLED BIOAUGMENTATION)

In the biomining of gold (Au), iron-oxidizing and sulfur-oxidizing microbes are widely used to increase the accessibility of Au by oxidizing sulfidic refractory gold ores. This biooxidation process will be followed by the chemical extraction of Au via the formation of a gold-cyanide complex under alkaline conditions. Cyanide is one of few compounds, aside from thiourea, chloride and other halide compounds, forming soluble complexes with Au in water with high extraction efficiency even with concentration as low as 0.25% (Smith & Mudder 1991; Syed 2012). The cyanidation process is a common practice for the recovery of Au in hydrometallurgy. Biotechnology methods using cyanogenic organisms to minimize the toxicity of this process have gained significant attention this past decade—some living microbes such as bacteria, fungi, and microalgae are naturally able to produce hydrogen cyanide (HCN). Through bioaugmentation, which is the addition of pre-cultured and pre-adapted microbes into a biological system, these cyanogenic microbes can extract Au from WEEE. The use of biogenic cyanide (bio-cyanidation) offers an alternative to conventional gold extraction technique (heterotrophic bioleaching under alkaline conditions) using cyanogenic microbes (Gadd 2000) or fungi (Sabra et al. 2011).

Chromobacterium violaceum is the most extensively studied bacterium for recovering Au from ores (Lawson et al. 1999). *C. violaceum* produces HCN as a secondary product from oxidative decarboxylation of glycine catabolism, which is the primary precursor of cyanide. HCN production occurs during the short period of bacterial growth phase; between the late logarithmic and stationary phase (for *C. violaceum*) and early stationary phase (for the genus *Pseudomonas*). Au solubilization with biocyanidation includes anodic and cathodic reactions as follows:

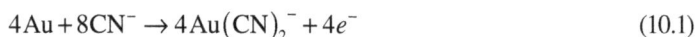

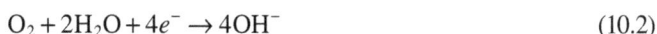

$$4Au + 8CN^- \rightarrow 4Au(CN)_2^- + 4e^- \tag{10.1}$$

$$O_2 + 2H_2O + 4e^- \rightarrow 4OH^- \tag{10.2}$$

The overall reaction below is known as Elsner's equation (Kita et al. 2006).

$$4Au + 8CN^- + O_2 + 2H_2O \rightarrow 4Au(CN)_2^- + 4OH^- \tag{10.3}$$

Other metals' cyanidation processes largely follow the above equations. Cyanogenic bacteria, including *C. violaceum,* have optimum pH ranges between 7 and 8 for cyanogenesis, even though HCN has a pKa value of 9.3 (Lear et al. 2010). Cyanide for metals complexation is less available at physiology pH due to a considerable loss of HCN by evaporation. Hence, at the same pH value or above pH 10.5, the available form is cyanide ions. Identifying the optimum operational conditions is a substantial challenge of biocyanidation application for WEEE. Another critical factor that has to be addressed is the existence of metals aside from Au in WEEE. Cu and other precious metals will compete with Au for the complexation with cyanide ions. Based on these reasons, some studies recommended pre-treatment of WEEE powder to remove the sizeable amount of undesired metals. Natarajan and Ting (2014) showed that nitric acid solution (6 M) was able to recover almost 80% of Cu, 55%–70% of Ag, Fe, Al, Zn, Pb, and Ni, and 90% of Sn. Studies on the effect of structure and composition of WEEE heterogeneity are also needed. To date, there are no studies on large-scale applications.

The biomineralization process by bacteria capable of forming minerals is used to precipitate the extracted soluble Au. *Delftia acidovorans* is one of the microbes able to mediate the biomineralization of Au (Au^{3+} to Au^0). In contact with Au^{3+}, this bacterium will produce nonribosomal extracellular peptide delftibactin, forming Au particles through complexation to protect itself from the metal's toxicity (Johnston et al. 2013). Apart from bacteria, fungi are also able to modify mineral substrate (Fomina et al. 2010). Previous studies on fungi showed their ability to immobilize Au through biomineralization. Recent studies have shown that fungi have the ability to recover Au through processes such as biosorption, bioleaching, biomineralization, and mechanical attack. Fungi explore their surrounding environment in search of food. If they grow on the surface of the gold layer, some fungal strains can form fractures. The fungi may carry this out to obtain the nutrition needed for them to grow. Some fungi which can form fractures and holes on the surface layer of Au include *Fusarium oxysporum, Aureobasidium pullulans, Verticillium* sp., and *Epicoccum* sp. Other fungi with the ability to form fractures with lower efficiency are *Acremonium* sp., *Aspergillus niger, Phoma* sp., *Trichoderma* sp., and *Ulocladium* sp. Observation by energy-dispersive X-ray spectroscopy and scanning electron micrograph (SEM) on mycelium grown over gold sheet revealed accumulation of Au related to the fungal hyphae.

Studies showed that fungi could mediate oxidation of Au inhabiting below the earth's surface as part of soil microbes. The study led by Bohu et al. (2019) showed that fungi could affect Au biogeochemical cycles. Superoxide and carbonyl compounds produced from fungal cell metabolism play a role in the redox reaction of Au, in which Au lowers the energy limitation of fungal metabolism in using the substrate and accelerates its growth in the environment. As a result, it generates alteration of biogeochemical cycles and affects mobilization and accumulation of Au.

10.3   RECOVERY OF COPPER (CU) AND OTHER METALS AS PRE-TREATMENT FOR GOLD EXTRACTION

Cyanide forms complex with Cu more readily and stable compared to that with Au. Thus, pre-treatment is necessary before performing bioleaching of Au using cyanogenic bacteria for WEEE containing a significant amount of Cu. Chemical leaching using HNO_3 or bioleaching with iron- and sulfur-oxidizing mixotrophic bacteria could be used to remove Cu from WEEE samples. Some kinds of direct mechanisms allow iron- and sulfur-oxidizing bacteria to mobilize metals from electronic waste. The bacteria cell will move toward the mineral surface and secrete extracellular polymeric substances (EPS), forming biofilm while the bulk of their mass will attach to the sulfide surfaces. The area between bacterial cell walls and mineral surfaces will be filled by a complex slime matrix from the EPS released by the bacterial intercedes with the adherence process.

In bioleaching with iron- and sulfur-oxidizing bacteria, metals solubilization happens through acid leaching and/or redox reaction. Generation and regeneration of leaching agents by bacterial leaching will occur mainly for Fe (III) ions and protons (H^+). Oxidation of insoluble Cu^0 into soluble Cu^{2+} will take place with biooxidation of Fe (II) into Fe (III), using iron as a source of energy for leaching, as shown in Equations (10.4) and (10.5).

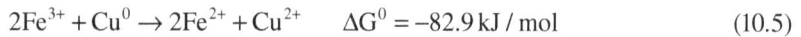

$$4Fe^{2+} + O_2 \rightarrow 4Fe^{3+} + H_2O \tag{10.4}$$

$$2Fe^{3+} + Cu^0 \rightarrow 2Fe^{2+} + Cu^{2+} \quad \Delta G^0 = -82.9\,kJ/mol \tag{10.5}$$

In the absence of iron, the leaching of Cu has also been observed with sulfur as the source of energy. Equation (10.6) explains the assisted solubilization of Cu^0 by proton and molecular oxygen.

$$2Cu^0 + 4H^+ + O_2 \rightarrow 2Cu^{2+} + 2H_2O \tag{10.6}$$

During Cu solubilization, Fe(III) ions were released from WEEE components. These ions will assist in producing new protons through hydrolysis and, in turn, increase Cu solubility. Consistent with their thermodynamic suitability reactions, solubilization of other metals such as Al, Ni, and Zn will go through the same mechanism as shown in Equations (10.7)–(10.9).

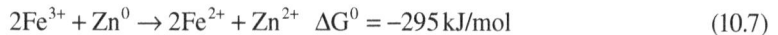

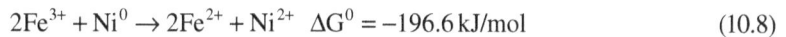

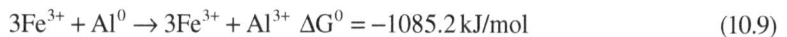

$$2Fe^{3+} + Zn^0 \rightarrow 2Fe^{2+} + Zn^{2+} \quad \Delta G^0 = -295\,kJ/mol \tag{10.7}$$

$$2Fe^{3+} + Ni^0 \rightarrow 2Fe^{2+} + Ni^{2+} \quad \Delta G^0 = -196.6\,kJ/mol \tag{10.8}$$

$$3Fe^{3+} + Al^0 \rightarrow 3Fe^{3+} + Al^{3+} \quad \Delta G^0 = -1085.2\,kJ/mol \tag{10.9}$$

In addition to bacteria (prokaryotes), fungi (eukaryotes) have been found to extract metals including Cu from WEEE. Under unfavorable conditions, many fungal species have been found to survive and grow, showing a high tolerance level for various toxic contaminants (Gadd 2010). In bioremediation, which uses living organisms to remove pollutants from the environment, fungi have also been studied extensively as a biological agent. For the extraction of metals from low-grade ores and mining tailings, industrial solid waste, and wastewater, fungal/heterotrophic leaching (metals bioleaching by fungi) has been widely explored.

Fungi offer a potential alternative in precious metals recovery from WEEE, which usually are in their zero-valent form, owning to their ability to modify metals speciation and mobility.

Fungi solubilize metals through several mechanisms; one of them is by way of weak organic acid production. Functional groups on the surface of fungal cell walls such as carboxyl, carbonyl, amine, amide, hydroxyl, and phosphate groups are bound to form complexation with metals (Baldrian 2003). *Aspergillus niger* and *Penicillium simplicissimum* were studied for their role in fungal leaching of WEEE, with the former showing the best efficiency to mobilize Cu (Brandl et al. 2001). Fungal leaching offers an operational advantage in acidic conditions compared to leaching by chemolithoautotrophic bacteria; H_2S gas produced by adding strong inorganic acid can be minimized (Sabra et al. 2012).

10.4 APPLICATION OF MICROORGANISMS IN RECOVERY OF GOLD AND OTHER METALS FROM E-WASTES

In general, microorganisms involved in the recovery of Au and other metals from ores and secondary sources such as electronic wastes can be classified into three groups: (i) chemolithoautotrophic bacteria and archaea such as iron- and sulfur-oxidizing bacteria, (ii) cyanogenic heterotrophic bacteria, and (iii) fungi. As mentioned previously, *Chromobacterium violaceum* is one of the most widely studied cyanogenic bacterial strains for the leaching of gold from ores and electronic wastes. Chi et al. (2011) reported *C. violaceum* for the leaching of Au and Cu from the used mobile phone PCBs in yeast extract and polypeptone medium supplemented with glycine. With a 15 g/L pulp density of PCB in an alkaline condition (pH 8-11) supplemented with oxygen and 0.004% (v/v) hydrogen peroxide (H_2O_2), the leaching of Au and Cu reached 11.31% and 24.6%, respectively, with the increase in pH (Chi et al. 2011).

The possibility of another bacterial strain generating cyanide ions, *Pseudomonas aeruginosa*, in solubilizing metals from electronic wastes was first studied by Pradhan and Kumar (2012). They conducted a study on the mobilization of various metals from electronic waste through two-step bioleaching by cyanogenic bacterial strains *C. violaceum*, *P. aeruginosa*, and *Pseudomonas fluorescens* as single and mixed cultures. As a single culture, *C. violaceum* was able to recover up to 79% of Cu, 69% of Au as well as other metals, while in a mixed culture with *P. aeruginosa*, the metals extracted were exceeding 83% and 73%, respectively, from 1% w/v electronic wastes concentration (Pradhan and Kumar 2012). Natarajan and Ting (2015) reported the leaching of Au from shredded electronic waste with a particle size of less than 100 μm using single and mixed cultures of *C. violaceum*, *P. aeruginosa*, and *P. fluorescens*, showing the highest cyanide production and Au recovery for a single culture of *C. violaceum*. The cell-free used medium of the cultures was utilized for the bioleaching process (the so-called spent medium bioleaching or indirect bioleaching) in this study. This bioleaching process offered higher recovery efficiency of 18%, which was improved with the intentionally increased pH to support cyanide ions production, suggesting better efficiency with pH modification than two-step bioleaching (Natarajan and Ting 2015).

The process of bioleaching depends on the production of leaching agents as a lixiviant secreted by microbes. Hence, industrial application of bioleaching has been limited by, for one reason, the abilities of bacteria or fungi to produce cyanide ions or other leaching agents continuously. One way to overcome this setback was to build a metabolically engineered strain of the bacterium *C. violaceum*. The bacterium was able to generate 70% more cyanide ions, which increased the recovery efficiency of Au from electronic waste more considerably compared to wild-type bacteria (Tay et al. 2013). Two genetically engineered *C. violaceum* strains, which could generate more cyanide, and a wild type were tested for the Au complexation ability from electronic waste scraps, resulting in one of the strains producing elevated cyanide levels. They exhibited 30% Au recovery at 0.5% pulp density compared with 11% efficiency of the wild-type bacteria (Natarajan et al. 2015). To reduce the amount of Cu forming stable complexes with cyanide and to increase Au mobilization in bioleaching, Das and Ting (2017) treated the electronic scrap materials with sulfuric acid (H_2SO_4) and hydrogen peroxide (H_2O_2) before the two-step leaching and spent medium bioleaching process using a metabolically engineered strain of *C. violaceum* were performed. Microbes were first cultured in the medium until a sufficient amount of cyanide was produced in the two-step bioleaching system, and shredded electronic waste was added to the cultures. The culture medium used to grow microbes containing secondary metabolites needed for leaching was filtered, and the electronic waste sample was then added to the cell-free medium for spent medium bioleaching.

The heterogeneity of metals composition in the electronic waste can lead to poor Au recovery efficiency due to the ease of these other metals to form cyanide complexation. Li et al. (2015) investigated a combined pre-treatment of electronic wastes with a chemolithoautotrophic acido-philic bacterium *Acidithiobacillus ferrooxidans* to recover Cu and other metals, followed by leaching of Au with *C. violaceum*. Applying optimum conditions with nutritive salts such as NaCl and $MgSO_4.7H_2O$ added to maximize bacterial cyanide generation and oxygen supplement along with pre-treatment of e-waste through biooxidation by the chemolithoautotrophic bacteria has proven to remarkably increase Au leaching efficiency up to 70%, accompanied by recovery of 80% Cu and other metals (Li et al. 2015).

Another study by Isildar et al. (2015) reported a two-step bioleaching process using chemo-lithoautotrophic bacteria *Acidithiobacillus ferrivorans* and *Acidithiobacillus thiooxidans* to recover Cu, followed by leaching of Au by cyanogenic heterotrophic bacteria *Pseudomonas putida* and *Pseudomonas fluorescens*. Shredded PCBs were added to the leaching medium containing grown cultures of *A. ferrivorans* and *A. thiooxidans*. Then, the filtered and sterilized residue from the previous step was added to the cultures of *P. putida* and *P. fluorescens*, which mobilized 44% of Au under alkaline conditions (Isildar et al. 2015). *P. fluorescens*, which was grown at pH 8 and 9 (under alkaline condition) in a medium supplemented with glycine and methionine to increase cya-nide production, was used to extract Au from PCBs and showed that the bioleaching efficiency was affected by the initial amount of the bacterial cells, pulp density, pre-adapted bacteria and addition of methionine (Li et al. 2020).

A bacterial strain identified as *Pseudomonas balearica* SAE1 isolated from a WEEE recycling facility was used to dissolve Au and Ag from the crushed waste of computer PCBs (Kumar et al. 2017). By optimizing factors affecting the leaching of metals such as pH level and temperature, gly-cine concentration, e-waste pulp density, the study showed that as much as 68.5% of Au and 33.8% of Ag were recovered. Kumar et al. (2017) also assessed *P. balearica* SAE1 tolerance to WEEE tox-icity along with the well-studied cyanogenic bacterium *C. violaceum* widely used for Au leaching.

A two-step bioleaching process for the extraction of base metals, precious metals, and rare earth elements (REE) from the shredded electronic waste dust intended for landfill disposal was inves-tigated by Marra et al. (2018). During the first step, *Acidithiobacillus thiooxidans* could leach out base metals such as Cu, Al, and Fe and REE such as cerium, lanthanum, and yttrium at a high recovery efficiency in 8 days. The shredded electronic waste dust from the first step, which has expe-rienced leaching by *At. thiooxidans*, was then subjected to treatment with the cyanogenic bacterium *P. putida*, allowing the mobilization of 48% Au in 3 hours (Marra et al. 2018).

Bacillus megaterium is another commonly studied bacterium able to generate cyanide to mobilize metals in addition to *Chromobacterium violaceum*, *Pseudomonas aeruginosa*, and *Pseudomonas fluorescens*. Arshadi and Mousavi (2015) investigated the optimum simultaneous leaching process of Cu and Au from computer PCBs utilizing a single culture of *B. megaterium*. They assessed the effect of e-waste pulp density, particle size, initial pH, and glycine addition on the efficiency of the bioleaching process. Pre-treatment using *Acidithiobacillus ferrooxidans* was first carried out to remove Cu to prevent interference in Au mobilization by cyanide complexation. The results showed that 63.8% of Au was leached from e-waste powder at the optimal condition of 2 g/L pulp density and a 0.5 g/L glycine concentration (Arshadi and Mousavi 2015). A more recent study constructed a metabolically engineered *B. megaterium* to generate excessive cyanide and improve the bioleach-ing process. The engineered *B. megaterium* strain grown in a specific medium contained glucose and cysteine, an amino acid containing sulfur, could recover 55% Au from PCB only in 4 days (Aminian-Dehkordi et al. 2020).

The bacteria *Frankia* sp. and *Frankia casuarinae* were used to leach metals such as Cu, Au, Ag, and Cu in electronic waste (i.e., PCB) shredded into fine particles (Marappa et al. 2020). This study reported up to 75% recovery of Au by *F. casuarinae* and 94% recovery of Cu by *Frankia* sp. from a lower e-waste pulp density (0.2%). *Frankia* is a genus of gram-positive actinobacteria with nitrogen-fixing ability either in free-living form or in symbiosis with actinorhizal plants.

Apart from the highly toxic cyanide, the possibility of using thiourea as a lixiviant or leaching agent used in hydrometallurgy to extract selected metals from ores has also been investigated. An earlier study reported two-step thiourea leaching preceded by a short period of biooxidation resulting in 95% of Au solubilization, although the role of Fe^{3+} regenerated by bacterial strains for the final recovery efficiency had not been clearly understood (Guo et al. 2017). A more recent study examined thiourea tolerance in several iron-oxidizing bacteria and archaea and the application in mobilizing Au from e-wastes. In the presence of PCB and 10 mM thiourea, an iron-oxidizing bacterial strain, *Acidiplasma* sp. Fv-Ap exhibited optimal Fe^{3+} regeneration, giving consistent Au solubilization up to 98% (Rizki et al. 2019). This study implied the potential application of thiourea bioleaching of precious metals from e-wastes under controlled Eh (redox potential) level and minimized reagent utilization with optimum recovery efficiency.

The majority of earlier studies on the recovery of metals from e-waste using bioagents were concentrated on the use of bacteria. Fungal leaching has been commonly studied and, in some consideration, more preferred than bacterial leaching due to its further tolerance of toxic materials and ability to form soluble metal complexes and precipitations. One of the widely studied fungi for bioleaching is the fungus *Aspergillus niger*. Two strains of *A. niger* were examined for their leaching efficiency of Cu, Au, and Ni from cellular phone PCBs and gold-layered integrated circuits computer motherboards. In the mixed culture of the two strains, the bioleaching of Au reached a recovery efficiency of 28% for computer motherboard parts and 87% for PCBs compared to Cu recovery, which gave higher efficiency for one of the strains instead of in combination (Madrigal-Arias et al. 2015). Horeh et al. (2016) conducted three modified leaching experiments of metals from spent lithium-ion battery powder with the addition of pre-cultured *Aspergillus niger*. At a pulp density of 1%, recovery of Cu reached 100% efficiency in the spent sucrose medium previously incubated with *A. niger* spores until reaching stationary phase and compared to chemical leaching; fungal leaching has given greater recovery efficiency (Horeh et al. 2016). To examine the possibility of bioleaching processes with reduced energy consumption, Argumedo-Delira et al. (2019) used two strains of *A. niger* in a single culture and in a consortium to recover Au from mobile phone PCB in a culture medium without agitation and with an added carbon source (glucose) which showed considerably higher Au leaching concentration up to 56% for the fungal consortium.

REFERENCES

Aminian-dehkordi, J., S. Mohammad Mousavi, and S.-a. Marashi. 2020. "A systems-based approach for cyanide overproduction by bacillus megaterium for gold bioleaching enhancement." *Frontiers in Bioengineering and Biotechnology* 8(June): 1–18. https://doi.org/10.3389/fbioe.2020.00528.

Argumedo-Delira, R., M.J. Gómez-Martínez, and B. Joan Soto. 2019. "Gold bioleaching from printed circuit boards of mobile phones by Aspergillus niger in a culture without agitation and with glucose as a carbon source." *Metals* 9(5). https://doi.org/10.3390/met9050521.

Arshadi, M., and S. M. Mousavi. 2015. "Enhancement of simultaneous gold and copper extraction from computer printed circuit boards using Bacillus megaterium." *Bioresource Technology* 175: 315–24. https://doi.org/10.1016/j.biortech.2014.10.083.

Baldrian, P. 2003. Interactions of heavy metals with white-rot fungi. *Enzyme and Microbial Technology* 32(1): 78–91.

Bigum, M., L. Brogaard and T. H. Christensen. 2012. Metal recovery from high-grade WEEE: A life cycle assessment. *Journal Hazardous Material* 207–208: 8–14.

Bohu, T., R. Anand, R. Noble, M. Lintern, A. H. Kaksonen, Y. Mei … M. Power. 2019. Evidence for fungi and gold redox interaction under Earth surface conditions. *Nature Communications* 10(1): 2290.

Brandl, H., R. Bosshard and M. Wegmann. 2001. Computer-munching microbes: Metal leaching from electronic scrap by bacteria and fungi. *Hydrometallurgy* 59(2): 319–326.

Brandl, H., S. Lehmann, M. A. Faramarzi and D. Martinelli. 2008. Biomobilization of silver, gold, and plati-num from solid waste materials by HCN-Forming microorganisms. *Hydrometallurgy* 94(1–4): 14–17.

Brierley, C. L. 2010. Biohydrometallurgical prospects. *Hydrometallurgy* 104(3–4): 324–328.

Chi, T. D., J. C. Lee, B. D. Pandey, K. Yoo, and J. Jeong. 2011. "Bioleaching of gold and copper from waste mobile phone pcbs by using a cyanogenic bacterium." *Minerals Engineering* 24(11): 1219–22. https://doi.org/10.1016/j.mineng.2011.05.009.

Crowe, M., A. Elser, B. Göpfert, L. Mertins, T. Meyer, J. Schmid, A. Spillner and R. Ströbel. 2003. *Waste from Electrical and Electronic Equipment (WEEE)-Quantities, Dangerous Substances and Treatment Methods.* D. Tsotsos, European Topic Centre on Waste, European Environmental Agency (EEA).

Das, S., and Y.-p. Ting. 2017. "Improving gold (bio) leaching efficiency through pretreatment using hydro-gen peroxide assisted sulphuric acid the copyediting, typesetting, pagination and proofreading process, which may lead to differences between," no. December 2015.

Das, S., S. H. Lee, P. Kumar, K.H. Kim, S.S. Lee, and S.S. Bhattacharya. 2019. Solid waste management: Scope and the challenge of sustainability. *Journal of Cleaner Production* 228: 658–678.

Ehrlich, H. L. 2001. Past, present and future of biohydrometallurgy. *Hydrometallurgy* 59(2–3): 127–134.

European Commission Eurostat. 2016. Guidance on municipal waste data collection. Unit E-2: Environmental statistics and accounts; sustainable development.

Fomina, M., E. P. Burford, S. Hillier, M. Kierans, & G. M Gadd. (2010). Rock-building fungi. *Geomicrobiology Journal* 27(6–7): 624–629. https://doi.org/10.1080/01490451003702974.

Gadd, G. M. 2010. Metals, minerals and microbes: Geomicrobiology and bioremediation. *Microbiology* 156(March): 609–643.

Guo, Y., X. Guo, H. Wu, S. Li, G. Wang, X. Liu, G. Qiu, D. Wang, 2017. A novel biooxidation and two-step thiourea leaching method applied to a refractory gold concentrate. *Hydrometallurgy* 171, 213–221. https://doi.org/10.1016/j.hydromet.2017.05.023.

He, W., G. Li, X. Ma, H. Wang, J. Huang, M. Xu and C. Huang. 2006. WEEE recovery strategies and the WEEE treatment status in China. *Journal of Hazardous Materials* 136(3): 502–512.

Hoornweg, D., P. Bhada-Tata, 2012. *What a Waste: a Global Review of Solid Waste Management. Urban Development Series – Knowledge Papers.* The World Bank.

Horeh, N. Bahaloo, S. M. Mousavi, and S. A. Shojaosadati. 2016. "Bioleaching of valuable metals from spent lithium-ion mobile phone batteries using Aspergillus niger." *Journal of Power Sources* 320: 257–66. https://doi.org/10.1016/j.jpowsour.2016.04.104.

Johnston, C.W., M.A. Wyatt, X. Li, A. Ibrahim, J. Shuster, G. Southam, N.A. Magarvey, 2013. Gold biominer-alization by a metallophore from a gold-associated microbe. *Nature Chemical Biology* 9: 241.

Kita, Y., H. Nishikawa and T. Takemoto. 2006. Effects of cyanide and dissolved oxygen concentration on biological Au recovery. *Journal of Biotechnology* 124(3): 545–551.

Korte, F., M. Spiteller and F. Coulston. 2000. The cyanide leaching gold recovery process is a nonsustain-able technology with unacceptable impacts on Ecosystems and Humans: The Disaster in Romania. *Ecotoxicology and Environmental Safety* 46(3): 241–245.

Kumar, A., H. Singh Saini, and S. Kumar. 2018. "Bioleaching of Gold and Silver from Waste Printed Circuit Boards by Pseudomonas Balearica SAE1 Isolated from an E-Waste Recycling Facility." *Current Microbiology* 75(2): 194–201. https://doi.org/10.1007/s00284-017-1365-0.

Lawson, E. N., M. Barkhuizen and D. W. Dew. 1999. Gold solubilisation by cyanide producing bacteria Chromobacterium violaceum. In *Biohydrometallurgy and the Environment Toward the Mining of the 21st Century.* eds. Amils R., Ballester A. 9A: 239–246. Elsevier, Amsterdam.

Lear, G., J. M. McBeth, C. Boothman, D. J. Gunning, B. L. Ellis, R. S. Lawson, K. Morris et al. 2010. Probing the biogeochemical behavior of technetium using a novel nuclear imaging approach. *Environmental Science & Technology* 44(1): 156–162.

Li, J., C. Liang, and C. Ma. 2015. "Bioleaching of gold from waste printed circuit boards by chromobacterium violaceum." *Journal of Material Cycles and Waste Management* 17(3): 529–39. https://doi.org/10.1007/s10163-014-0276-4.

Marappa, N., L. Ramachandran, D. Dharumadurai, and T. Nooruddin. 2020. "Recovery of gold and other precious metal resources from environmental polluted e-waste printed circuit board by bioleaching frankia." *International Journal of Environmental Research* 14(2): 165–76. https://doi.org/10.1007/s41742-020-00254-5.

Marra, A., A. Cesaro, E.R. Rene, V. Belgiorno, and P.N.L. Lens. 2018. "Bioleaching of metals from WEEE shredding dust." *Journal of Environmental Management* 210: 180–90. https://doi.org/10.1016/j.jenvman.2017.12.066.

Mecucci, A. and K. Scott. 2002. Leaching and electrochemical recovery of copper, lead and tin from scrap printed circuit boards. *Journal of Chemical Technology & Biotechnology* 77(4): 449–457.

Natarajan, G., S. B. Tay, W. S. Yew, and Y. P. Ting. 2015. "Engineered Strains Enhance Gold Biorecovery from Electronic Scrap." *Minerals Engineering* 75: 32–37. https://doi.org/10.1016/j.mineng.2015.01.002.

Natarajan, G. and Y.-P. Ting. 2014. Pretreatment of e-waste and mutation of alkali-tolerant cyanogenic bacteria promote gold biorecovery. *Bioresource Technology* 152: 80–85.

Natarajan, G., and Y. P. Ting. 2015. "Gold biorecovery from e-waste: an improved strategy through spent medium leaching with pH modification." *Chemosphere* 136: 232–38. https://doi.org/10.1016/j.chemosphere.2015.05.046.

Oguchi, M., H. Sakanakura and A. Terazono. 2013. Toxic metals in WEEE: Characterization and substance flow analysis in waste treatment processes. *The Science of the Total Environment* 463–464: 1124–1132.

Ongondo, F. O., I. D. Williams and T. J. Cherrett. 2011. How are WEEE doing? A global review of the management of electrical and electronic wastes. *Waste Management* 31(4): 714–730.

Pradhan, J. K., and S. Kumar. 2012. "Metals Bioleaching from Electronic Waste by Chromobacterium Violaceum and Pseudomonads sp." *Waste Management and Research* 30(11): 1151–59. https://doi.org/10.1177/0734242X12437565.

Robinson, B. H. 2009. E-waste: An assessment of global production and environmental impacts. *The Science of the Total Environment* 408 (2): 183–191.

Rocchetti, L., F. Vegliò, B. Kopacek, and F. Beolchini. 2013. Environmental impact assessment of hydrometallurgical processes for metal recovery from WEEE residues using a portable prototype plant. *Environmental Science & Technology* 47(3), 1581–1588.

Sabra, N., H. C Dubourguier, M. N. Duval and T. Hamieh. 2011. Study of canal sediments contaminated with heavy metals: Fungal versus bacterial bioleaching techniques. *Environmental Technology* 32: 1307–1324.

Sabra, N., H.-C. Dubourguier and T. Hamieh. 2012. Fungal leaching of heavy metals from sediments dredged from the deûle canal, France. *Advances in Chemical Engineering and Science* 02(01): 1–8.

Schluep, M., C. Hagelueken, R. Kuehr, F. Magalini, C. Maurer, C. Meskers, E. Mueller and F. Wang. 2009. *Sustainable Innovation and Technology Transfer Industrial Sector Studies: Recycling–From E-Waste to Resources*. United Nations Environment Programme & United Nations University, Bonn, Germany.

Smith, A. and T. Mudder. 1991. *Chemistry and Treatment of Cyanidation Wastes*. Mining Journal Books Ltd. (UK), 345.

Syed, S. 2012. Recovery of gold from secondary sources—A review. *Hydrometallurgy* 115–116: 30–51.

Tsydenova, O. and M. Bengtsson. 2011. Chemical hazards associated with treatment of waste electrical and electronic equipment. *Waste Management* 31(1): 45–58.

11 Bioaugmentation in the Bioremediation of the Heavy Metals and Radionuclides

Bhagwan Toksha, Saurabh Tayde, Ajinkya Satdive,
Shyam Tonde, and Aniruddha Chatterjee
Maharashtra Institute of Technology

CONTENTS

11.1 INTRODUCTION

The notion of ecosystem services has been a significant embodiment for linking the functioning of the ecosystem services to human well-being (Fisher et al. 2007; Summers et al. 2018). In the human-dominated environment, humans in the pursuit of well-being have either eroded or overexploited the capacity of the ecosystem services subsequently altering the ecosystem. The omnipresence of mining activities has been found to be deteriorating the environmental system resulting in physical disturbance and chemical contamination of soil and water (Haddaway et al. 2019).

DOI: 10.1201/9781003187622-11

Soil is the only habitat that contains biota providing nutritional and mechanical support to the higher plants. The heavy discharges of heavy metals and radionuclides from the industrial areas, mine tailings, pesticides, coal combustion residues, oil and petrochemical spillages, etc., lead to soil contamination (Wuana and Okieimen 2011; Zwolak and Sarzy 2019). Heavy metals are natural constituents of the earth's crust which are generally referred to as a group of metals and metalloids that have a relatively high density (>4,000 kg/m^3). Some heavy metals (Fe, Mn, Cu, Zn, Mo) also act as micronutrients which are essential for the metabolic processes in plants but might become harmful when used in excessive quantities. Other elements like As, Hg, Pb, or Cd affect plant growth and development even at low concentrations and are toxic to human and animal health. Based on their toxicity heavy metals can be put in order as Hg>Cu>Zn>Ni>Pb>Cd>Cr>Sn>Fe>Mn> Al (Filipiak-Szok, Kurzawa, and Szłyk 2015; Zwolak and Sarzy 2019). Radioactive elements are physically unstable that undergo radioactive decay by emitting energy in the form of ionizing radiation. This ionizing radiation emanates the alpha particles, beta particles, and/or gamma rays. Of these, gamma rays are the most powerful ionizing radiation that can knock electrons out of atoms. These gamma rays pose a health risk and can damage various human tissue, DNA, and protein. Among the naturally existing radionuclides, uranium (U) and radium (Ra) are found in significant amounts. Most radionuclides are either produced in particle accelerators, nuclear reactors, or radioactive decay of uranium or other radionuclides (Palmisano and Hazen 2003). The release of such heavy metals and radionuclides in the soil by either natural or anthropogenic activity disturbs the soil ecosystem. Hence to restore the soil ecosystem one would require to perform characterization and bioremediation. The soil characterization would provide an insight into the heavy metal type, while remediation entails the knowledge of contamination source, environmental and associated health risks, and basic chemistry.

This chapter focuses on the role of bioaugmentation in the bioremediation of the ecosystem. The effects of heavy metals and radionuclides on the biological system will be discussed. We will also go through the bioaugmentation mechanism and the factors affecting it. Ultimately, we will see how nanotechnology can be exploited in the process of bioaugmentation.

11.2 BIOREMEDIATION TECHNIQUES

Bioremediation is the technique in which pollutants present in the environment are treated by microorganisms naturally to reduce their toxic effect. Bioremediation is an effective and efficient eco-friendly tool in removing heavy metal pollutants and radionuclides. It is also known as the eco-friendly waste treatment method. Bioremediation can be accomplished by in situ and ex situ techniques. The basic principle involved in bioremediation is to convert the toxic pollutant into a low or non-hazardous form. Biotic and abiotic components play a vital role in deciding the rate of degradation (Sharma 2020). Bioremediation consists of reduction, detoxification, biodegradation, immobilization, eradication of hazardous toxic chemical pollutants to their reduced form or low toxic compound. Bioremediation is a low-cost remediation technique as compared to physical and chemical techniques while treating heavy metals and nuclides. For microbial degradation activity, the presence of carbon in the pollutant is the basic requirement as it may be the energy source for them. Bioremediation is majorly accomplished by detoxification of hazardous chemicals and their waste through degradation, eradication, immobilization process by the action of microorganisms. Various methods and techniques are used for the bioremediation of pollutants. It involves the treatment of contaminated materials through the action of bacterial microbes, fungi, algae, and yeast. Microbes can grow at different temperature conditions over the pollutants. Various microorganisms participate in bioremediation such as *Alcaligenes, Achromobacter, Arthrobacter, Bacillus, Corynebacterium, Flavobacterium,* Pseudomonas, Mycobacterium, *Nitrosomonas,* and *Xanthobacter* (Singh et al. 2014).

The microorganisms can do bioremediation in aerobic or anaerobic conditions. The aerobic bacterium such as *Acinetobacter, Flavobacterium, Mycobacterium, Nocardia, Pseudomonas, Rhodococcus,*

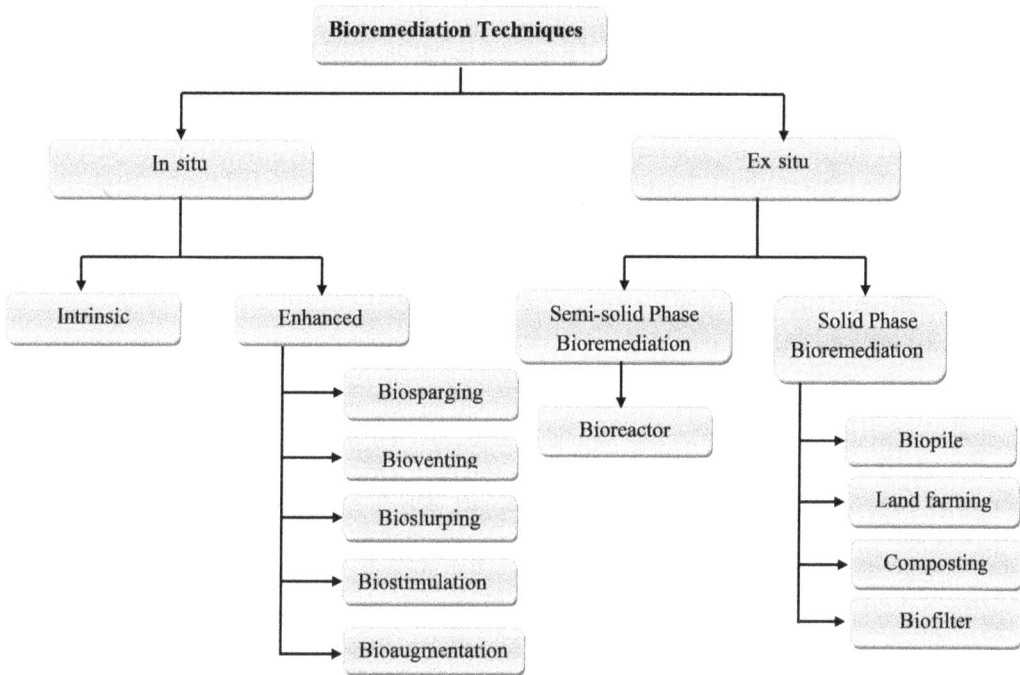

FIGURE 11.1 Classification of bioremediation technique.

and *Sphingomonas* are used for bioremediation of pesticides, aliphatic and aromatic hydrocarbons. While the anaerobic bacterium is used for the biological degradation of complex compounds like chloroform and chlorinated aromatic compounds. The concentration, structural complexity, physical and chemical properties of compounds as well as other factors such as temperature, pH, atmospheric oxygen, and soil containing nutrients affect the extent of bioremediation of pollutants. The bioremediation technique can be broadly categorized into in situ and ex situ type and their sub-types are given in Figure 11.1. They are discussed in detail in the following sections.

11.2.1 IN SITU TECHNOLOGY

In this method, the soil containing pollutants is treated itself at the polluted site and bioremediation does not require excavation. As no excavation, the soil structure remains unaffected. The cost of bioremediation of in situ technology is low as compared to ex situ technology. It is divided into intrinsic (natural) and engineered bioremediation. The engineered bioremediation involves the reduction of pollutants via biosparging, bioventing, bioslurping, biostimulation, and bioaugmentation. This technique is used to treat complex pollutants such as aliphatic and aromatic hydrocarbon, dyes, and heavy metals (Frascari et al. 2015; Roy et al. 2015).

11.2.1.1 Biosparging

Biosparging is quite the same as that of bioventing. In this technique, the microbial activities are enhanced by introducing (injecting) air through the soil subsurface. This leads to promoting microbial growth to treat the pollutants. Porosity and permeability of soil and biodegradability of toxic compounds affect the efficiency of biosparging (Philp and Atlas 2005). Biosparging helps to stimulate the rate of biodegradation. It is similar to in situ air sparging. Water contaminated by benzene, toluene, ethylbenzene, and xylene (BTEX) could be bioremediated using biosparging (Kao et al. 2008). It is also used to treat the water contaminated with kerosene and diesel.

11.2.1.2 Bioventing

It consists of air stimulation by incorporating oxygen to the unsaturated portion of the polluted site under controlled conditions. Delivery of oxygen under controlled conditions improves bioremediation rate through increased microbial activities. The rate of bioremediation can be stimulated by the incorporation of nutrients and moisture at the contaminated site. This is a widely used technique as compared to the other in situ technology (Philp and Atlas 2005; Höhener and Ponsin 2014). Frutos et al. (2010) investigated that the use of bioventing in remediation of soil contaminated with phenanthrene has removed about 93% of the contaminants after 7 months.

11.2.1.3 Bioslurping

Bioslurping includes vacuum pumping, extraction of vapor evolved from the soil along bioventing to accelerate polluted site bioremediation. Availing oxygen through bioventing stimulates bioremediation. This technique is helpful to remediate saturated and unsaturated portions and capillaries of light non-aqueous liquids and volatile organic components (Gidarakos and Aivalioti 2007). Slurp means to draw the liquid, soil gas upward from the soil layer which is very similar to the working of normal straw used to draw the liquid from the container. The slurping mechanism helps to separate light non-aqueous liquid from air and water. The excessive moisture present in soil resists the diffusion of oxygen through it and affects bioremediation. It is not only a cost-effective technique but also applicable to soil containing low oxygen permeability.

11.2.1.4 Biostimulation

Biostimulation is the bioremediation technique through which the microorganism's degradation activities are stimulated through modification of the environment at the contaminated site. By this technique, the existing degradation microorganisms are stimulated to enhance their physiological activities. It can be worked out for both polluted soil and water in their ex situ and in situ techniques. For biostimulation the environment at the polluted site can be achieved by adding suitable nutrients, bulking agents, and electron acceptors (nitrogen, oxygen, phosphorous). The bioremediation of halogenated pollutants is obtained by adding electron donors which stimulate degradation activities. It can be suitable to encounter pesticides and herbicides (Kanissery and Sims 2011). Biostimulation can enhance through bioaugmentation and enhance diversified microorganism groups. The induced nutrients help to provide acceleration and driving force to the bioremediation (Roy et al. 2018). The important benefit of this technique is the use of autochthonous microorganisms as they are more adaptable for this technique (Speight 2016).

11.2.1.5 Bioaugmentation

Bioaugmentation is an in situ process that involves the incorporation of bacterial culture, for example; archaea to the pollutant-containing site in order to accelerate the degradation rate of pollutants. This technique is found to be very effective when the native bacterial colonies are not capable of doing degradation of toxic compounds (Woźniak-Karczewska et al. 2019). The bacterial culture may be of the same type of bacterial colony or mixed cultures or in the form of a particular strain (Huang and Ye 2020). The bacteria which are generated from the native polluted site may be able to degrade the pollutants, but with slow and inefficient degradation speed. Bioaugmentation studies the indigenous species of bacteria found in the native to investigate whether biostimulation is carried or not by them. If the indigenous bacteria can degrade the pollutants, implementation of more indigenous bacterial culture will be done over the contaminants to enhance its degradation. If indigenous bacterial species are unable to degrade the pollutants, exogenous bacterial species with suitable degradation pathways are incorporated. Bioaugmentation has given its contribution to immobilization, biology, and bioreactor design (Herrero and Stuckey 2015). These techniques are found effective in the starting up of bioreactor and boosting its efficiency (Pandey et al. 2011).

11.2.2   Ex Situ Technology

In this technique, the hazardous pollutants are taken out by excavation from the polluted site and accordingly transported to the other site for their safe and effective treatment. Factors such as the geographical location of pollutant site, its geology, extent, and level of pollution, category of hazardous pollutant, and treatment expenses are taken into consideration. The ex situ bioremediation has been classified into slurry-phase and solid-phase bioremediation (Philp and Atlas 2005).

11.2.2.1   Solid-Phase Bioremediation

This technique involves the removal of solid organic waste such as agricultural waste, plant leaves, animal excreta, domestic-municipal and industrial waste and placing it into piles. Suitable conditions are made using a pipe to develop bacterial growth and the necessary air supply for microbial activity into the pile. It consists of different sub-techniques such as biopiling, land farming, composting, and land filter.

11.2.2.1.1   Biopile

It follows the above-ground piling bioremediation of excavated soil containing pollutants. After this sufficient air and nutrients are supplied to enhance microbial activities. Biopiling consists of aeration, irrigation, nutrient, leachate collection, and treatment bed. It is low-cost solid-phase bioremediation with the provision of controlled conditions of pH, temperature, oxygen, nutrients, and moisture content. Biopile is effective in the biodegradation of low molecular weight volatile pollutants. It is also effective at lower temperature conditions (Dias et al. 2015; Whelan et al. 2015). Biopiles are constructed in the form of a slab using concrete material to make it waterproof and reduce leaching over the surface which has a covering of waterproof membrane. This arrangement helps to inhibit the spreading of pollutants to the surrounding during rain (Jørgensen et al. 2000). This process has some limitations such as the requirement of large space and excess air heat which can restrict microbial growth and activities and causes an emission of volatile pollutants.

11.2.2.1.2   Land Farming

Land farming is the bioremediation process which comes with low-cost and easy operational requirement. Majorly, it is known as ex situ bioremediation, but in some instances, it is considered an in situ bioremediation technique. The depth of the pollutant decides whether the land farming should be carried out by in situ or ex situ technique (Azubuike, Chikere, and Okpokwasili 2016). If the pollutants are present at less than 1 m of depth from the soil surface, bioremediation can be carried out without digging. But if the pollutants are present at more than 1.7 m of depth below the soil surface, it requires excavation and transporting the polluted soil over the ground for its efficient remediation (Nikolopoulou et al. 2013). If the dig contaminated soil is treated at its site only, it is known as in situ bioremediation otherwise it will be an ex situ kind of bioremediation. The polluted soil is treated at the surface by proper aeration, nutrient supply, and irrigation using autochthonous microbes (Silva-Castro et al. 2015). Land farming is widely used for bioremediation soil contaminated with hydrocarbons to enhance soil structural properties (Macaulay and Rees 2014). This technique has some limitations of large space requirement, limited microbial activity, low efficiency in treating inorganic waste, and additional cost of digging (Ortega et al. 2018).

11.2.2.1.3   Composting

It is an ex situ technique following aerobic biodegradation. In this technique, the organic waste is decomposed to form compost which is used as organic fertilizer to improve soil fertility (Chadar 2018). For proper compositing environmental conditions always matters, it includes temperature (40°C–70°C), neutral pH, adequate oxygen, and nutrients availability. In compositing along with recycling of organic waste, the other contaminants present in the soil get bioremediated. This microbe degrades the organic matter but reduces metal bioavailability (Aguelmous et al. 2019).

The factors such as temperature, oxygen, moisture, and nutrient always affect the efficacy of composting although it is a low-cost technique. Currently, compositing is receiving widespread acceptance as it is found to be efficient in degrading complex organic compounds like pesticides, hydrocarbons, and chlorophenols (Sayara and Sánchez 2020).

11.2.2.1.4 Biofilter

Biofilters (BFs) involve bioreactor which uses microorganisms to treat biologically degradable pollutants. It processes wastewater and air containing toxic chemicals. BF which purifies air uses microorganisms for its bioremediation. It consists of a packed bed through which polluted air is passed and thin biofilm is formed on its surface. Immobilized bacteria and fungi are used to degrade the harmful polluted air. It finds its wide application in the separation of water-soluble volatile organic compounds from the air. BFs can be used even to treat polluted gases emitted from manufacturing industries and clean them up to 98%. Trickling BFs are generally used to treat water from different sources. Daily backwashing is plasticized in a trickling filter to maintain its efficiency (Chaudhary et al. 2003). BFs have low maintenance and operating cost, easy operation, and high surface area. It has some disadvantages such as poor process control, a large requirement of area, sensitivity towards environmental conditions, and drainage discharge from filter beds (Cabrera, Ramírez, and Cantero 2011). After the continuous operation, the BF media have an accumulation of biomass. Such a deteriorated bed can build pressure and improper distribution of contaminated air and water through the BF (Qamaruz-Zaman, Yaacof, and Kamarzaman 2020).

11.2.2.2 Semi-Solid (Slurry) Phase Bioremediation

In slurry-phase bioremediation, the soil containing polluted compounds is excavated and added with nutrients, oxygen, and water. This mixture is processed in the bioreactor for the creation of desired conditions for microbial activities for degradation of pollutants present in the soil rapidly. The addition of water and nutrients is decided by the type and concentration of pollutants along with the physico-chemical properties of soil. After completion of this process, pressure filters or centrifuges are used to dry the soil.

11.2.2.2.1 Bioreactor

The bioreactor is a reaction vessel in which a biochemical process is worked out using microorganisms. It is a manufacturing system that is supported by biologically active reaction conditions may be aerobic or anaerobic. The sludge bioreactors are used for the bioremediation of hydrocarbon pollutants. Generally, they are made up of stainless steel having a cylindrical shape. The bioremediation of biomass and pollutants is stimulated through biofilm formation (Salehi and Hakimghiasi 2017). The pollutants can be charged into the bioreactor in dry or suspension form. Operationally, bioreactors are classified as batch or continuous. Treating pollutants in a bioreactor has many advantages like a controlled process in terms of temperature, mixing, oxygen supply, pH, and proper inoculum level. Bioreactors can effectively bioremediate pesticides containing heavy metals, organic pollutants, and solvents. The microorganisms in bioreactors are submerged in the liquid or mounted on the surface of a solid medium. The introduced culture may remain in suspended or immobilized form (López, Lázaro, and Marqués 1997). Bioreactors are widely used for industrial biotechnology, healthcare, and waste treatment and environmental protection applications. In advanced bioreactors, sensors and actuators are used to improve their techno-economic efficiency (Wang et al. 2020).

11.3 EFFECT OF HEAVY METALS AND RADIONUCLIDES ON THE BIOLOGICAL SYSTEM

Heavy metals can be regarded as essential and non-essential exhibiting harmful effects at a higher concentration. The heavy metal threat to plants, animals, and human health is aggravated by their long-term persistence in the environment. This can be explained as metals tend to form covalent bonds with the organic molecules ending up in forming lipophilic ions and compounds. These ions and compounds become more toxic and their toxicity varies from the action of the simple ionic form of the same element. For example, the methylated form of arsenic is highly toxic (Briffa et al. 2020). Heavy metals can generate toxic effects when they bind to non-metallic elements of cellular macromolecules. For example, binding of lead and mercury to sulfhydryl groups of the protein. Heavy metals may enter into humans and animals by ingesting contaminated food, drinking contaminated water, inhaling through the atmosphere, or coming in contact with agricultural, pharmaceutical products (Masindi and Muedi 2018). Different cell components like mitochondria, organelles, nuclei, lysosomes, enzymes, and cell membranes are affected due to heavy metals. The heavy metals interact with the nuclear proteins and DNA causing damage to it which may lead to cell cycle modulation, carcinogenesis, or apoptosis (Tchounwou et al. 2012). An almost similar phenomenon is observed in the case of plants, where cellular damage takes place occurring as a result of oxidative stress generated due to heavy metal exposure. Unlike humans, plants have evolved detoxification mechanisms based on subcellular compartmentalization and chelation. An example of heavy metal chelator in the case of plants is phytochelatins (PCs) (Singh et al. 2011). Figure 11.2 represents the scheme of radionuclides interoperability with various bacterial processes.

Exposure to radionuclides induces the physical, chemical, and biological transformations in the human and animal body. When radiation is controlled it can be used to treat cancer but uncontrolled exposure or leak will lead to the generation of difficulties like kidney damage, genetic transformation, and leukemia existing through the ecosystem. Upon exposure, acute health effects can be experienced like nausea, vomiting, and headaches. The increased exposure can lead to hair loss, weakness, fever, diarrhea, blood in stool, disorientation, low blood pressure leading to death. Damages at the cellular level occur in the case of fetuses resulting in poorly formed eyes, mental retardation abnormal brain growth, and distorted head or brain size.

The fact that heavy metals and radionuclides are entering our food chain is ruining our ecosystem. These are absorbed by the plants which make them hazardous both to the plant and the food chain that eats the plant (Caplin and Willey 2018; Briffa et al. 2020). This will alter the natural chemistry and other properties like porosity, pH, and the color of the soil, thus deteriorating the quality of the soil and water (Zhu and Shaw 2000; Yadav et al. 2014). The radionuclides interoperability with various bacterial processes is depicted schematically in Figure 11.3. The radionuclides are seen to be attached with the precipitate and immobilized accumulation inside the bacterial cell (Shukla et al. 2017).

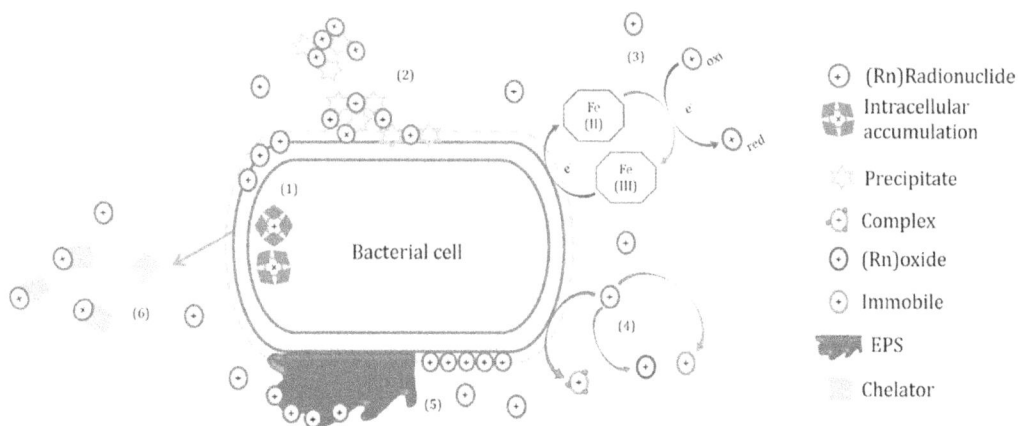

FIGURE 11.2 The schematic depicting radio nuclides interoperability with various bacterial processes (Source: Shukla et al. 2017).

11.4   MICROBIAL ACTIVITY AND BIOAUGMENTATION

Microbial activity is an important parameter for understanding the functioning of different environments. This parameter helps in understanding how soil microbes, for example, fungi and bacteria, play a role in carbon recycling. Microbial respiration or soil respiration is the measure of microbial activity that provides an insight into the propensity of soil microbes to welcome the residues or amendments, mineralize and generate the nutrients for other plants and microorganisms, preserve the nutrients and safeguard their availability in the long run, and improve the soil structure (Stevenson et al. 2011). Microbial activity is affected by the change in temperature. For example, when the temperature falls below 5°C, the activity becomes very slow, and if the temperature is increased to 25°C–30°C, the microbial activity starts increasing linearly to a maximum value. If you continue to increase the temperature, the activity starts decreasing to the point where the enzyme denaturation occurs and growth stops. Microbial activity involving oxygen requirement as in the case of fungal and aerobic type causes complete decay of plant (Dombrowski et al. 2016). In the ecosystem where the plants provide the organic substrate, microbial activity plays a vital role in nutrient recycling. Different factors such as the chemical nature of the soil and physical conditions also influence microbial activities (Ma et al. 2000).

Inoculating the microorganism population through the introduction of genetically modified microorganisms having specific catabolic activities with a view to enhance the performance and degradation rate is a technique termed bioaugmentation (Kim et al. 2018).

11.4.1   Basic Metabolic Activity of Microbes

Metabolic activity is the means through which microorganisms obtain the required nutrients and energy via chemical pathways, sequences of biochemical reactions resulting in the activity, growth, and reproduction (Zheng et al. 2006). Larger molecules are broken down into simpler compounds during the catabolic activity producing the energy required for growth and reproduction. Harmful compounds are transformed into a less harmful form or are degraded completely through such catabolic processes.

11.4.2   Bioaugmentation in Heavy Metal Transformation

Bioaugmentation involves the enhancement of the degradation capabilities of microorganisms by inoculating the contaminated site with genetically engineered microorganisms. The microbes used in bioaugmentation can be native or non-native types. Though this technique manifested the cleaning up of the contaminated site containing aromatic compounds, still some environmental challenges lie ahead. The survival of the strains, when introduced into the soil, is the most difficult problem. The number of exogenous microorganisms present in the soil tends to decrease soon after the inoculation. Thus, the effectiveness of bioaugmentation is influenced by both biotic and abiotic factors (Diep, Mahadevan, and Yakunin 2018). Bioaugmentation technique should be employed in the soil having a lower or non-detectable number of microbes capable of degrading the contaminants, where the process is harmful or toxic to the microbes, where cost-reduction is required, for example, replacing the non-biological methods with bioaugmentation (Haluška et al. 1995). However, given that bioaugmentation helps in removing the aromatic compounds present in the soil, it may also be considered useful in cleaning up the diverse biota containing heavy metals (Ogino et al. 2001).

Some advanced molecular techniques can be employed to improve bioremediation and analyze the diverse microorganisms capable of degrading the pollutant. Utilizing some engineering steps to modify the DNA of microorganisms making them capable of removing or converting heavy metals has become a way of increasing the bioremediation efficiency. Engineering steps include genetic engineering, sequence alteration of existing coding and regulatory genes, and pathway construction (Fredrickson et al. 2004). Such applications can further be utilized to control greenhouse gas emissions, waste conversion into value-added eco-friendly products, and carbon sequestration. Still, to make such potentials a reality, there arises a need for applying cost benefit driving force and regulatory safety. However, employing genetically engineered microbes would also cause some environmental problems such as genetic pollution that must also be accounted for before using the bioaugmentation technique (Hildén and Mäkelä 2018).

11.4.3 BIOAUGMENTATION IN RADIONUCLIDE TRANSFORMATION

The processing of radionuclides to mitigate its effects on our ecosystem has been thought of by researchers in recent years (Ahier and Tracy 1995). Different cases of groundwater and soil getting contaminated or affected due to the radioactive wastes are observed. These radionuclides have become a major concern for the environment and human health. When the general population comes into contact with such radioactive waste, it causes serious life-threatening diseases. To eliminate the radioactive materials from the environment, bioremediation has been looked at as an ecological alternative (Ganesh et al. 1997). The endogenous genetic, physiological, and biochemical properties of microorganisms make them ideal agents for the bioremediation of soil and groundwater. The biotransformation of radionuclides can be carried out by microorganism-mediated mechanisms using different environmental conditions (Lu et al. 2009). Using microorganism-mediated bioremediation will alter the bioavailability, mobility, and solubility of the radionuclides.

11.5 FACTORS AFFECTING BIOAUGMENTATION IN BIOREMEDIATION

Various biotic (biological) and abiotic (environmental) factors highly influence bioaugmentation in bioremediation. Biotic factors include microbial activity in the biological degradation of complex and toxic compounds having insufficient carbon. Bioaugmentation is the process of incorporation of bacterial culture over the pollutants in order to enhance their rate of degradation. The enzyme released, gene mutation, and increase in the number of colonies by the microbes are some factors that supports the bioremediation. (Boopathy 2000). The abiotic or environmental conditions such as temperature, pH, soil quality, nutrients, oxygen and moisture contents, conditions at the site, concentration, structural complexity, and physical and chemical properties of compounds affect the bioremediation. This abiotic factor influences the metabolic activities of microorganisms and affects their kinetics of bioremediation. Some of the important factors that affect bioaugmentation are discussed below.

11.5.1 PH

The increase or decrease of pH affects the physiological properties of microorganisms. The acidic or basic pH of the compound, as well as the pH of the polluted site, decides the rate of bioremediation. It has been investigated that the biodegradation of pollutants occurs only at optimal pH ranging from 6.5 to 8.5 for the terrestrial as well as aquatic environmental conditions (Cases and de Lorenzo 2005). The metabolic activities of the microorganisms are highly prone to a small change in the pH value. Hence, measuring the pH and maintaining it at the required value can lead to a suitable condition for rapid microbial growth and degradation (Wang et al. 2011).

11.5.2 TEMPERATURE

It is one of the most vital factors in deciding the endurance of microorganisms. As the microbial activities are very slow at a lower temperature, the degradation occurs faster at high temperatures than that at low temperatures (Das and Chandran 2011). But the degradation of a specific pollutant occurs at a particular temperature only. Thus the rate of bacterial degradation rises with temperature and is found to be highest at an appropriate optimum temperature. If the temperature increases beyond its optimum value, the rate of degradation decreases or sometimes no degradation occurs after attaining a particular temperature (Macaulay 2015).

11.5.3 OXYGEN AND MOISTURE CONTENTS

Bioremediation occurs at both aerobic and anaerobic conditions. Hence the concentration of oxygen plays an important role in bioremediation. The sufficient oxygen availability is the requirement of most of the microorganisms in improving biodegradation in the case of hydrocarbons (Macaulay 2015). Also, the sufficient moisture content supports the microbial growth in many cases. The moisture content of the soil is a crucial regulatory property in the case of degradation of chlorophenol by bioremediation (Cho, Rhee, and Lee 2000). Collectively, the moisture content of soil can affect the oxygen requirement for the biodegradation of pollutants. The proper provision of moisture content at the polluted site is needed for its successful bioremediation. The hydrocarbon-degrading microorganism consumes its carbon as a source of energy in the presence of oxygen (electron acceptor) and suitable moisture with nutrients (Hesnawi and Adbeib 2013). A sufficient amount of moisture content should be available to promote microbial activity at the contaminated site. The insufficient moisture content or water availability may inhibit the microbial activity and can stop the bioremediation of soil, whereas the excess water content fills all pores present in the soil and resists the diffusional transfer of oxygen through it. This may also reduce the rate of soil bioremediation. Hence optimum oxygen and moisture content at contaminated sites support the soil bioremediation.

11.6 BIOAUGMENTATION USING NANOTECHNOLOGY

The basis of the bioremediation process revolves around the living organisms and incorporating the nanomaterials in their close vicinity requires a thorough understanding of the interaction between nanomaterials and living organisms. The demand for bioremediation strategies being less toxic on microorganisms achieves an improved microbial activity in the specific waste and toxic material, reducing the time taken to complete the process at reduced cost could be one of the promises of nanotechnology. There are various facets to the dynamics of how the microorganisms, contaminants interaction gets modified with the presence of nanomaterials. The contribution of the presence of nanoparticle-based methodologies may have significant positive and negative effects. The nanomaterials are known to be stimulants as well as toxic for microorganisms. Hence the selection of nanomaterials for inclusion in bioremediation becomes a critically important task.

The bioremediation strategies based on nanomaterials integrate the biological processes to accelerate and promote the removal of toxic compounds from the environment (Kumari and Singh 2016). The term nano-bioremediation is coined with the meaning of processes involving the nanomaterials are added in the presence of microorganisms and plants to remove contaminants (Koul and Pooja 2018). The variations in such approaches were labeled as phyto-nano-remediation, microbial nano-remediation, and zoo-nano-remediation depicting the nature of the organism utilized for the remediation of contaminants (Ojuederie and Babalola 2017). The physio-chemical exchange between contaminants, nanomaterials, flora, and fauna depends on nanomaterials' inherent virtues such as size and shape, surface coating, chemical nature of the nanomaterials, type of microorganism, and ambient conditions such as media, pH, and temperature. Furthermore, the responses these parameters create once they interact with the contaminant (Tan et al. 2018). Depending upon the nature of the contaminant, the class of nanomaterials such as iron-based, carbon-based, other metals, and other engineered polymorphs are selected (Kiciński and Dyjak 2020). The magnetic nanoparticle, SiO_2 core-shell nanoparticles with synthetic bacteria were co-assembled to separate the Cd^{2+} and Pb^{2+} contaminants, and in another example, the magnetite was utilized to separate heavy metals in soils or water (Zhu et al. 2020; Vázquez-Núñez et al. 2020). Figure 11.3 shows schematically the genetic and chemical modifications brought in *Escherichia coli* cells via protein SynHMB encoding developing synthetic bacteria and magnetic nanoparticles assembly (Zhu et al. 2020).

The functional nanomaterials when synthesized and integrated with living microorganisms exhibit the newer augmented properties like enhanced tolerance against stress, programmed metabolism and proliferation, conductivity, and artificial photosynthesis. In the study of soil bioremediation iron

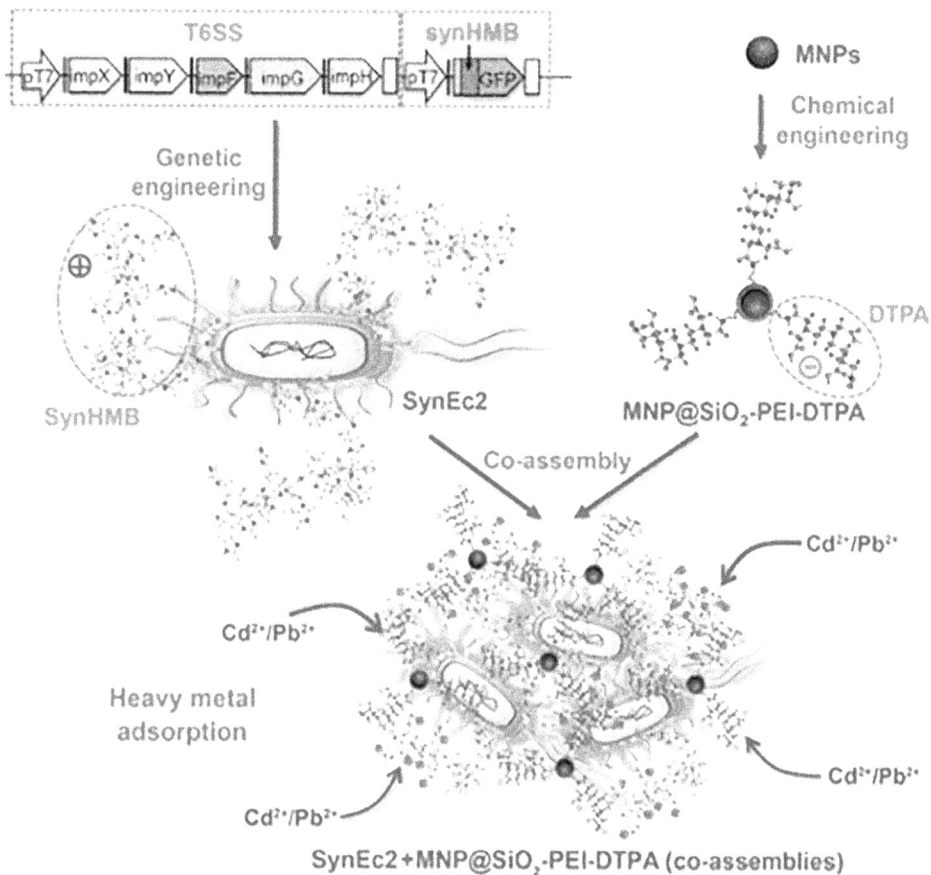

FIGURE 11.3 The genetic and chemical modifications brought in *Escherichia coli* cells via protein SynHMB encoding developing synthetic bacteria and magnetic nanoparticles assembly (Source: Zhu et al. 2020).

nanoparticles and rhizosphere microorganisms on the phytoremediation, the stress tolerance of white willow is reported by Kashtiban et al. (2019). The bioremediation of soil contaminated with hydrocarbons and heavy metals through programmed metabolism was reported to be an effective approach (Alisi et al. 2009). Abhishek Panchal et al carried out the study and reported that the performance of Halloysite clay nanotubes is more effective as compared to surfactant emulsions for metabolism bacteria for bioremediation of the oil spill (Panchal et al. 2018). Nanoparticles self-assembly for lead, copper, and cadmium bioremediation are the instances of artificial photosynthesis achieved through functionalization of nanomaterials (Li et al. 2017; Yeh et al. 2018). Figure 11.3 represents the genetic and chemical modifications brought in *Escherichia coli* cells via protein SynHMB encoding developing synthetic bacteria and magnetic nanoparticles assembly.

The production of nanomaterials with simultaneous incorporation of microbes is a cost-effective sustainable and eco-friendly way to achieve bioremediation of heavy metals. The synthesis through green routes has certain advantages over chemical routes of producing nanoparticles in terms of toxic effects due to chemicals and agglomeration tendencies of synthesized nanoparticles (Mahanty et al. 2020). The nanoparticles extracted from the plants such as Noaea mucronata and other close variants are found to be effective for Cu, Zn, and Pb metals in the soil (Mohsenzadeh and Rad 2012). The enzyme-based Maize plant growth under the influence of Ag-nano particle in the bioremediation of heavy metals under wastewater irrigation was tested by Khan and Bano (2016). The properties of nanomaterials being highly reactive and catalytic could be exploited to produce effective enzymes. The sewage micro-pollutants were degraded by immobilizing laccase enzymes onto the surface of functionalized TiO_2 nanoparticles (Ji et al. 2017). The research in this direction leads to bioremediation with the inclusion of microbes, plants, and enzymes along with the nanomaterials.

11.7 CONCLUSION

The use of bacterial cultures to enhance bioremediation for heavy metal degradation using nanomaterials via in situ and ex situ technologies is a current topic of research. Bioaugmentation is effective with minimal side effects in heavy metal and radionuclide transformation by precipitation, immobilization, and creating environmental partitioning. The research in the direction of mutual effects and interdependence of the factors pH, ambient temperature, substrate, and concentration needs more focus for scientific understanding. The nanomaterials synthesis directly from the microorganisms and use of nanomaterials supplementing the accelerated growth of bacterial cultures for the heavy metal effluent elimination/accumulation are found to be among the several other ways in which microorganisms and nanotechnology can complement each other. The bioremediation routes with the use of bioaugmentation without any long-term imperishable effects on microorganisms improving the microbial activity for the heavy metals and radionuclide treatments and still time/cost-effectiveness is the need of the hour.

ACKNOWLEDGMENTS

AC is sincerely thankful to the Department of Science and Technology (DST), Government of India, for using the facility provided under the Nano Mission Scheme (File No.: SR/NM/NT-1106/2016, 05/09/2018).

REFERENCES

Aguelmous A., L. El Fels, S. Souabi, M. Zamama, and M. Hafidi. 2019. "The Fate of Total Petroleum Hydrocarbons during Oily Sludge Composting: A Critical Review." *Reviews in Environmental Science and Bio/Technology* 18 (3): 473–93.

Ahier B A and Tracy B L. "Radionuclides in the Great Lakes Basin. 1995." *Environmental Health Perspectives* 103 (suppl 9): 89–101.

Alisi C., R. Musella, F. Tasso, C. Ubaldi, S. Manzo, C. Cremisini, and A. Rosa Sprocati. 2009. "Bioremediation of Diesel Oil in a Co-Contaminated Soil by Bioaugmentation with a Microbial Formula Tailored with Native Strains Selected for Heavy Metals Resistance." *Science of the Total Environment* 407 (8): 3024–32.

Azubuike C. C., C. B. Chikere, and G. C. Okpokwasili. 2016. "Bioremediation Techniques–Classification Based on Site of Application: Principles, Advantages, Limitations and Prospects." *World Journal of Microbiology and Biotechnology* 32 (11): 180.

Boopathy R. 2000. "Factors Limiting Bioremediation Technologies." *Bioresource Technology* 74 (1): 63–67.

Briffa J., E. Sinagra, and R. Blundell. 2020. "Heavy Metal Pollution in the Environment and Their Toxicological Effects on Humans." *Heliyon* 6 (9).

Cabrera G., Ramírez M., and Cantero D. 2011. "Biofilters." In *Comprehensive Biotechnology*, Elsevier. Murray Moo-Young 303–18.

Caplin N., and N. Willey. 2018. "Ionizing Radiation, Higher Plants, and Radio protection: From Acute High Doses to Chronic Low Doses." In *Frontiers in Plant Science*. Frontiers Media S.A. Vicent Arbona 9: 847

Cases I., and V. de Lorenzo. 2005. "Genetically Modified Organisms for the Environment: Stories of Success and Failure and What We Have Learned from Them". *International Microbiology* 8 (3): 213–22.

Chadar S. 2018. "Composting as an Eco-Friendly Method to Recycle Organic Waste." *Progress in Petrochemical Science* 2: 252–54.

Chaudhary D. S., S. Vigneswaran, H.-H. Ngo, W. G. Shim, and H. Moon. 2003. "Biofilter in Water and Wastewater Treatment." *Korean Journal of Chemical Engineering* 20 (6): 1054.

Cho Y.-G., S.-K. Rhee, and S.-T. Lee. 2000. "Effect of Soil Moisture on Bioremediation of Chlorophenol-Contaminated Soil." *Biotechnology Letters* 22 (11): 915–19.

Das N., and P. Chandran. 2011. "Microbial Degradation of Petroleum Hydrocarbon Contaminants: An Overview." *Biotechnology Research International* 2011: 1–13.

Dias Romina L., L. Ruberto, A. Calabró, A. Lo Balbo, M. T. Del Panno, and W. P. Mac Cormack. 2015. "Hydrocarbon Removal and Bacterial Community Structure in On-Site Biostimulated Biopile Systems Designed for Bioremediation of Diesel-Contaminated Antarctic Soil." *Polar Biology* 38 (5): 677–87.

Diep P., R. Mahadevan, and A. F. Yakunin. 2018. "Heavy Metal Removal by Bioaccumulation Using Genetically Engineered Microorganisms." *Frontiers in Bioengineering and Biotechnology* 6: 157.

Dombrowski N., J. A. Donaho, T. Gutierrez, K. W. Seitz, A. P. Teske, and B. J. Baker. 2016. "Reconstructing Metabolic Pathways of Hydrocarbon-Degrading Bacteria from the Deepwater Horizon Oil Spill." *Nature Microbiology* 1 (7): 1–7.

Filipiak-Szok, A., M. Kurzawa, and E. Szłyk. 2015. "Determination of Toxic Metals by ICP-MS in Asiatic and European Medicinal Plants and Dietary Supplements." *Journal of Trace Elements in Medicine and Biology* 30: 54–58.

Fisher B., R. Costanza, R. K., Turner, and M. Paul. 2007. "Defining and Classifying Ecosystem Services for Decision Making." CSERGE Working Paper EDM 07–04:20.

Frascari D., G. Zanaroli, and A. S. Danko. 2015. "In Situ Aerobic Cometabolism of Chlorinated Solvents: A Review." *Journal of Hazardous Materials* 283: 382–99.

Fredrickson J. K., J. M. Zachara, D. L. Balkwill, D. Kennedy, S.-m. W. Li, H. M. Kost, M. J. Daly, et al. 2004. "Geomicrobiology of High-Level Nuclear Waste-Contaminated Vadose Sediments at the Hanford Site." *Applied Environmental Microbiology*. https://doi.org/10.1128/AEM.70.7.4230-4241.2004.

Frutos F. J. G., O. Escolano, S. García, M. Babín, and M. Dolores Fernández. 2010. "Bioventing Remediation and Ecotoxicity Evaluation of Phenanthrene-Contaminated Soil." *Journal of Hazardous Materials* 183 (1): 806–13.

Ganesh R., Robinson K. G., Reed G. D., and Sayler G. S. 1997. "Reduction of Hexavalent Uranium from Organic Complexes by Sulfate- and Iron-Reducing Bacteria." *Applied and Environmental Microbiology* 63 (11): 4385–91.

Gidarakos E., and M. Aivalioti. 2007. "Large Scale and Long Term Application of Bioslurping: The Case of a Greek Petroleum Refinery Site." *Journal of Hazardous Materials, Pollution Prevention and Restoration of the Environment* 149 (3): 574–81.

Haddaway Neal R., S. J. Cooke, P. Lesser, B. Macura, A. E. Nilsson, J. J. Taylor, and K. Raito. 2019. "Evidence of the Impacts of Metal Mining and the Effectiveness of Mining Mitigation Measures on Social-Ecological Systems in Arctic and Boreal Regions: A Systematic Map Protocol." *Environmental Evidence* 8 (1): 1–11.

Haluška L., G. Barančíková, Š. Baláž, K. Dercová, B. Vrana, M. Paz-Weisshaar, E. Furčiová, and P. Bielek. 1995. "Degradation of PCB in Different Soils by Inoculated Alcaligenes Xylosoxidans." *Science of the Total Environment* 175 (3): 275–85.

Herrero M., and D. C. Stuckey. 2015. "Bioaugmentation and Its Application in Wastewater Treatment: A Review." *Chemosphere, Wastewater-Energy Nexus: Towards Sustainable Wastewater Reclamation*, 140: 119–28.

Hesnawi R. M., and M. M. Adbeib. 2013. "Effect of Nutrient Source on Indigenous Biodegradation of Diesel Fuel Contaminated Soil." *APCBEE Procedia, 4th International Conference on Environmental Science and Development- ICESD* 2013 (5): 557–61.

Hildén K., and M. R. Mäkelä. 2018. "Role of Fungi in Wood Decay." *Reference Module in Life Sciences*.

Höhener P., and V. Ponsin. 2014. "In Situ Vadose Zone Bioremediation." *Current Opinion in Biotechnology, Energy Biotechnology, Environmental Biotechnology* 27: 1–7.

Huang H., and L. Ye. 2020. "Chapter 9-Biological Technologies for CHRPs and Risk Control." In *High-Risk Pollutants in Wastewater*, H. Ren and X. Zhang (ed.). Elsevier. 209–36.

Ji C., L. N. Nguyen, J. Hou, F. I. Hai, and V. Chen. 2017. "Direct Immobilization of Laccase on Titania Nanoparticles from Crude Enzyme Extracts of P. Ostreatus Culture for Micro-Pollutant Degradation." *Separation and Purification Technology* 178: 215–23.

Jørgensen K. S., Puustinen J., and Suortti A. -M. 2000. "Bioremediation of Petroleum Hydrocarbon-Contaminated Soil by Composting in Biopiles." *Environmental Pollution* 107 (2): 245–54.

Kanissery R. G., and G. K. Sims. 2011. "Biostimulation for the Enhanced Degradation of Herbicides in Soil." *Applied and Environmental Soil Science* 2011: 1–10.

Kao C. M., Chen C. Y., Chen S. C., Chien H. Y., and Chen Y. L. 2008. "Application of in Situ Biosparging to Remediate a Petroleum-Hydrocarbon Spill Site: Field and Microbial Evaluation." *Chemosphere* 70 (8): 1492–99.

Khan N., and A. Bano. 2016. "Role of Plant Growth Promoting Rhizobacteria and Ag-Nano Particle in the Bioremediation of Heavy Metals and Maize Growth under Municipal Wastewater Irrigation." *International Journal of Phytoremediation* 18 (3): 211–21.

Kiciński W., and S. Dyjak. 2020. "Transition Metal Impurities in Carbon-Based Materials: Pitfalls, Artifacts and Deleterious Effects." *Carbon* 168: 748–845.

Kim S., K. H. Chu, Y. A. J. Al-Hamadani, C. M. Park, M. Jang, D.-H. Kim, M. Yu, J. Heo, and Y. Yoon. 2018. "Removal of Contaminants of Emerging Concern by Membranes in Water and Wastewater: A Review." *Chemical Engineering Journal* 335: 896–914.

Koul B., and T. Pooja. 2018. *Biotechnological Strategies for Effective Remediation of Polluted Soils.* Springer, Singapore.

Kumari B., and D. P. Singh. 2016. "A Review on Multifaceted Application of Nanoparticles in the Field of Bioremediation of Petroleum Hydrocarbons." *Ecological Engineering* 97: 98–105.

Li C., L. Zhou, H. Yang, R. Lv, P. Tian, X. Li, Y. Zhang, Z. Chen, and F. Lin. 2017. "Self-Assembled Exopolysaccharide Nanoparticles for Bioremediation and Green Synthesis of Noble Metal Nanoparticles." *ACS Applied Materials & Interfaces* 9 (27): 22808–18.

López A., N. Lázaro, and A. M. Marqués. 1997. "The Interphase Technique: A Simple Method of Cell Immobilization in Gel-Beads." *Journal of Microbiological Methods* 30 (3): 231–34.

Lu H., G. Gao, G. Xu, L. Fan, L. Yin, B. Shen, and Y. Hua. 2009. "Deinococcus Radiodurans PprI Switches on DNA Damage Response and Cellular Survival Networks after Radiation Damage." *Molecular & Cellular Proteomics* 8 (3): 481–94.

Ma L.-J., S. O. Rogers, C. M. Catranis, and W. T. Starmer. 2000. "Detection and Characterization of Ancient Fungi Entrapped in Glacial Ice." *Mycologia* 92 (2): 286–95.

Macaulay B. 2015. "Understanding the Behaviour of Oil-Degrading Micro-Organisms to Enhance the Microbial Remediation of Spilled Petroleum." *Applied Ecology and Environmental Research* 13: 247–62.

Macaulay B. M., and D. Rees. 2014. "Bioremediation of Oil Spills: A Review of Challenges for Research Advancement." *Annals of Environmental Science* 8: 9–37.

Mahanty S., S. Chatterjee, S. Ghosh, P. Tudu, T. Gaine, M. Bakshi, S. Das. 2020 "Synergistic Approach towards the Sustainable Management of Heavy Metals in Wastewater Using Mycosynthesized Iron Oxide Nanoparticles: Biofabrication, Adsorptive Dynamics and Chemometric Modeling Study." *Journal of Water Process Engineering* 37: 101426.

Masindi V, and Muedi KL. 2018. *Environmental Contamination by Heavy Metals in: Heavy Metals.* InTech.

Mohsenzadeh F., and A. C. Rad. 2012. "Bioremediation of Heavy Metal Pollution by Nano-Particles of Noaea Mucronata." *International Journal of Bioscience, Biochemistry and Bioinformatics* 85–89.

Mokarram-Kashtiban S., S. M. Hosseini, M. T. Kouchaksaraei, and H. Younesi. 2019. "The Impact of Nanoparticles Zero-Valent Iron (NZVI) and Rhizosphere Microorganisms on the Phytoremediation Ability of White Willow and Its Response." *Environmental Science and Pollution Research* 26 (11): 10776–89.

Nikolopoulou M., N. Pasadakis, H. Norf, N. Kalogerakis. 2013. "Enhanced Ex Situ Bioremediation of Crude Oil Contaminated Beach Sand by Supplementation with Nutrients and Rhamnolipids." *Marine Pollution Bulletin* 77 (1): 37–44.

Ogino A., H. Koshikawa, T. Nakahara, H. Uchiyama. 2001. "Succession of Microbial Communities during a Biostimulation Process as Evaluated by DGGE and Clone Library Analyses." *Journal of Applied Microbiology* 91 (4): 625–35.

Ojuederie O. B., and O. O. Babalola. 2017. "Microbial and Plant-Assisted Bioremediation of Heavy Metal Polluted Environments: A Review." *International Journal of Environmental Research and Public Health* 14(12).

Ortega Marcelo F., D. E. Guerrero, M. J. García-Martínez, D. Bolonio, J. F. Llamas, L. Canoira, and J. L. R. Gallego. 2018. "Optimization of Landfarming Amendments Based on Soil Texture and Crude Oil Concentration." *Water, Air, & Soil Pollution* 229 (7): 234.

Palmisano A., and T. Hazen. 2003. *Bioremediation of Metals and Radionuclides: What It Is and How It Works* (2nd Edition).

Panchal A., L. T. Swientoniewski, M. Omarova, T. Yu, D. Zhang, D. A. Blake, V. John, and Y. M. Lvov. 2018. "Bacterial Proliferation on Clay Nanotube Pickering Emulsions for Oil Spill Bioremediation." *Colloids and Surfaces B: Biointerfaces* 164: 27–33.

Pandey A., C. Larroche, and S. C. Ricke. 2011. *Biofuels: Alternative Feedstocks and Conversion Processes.* Academic Press.

Philp Jim C., and R. M. Atlas. 2005. "Bioremediation of Contaminated Soils and Aquifers." In *Bioremediation,* John Wiley & Sons. 139–236

Qamaruz-Zaman, N., N. Yaacof, and F. F. Kamarzaman. 2020. "9- Control of Odors in the Food Industry." In *The Interaction of Food Industry and Environment,* C. Galanakis, (ed.) Academic Press. 281–313.

Roy A., A. Dutta, S. Pal, A. Gupta, J. Sarkar, A. Chatterjee, A. Saha, P. Sarkar, P. Sar, and S. K. Kazy. 2018. "Biostimulation and Bioaugmentation of Native Microbial Community Accelerated Bioremediation of Oil Refinery Sludge." *Bioresource Technology* 253: 22–32.

Roy M., A. K. Giri, S. Dutta, and P. Mukherjee. 2015. "Integrated Phytobial Remediation for Sustainable Management of Arsenic in Soil and Water." *Environment International* 75: 180–98.

Salehi M. A., and N. Hakimghiasi. 2017. "Hydrodynamics and Mass Transfer Inthree-Phase Airlift Reactors for Activated Carbon and Sludge Filtration." *Advances in Environmental Technology* 2 (4): 179–84.

Sayara T., and A. Sánchez. 2020. "Bioremediation of PAH-Contaminated Soils: Process Enhancement through Composting/Compost." *Applied Sciences* 10 (11): 3684.

Sharma I. 2020. Bioremediation Techniques for Polluted Environment: Concept, Advantages, Limitations, and Prospects. *Trace Metals in the Environment - New Approaches and Recent Advances.* InTechOpen.

Shukla, A., P. Parmar, and M. Saraf. 2017. "Radiation, Radionuclides and Bacteria: An in-Perspective Review." *Journal of Environmental Radioactivity* 180: 27–35.

Silva-Castro, G. A., I. Uad, A. Rodríguez-Calvo, J. González-López, and C. Calvo. 2015. "Response of Autochthonous Microbiota of Diesel Polluted Soils to Land-Farming Treatments." *Environmental Research* 137: 49–58.

Singh R., N. Gautam, A. Mishra, and R. Gupta. 2011. "Heavy Metals and Living Systems: An Overview." *Indian Journal of Pharmacology* 43 (3): 246–53.

Singh R., P. Singh, and R. Sharma. 2014. "Microorganism as a Tool of Bioremediation Technology for Cleaning Environment: A Review," 6.

Speight J. G. 2016. *Environmental Organic Chemistry for Engineers.* Butterworth-Heinemann.

Stevenson Bradley S., H. S. Drilling, P. A. Lawson, K. E. Duncan, V. A. Parisi, and J. M. Suflita. 2011. "Microbial Communities in Bulk Fluids and Biofilms of an Oil Facility Have Similar Composition but Different Structure." *Environmental Microbiology* 13 (4): 1078–90.

Summers J. K., L. M. Smith, R. S. Fulford, and R. de Jesus Crespo. 2018. "The Role of Ecosystem Services in Community Well-Being." *Ecosystem Services and Global Ecology.*

Tan W., J. R. Peralta-Videa, and J. L. Gardea-Torresdey. 2018. "Interaction of Titanium Dioxide Nanoparticles with Soil Components and Plants: Current Knowledge and Future Research Needs – A Critical Review." *Environmental Science: Nano* 5 (2): 257–78.

Tchounwou, P. B., C. G. Yedjou, A. K. Patlolla, and D. J. Sutton. 2012. Heavy Metal Toxicity and the Environment. EXS. Vol. 101. Springer, Basel.

Vázquez-Núñez, E., C. E. Molina-Guerrero, J. M. Peña-Castro, F. Fernández-Luqueño, and M. G. de la Rosa-Álvarez. 2020. "Use of Nanotechnology for the Bioremediation of Contaminants: A Review." *Processes* 8 (7): 826.

Wang, B., Z. Wang, T. Chen, and X. Zhao. 2020. "Development of Novel Bioreactor Control Systems Based on Smart Sensors and Actuators." *Frontiers in Bioengineering and Biotechnology* 8.

Wang, Q., S. Zhang, Y. Li, and W. Klassen. 2011. "Potential Approaches to Improving Biodegradation of Hydrocarbons for Bioremediation of Crude Oil Pollution." *Journal of Environmental Protection* 2 (1): 47.

Whelan, M. J., F. Coulon, G. Hince, J. Rayner, R. McWatters, T. Spedding, and I. Snape. 2015. "Fate and Transport of Petroleum Hydrocarbons in Engineered Biopiles in Polar Regions." *Chemosphere* 131: 232–40.

Woźniak-Karczewska, M., P. Lisiecki, W. Białas, M. Owsianiak, A. Piotrowska-Cyplik, Ł. Wolko, Ł. Ławniczak, H. J. Heipieper, T. Gutierrez, and Ł. Chrzanowski. 2019. "Effect of Bioaugmentation on Long-Term Biodegradation of Diesel/Biodiesel Blends in Soil Microcosms." *Science of the Total Environment* 671: 948–58.

Wuana Raymond A., and F. E. Okieimen. 2011. "Heavy Metals in Contaminated Soils: A Review of Sources, Chemistry, Risks and Best Available Strategies for Remediation." *ISRN Ecology* 2011: 1–20.

Yadav M., M. Rawat, A. Dangwal, M. Prasad, G. S. Gusain, and R. C. Ramola. 2014. "Levels and Effects of Natural Radionuclides in Soil Samples of Garhwal Himalaya." *Journal of Radioanalytical and Nuclear Chemistry* 302 (2): 869–73.

Yeh Chia-Shen, R. W., W.-C. Chang, and J.-h. Shih. 2018. "Synthesis and Characterization of Stabilized Oxygen-Releasing CaO2 Nanoparticles for Bioremediation." *Journal of Environmental Management* 212: 17–22.

Zheng D., Y. Liu, Y. Zhang, X.-J. Chen, and Y.-F. Shen. 2006. "Microcalorimetric Investigation of the Toxic Action of Cr(VI) on the Metabolism of Tetrahymena Thermophila BF5 during Growth." *Environmental Toxicology and Pharmacology* 22 (2): 121–27.

Zhu N., B. Zhang, and Q. Yu. 2020. "Genetic Engineering-Facilitated Coassembly of Synthetic Bacterial Cells and Magnetic Nanoparticles for Efficient Heavy Metal Removal." *ACS Applied Materials & Interfaces* 12 (20): 22948–57.

Zhu Y. G., and Shaw G. 2000. "Soil Contamination with Radionuclides and Potential Remediation." *Chemosphere* 41 (1–2): 121–28.

Zwolak A., and M. Sarzy. 2019. "Sources of Soil Pollution by Heavy Metals and Their Accumulation in Vegetables : A Review". *Water, Air, & Soil Pollution* 230: 164. https://doi.org/10.1007/s11270-019-4221-y.

Index

Note: **Bold** page numbers refer to tables; *italic* page numbers refer to figures.

For Product Safety Concerns and Information please contact our EU
representative GPSR@taylorandfrancis.com
Taylor & Francis Verlag GmbH, Kaufingerstraße 24, 80331 München, Germany

9 781032 035017